SOP

Principles of
Fluid Mechanics

Principles of
Fluid Mechanics

by

WEN-HSIUNG LI

Department of Civil Engineering
Syracuse University

SAU-HAI LAM

Department of Aerospace and Mechanical Sciences
Princeton University

ADDISON-WESLEY PUBLISHING COMPANY, INC.

READING, MASSACHUSETTS · PALO ALTO · LONDON

Preface

The study of fluid mechanics is basic to many branches of engineering. The purpose of this book is to present the fundamental principles of fluid mechanics and their application in a variety of engineering problems. The book is intended for use as a first course in this subject, probably in the junior year. The student is expected to have had about two years of college mathematics, including such topics as partial differentiation, multiple integrals, and Taylor's series.

In the early chapters, the pace of presentation is purposely kept slow, and the student is led carefully through the analyses. A large number of illustrative examples and problems are provided to emphasize the practical implications of the newly acquired concepts. In the later chapters, the presentation is less detailed in anticipation of the increased sophistication of the student.

Vector notations are introduced and used wherever it is convenient, but the more advanced methods of vector analysis are avoided in the belief that undergraduate students who are not familiar with advanced vector analysis should not be overly penalized in the study of fluid mechanics. Cartesian tensor analysis is introduced and used only in Chapter 11 on viscous fluids for the expressed purpose of deriving the relation between stress and rate of strain.

The book covers a wide range of topics spanning the interests of aeronautical, chemical, civil, and mechanical engineers. Each chapter is written in a reasonably self-contained manner, so as to allow the instructor flexibility in organizing his course to suit the needs of his class. The materials in this book can be covered in a one-year intensive course, and students so prepared should be able to adequately handle graduate-level courses in fluid mechanics and other related subjects.

The book can also be used in a one-term terminal course. The sections recommended for such a course are indicated with asterisks in the Contents.

W. H. L. and S. H. L.

May 1964

v

Contents

*Sections recommended for a short terminal course.

1 Introduction

1–1. Fluids

Fluids may be defined as materials which continue to deform in the presence of any shearing stress. When the space between two plates is filled with a fluid as shown in Fig. 1–1, the plates can be kept moving relative to each other by a force, however small. Generally, the larger the force, the higher the rate of the relative motion. As a contrast, a *solid* under shearing stresses can remain in a deformed position if the stresses do not exceed a certain limit. There are solids that will continue to deform like a fluid when the stress exceeds a certain value. The solid is then said to be in a *plastic state*. The study of plastic flow is outside the domain of fluid mechanics.

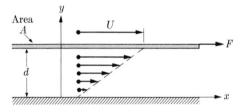

FIGURE 1–1

In fluid mechanics, fluids are considered to be *continuous* although they, like any substance, consist of discrete molecules. This approach is taken not only for the resultant simplicity in analysis, but also because the behavior of the in- dividual molecules is not usually of primary interest in technology. The average properties of the molecules in a small parcel of fluid are used as the properties of the continuous material. For example, the mass of all the molecules per unit volume of the parcel is called the *density* of the fluid. For this approach to be successful, the size of the flow system must be much larger than the mean free path of the molecules, so that the properties, such as density, of the fluid can be meaningfully computed. Ordinarily, this requirement presents no difficulties. For example, there are about 2.7×10^{16} molecules in one cubic millimeter of air under atmospheric conditions. However, there are cases where this requirement is not satisfied. For example, in the upper atmosphere where air molecules may be, on the average, several feet apart, the term *density* becomes meaningless if one considers the flow around a satellite of a foot in size.

1

Fluids may be classified into two groups, namely, *liquids* and *gases*. While a volume of liquid will not fill a container with a larger volume, a gas will always do so. Except for the fact that a liquid may thus exhibit a distinct surface, the two, as continuous materials, behave mechanically in a similar manner, and can therefore be studied together.

1–2. Mechanics

Fluid mechanics is to be studied according to Newton's laws of motion, the laws of thermodynamics, the principle of conservation of mass, and the mechanical and thermodynamic properties of fluids. Newton's laws of motion are reviewed in this section. The laws of thermodynamics will be presented in Chapter 14.

Newton's laws of motion may be summarized in the second law and the third law. The *second law* states that the motion of a particle of mass m, subject to forces whose vector sum is $\mathbf{F}$, is governed by

$$\mathbf{F} = k\,\frac{d(m\mathbf{q})}{dt}, \tag{1–1}$$

where t is time, and $\mathbf{q}$ is the velocity of the particle. (In this book, a boldface letter, such as $\mathbf{F}$, is used to represent a *vector* quantity which has a direction as well as a magnitude. The *scalar magnitude* of $\mathbf{F}$ is represented by F.) The velocity $\mathbf{q}$ is referred to an inertial system. In fluid mechanics problems, $\mathbf{q}$ relative to the earth can be used unless the large-scale motion in the atmosphere or the oceans is being studied. The value of the constant k depends on the units used; for example, $k = 1$ in the *English system*, where force is expressed in pounds, time in seconds, length in feet, and mass in slugs; and in the *cgs system*, where force is expressed in dynes, time in seconds, length in centimeters, and mass in grams. With such units, Eq. 1–1 can be written as

$$\mathbf{F} = \frac{d(m\mathbf{q})}{dt}. \tag{1–2}$$

Note that 1 lb = 1 slug-ft/sec², and 1 dyne = 1 gm-cm/sec².

Newton's *third law* of motion is the law of actions and reactions: the forces two particles exert on each other are equal and oppositely directed along the line joining them.

1–3. Mechanical properties of fluids

To study the mechanical behavior of a fluid system, Eq. 1–2 is to be applied to each infinitesimal fluid element. The following mechanical properties of the fluid are involved.

The *density* ρ of a material is defined as the mass per unit volume of the material. The unit of ρ may be slug/ft³, gm/cm³, etc. For an element of volume dV, the mass is $\rho\,dV$. The density of a fluid is related to the temperature and the pressure exerted on the element. The density of gases varies much more readily than that of liquids. In many cases, however, the effects of the variation of density are very small and the density of a fluid element can be assumed to remain constant as an approximation. Under this assumed condition, the flow is called an *incompressible flow*.

The forces acting on an infinitesimal fluid element may be classified into two groups: body forces and surface forces. A *body force* is one whose magnitude is proportional to the volume dV of the fluid element, e.g., gravitational and magnetic forces. The only body force to be considered in this book is the weight due to gravity. The weight per unit volume is called *unit weight* and, according to Eq. 1–2, can be written as ρg, where g is the magnitude of the gravitational acceleration. The value of g is approximately equal to 32.2 ft/sec² or 981 cm/sec² at sea level.

The *surface forces* are those forces acting on the surface bounding the element exerted by the adjacent fluid elements or solid boundaries. Consider a portion dA of the surface shown in Fig. 1–2. The force on dA does not in general lie along the normal of dA. This force may be resolved into two components, one normal and the other tangential to the area dA. The normal force per unit area is called the *normal stress*. This stress is compressive and is often called *pressure*. (A precise definition of pressure will be given in Section 11–7.) The tangential force per unit area is called the *shearing stress*.

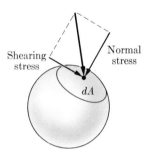

FIGURE 1–2

According to the definition of a fluid, a fluid element will continue to deform when there are shearing stresses acting on its surface. The relationship between the rate of deformation and the shearing stresses defines a mechanical property of the fluid. Most common fluids, when sheared as shown in Fig. 1–1, deform at a rate U/d proportional to the shearing stress F/A; that is $\mu U/d = F/A$, where the coefficient μ is independent of the speed U. Such fluids, including air and water, are called *Newtonian fluids*. The coefficient μ is called the *absolute viscosity* or *dynamic viscosity*. Under ordinary pressure, μ of a Newtonian fluid varies only with temperature. Fluids which do not behave in this manner are called *non-Newtonian fluids*. In any case, all real fluids offer some resistance to a finite rate of deformation. However, in many cases, the shearing stresses in most parts of the flow are unimportant and can be neglected in an approximate analysis. Such an idealized fluid is called a *frictionless fluid*.

All fluid flows are governed by the same principles. The behavior of a given flow must also depend on the conditions at the boundaries confining the flow. Two common boundary conditions are the following.

(a) At a solid boundary such as the wall of a pipe and the surface of an airfoil, all fluids immediately in contact with the solid have been observed to have no velocity relative to the solid except in the case of highly rarefied gases. Thus, in Fig. 1–1, the fluid particles in contact with the moving upper plate move with the plate at speed U, while those in contact with the fixed lower plate are stationary. (However, as will be seen later, we allow for mathematical reasons a relative tangential velocity in the case of the fictitious frictionless fluid.)

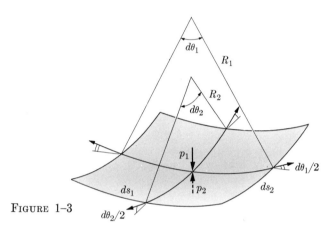

FIGURE 1–3

(b) The interface between two immiscible fluids has been observed to behave as if it were under tension (e.g., small volumes of mercury form globules). The magnitude of this tensile force per unit length of a line on the surface is called the *surface tension* σ of the interface. The value of σ varies with the fluids in contact and the temperature. As a result of surface tension, the pressure must be different on the two sides of a curved interface. Consider a small area of the interface with sides ds_1 and ds_2 as shown in Fig. 1–3. Since the mass of the interface is zero, the total force acting on it must be zero, in accordance with Eq. 1–2. Thus, neglecting the shearing stresses, if any,

$$p_2\, ds_1\, ds_2 - p_1\, ds_1\, ds_2 + 2\sigma\, ds_1 \sin \frac{d\theta_2}{2} + 2\sigma\, ds_2 \sin \frac{d\theta_1}{2} = 0.$$

With $d\theta_1 = ds_1/R_1$ and $d\theta_2 = ds_2/R_2$, $\sin(d\theta_1/2) = d\theta_1/2$, $\sin(d\theta_2/2) = d\theta_2/2$ for small angles, we have

$$p_1 - p_2 = \sigma \left(\frac{1}{R_1} + \frac{1}{R_2} \right). \tag{1–3}$$

However, except for interfaces with large curvature, this pressure difference is usually negligible.

Another property which has to be considered in a liquid is the *vapor pressure*. When the pressure at a point in a flow drops to the vapor pressure of the liquid,

TABLE 1–1

Property	English units	cgs units
Density ρ	slug/ft^3	gm/cm^3
Unit weight ρg	lb/ft^3	dynes/cm^3
Dynamic viscosity μ	slug/ft-sec	poise $=$ gm/cm-sec
	$=$ lb-sec/ft^2	$=$ dynes-sec/cm^2
Kinematic viscosity	ft^2/sec	stoke $=$ cm^2/sec
$\nu(=\mu/\rho)$		
Surface tension σ	lb/ft	dynes/cm

vapor bubbles will form. This phenomenon is called *cavitation*. Sometimes a vapor pocket may form and stay at the point of low pressure adjacent to a solid body; e.g., at the end of a fast-moving torpedo. This phenomenon is called *supercavitation*. In this case, the liquid does not flow along the whole solid boundary, but departs from it at the vapor pocket.

The fluid properties discussed above are involved in the study of fluid mechanics. Attempts have been made to relate these bulk properties to the molecular properties. Such attempts have met with considerable success for gases, and to a lesser extent for liquids. However, we shall not be concerned with the discussion of these relations, but shall take the phenomenological point of view.

For the mechanical properties of water, air, and mercury, see the Appendix. The units of these properties in the English and cgs systems are as shown in Table 1–1. The ratio μ/ρ will be encountered very often and is therefore given the name *kinematic viscosity* ν.

The unit of pressure in the English and cgs systems is lb/ft^2 and dynes/cm^2, respectively. Pressure may be expressed with reference to any arbitrary datum. The common ones are the complete vacuum and the local atmospheric pressure. When the former is used, the pressure is called *absolute pressure*; and when the latter is used, the pressure is called *gage pressure*. For example, a pressure of 100 lb/ft^2 higher than a local atmospheric pressure of 2100 lb/ft^2 may be said to be 100 lb/ft^2 gage or 2200 lb/ft^2 absolute. A pressure of 200 lb/ft^2 below the atmospheric pressure is called 200 lb/ft^2 vacuum or 1900 lb/ft^2 absolute. (There are many other units of pressure used in practice, such as psi or lb/in^2, and the standard atmosphere, which is 14.70 psi or 1.013×10^6 dynes/cm^2.)

PROBLEMS

1–1. Water weighs about 62.4 lb/ft^3. What is its density? What is its unit weight on the moon where the gravitational acceleration is 5.47 ft/sec^2?

1–2. A soap film of thickness t forms a bubble of outside diameter $2R$. The surface tension at the interfaces is σ. The pressure outside the bubble is atmospheric. What is the pressure inside the bubble?

1–4. Thermodynamic properties of fluids

All materials are to a certain extent compressible. The density of a fluid may therefore vary in a flow. The density is related to the pressure and the temperature. In the study of fluid systems where the effects of compressibility are important, thermodynamic properties of the fluid must be considered. The laws of thermodynamics and thermodynamic properties of fluids will be discussed in Chapter 14. For our present purpose, we introduce here only the equation of state and the isentropic relation for a perfect gas.

A *perfect gas* is one which obeys the following *equation of state:*

$$\rho = \frac{p}{RT},\qquad(1\text{–}4)$$

where R is a constant for a gas, and p and T are the absolute pressure and absolute temperature, respectively. (Absolute temperature may be expressed in degrees rankine or degrees kelvin: $°R = °F + 459.67$, and $°K = °C + 273.15$.) As density is lowered, all gases tend to behave as a perfect gas. Under ordinary conditions, air and many other gases are closely described by Eq. 1–4. In the technical literature, the equation $\rho g = p/(R'T)$ is sometimes used. For air, $R' = R/g = 53.35$ ft/°R or 2927 cm/°K.

In a flow, heat may be conducted from one fluid element to another. The rate of conduction is described by a coefficient of heat conduction. Heat may be generated by viscous stresses at the expense of the mechanical energy of the fluid. In many flow systems, however, heat conduction and viscous dissipation are negligible, and the flow is considered to be isentropic. Under this condition, a perfect gas with a constant specific heat C_v will obey the isentropic relation

$$\frac{p}{\rho^\gamma} = \text{constant for a fluid particle,}\qquad(1\text{–}5)$$

where $\gamma = 1 + (R/C_v)$ is a constant (see Section 14–7). For air under ordinary conditions, γ is approximately equal to 1.40. With Eqs. 1–4 and 1–5, we can determine two of the variables p, ρ, and T when the other one is known.

The compressibility of a fluid is often described with the *bulk modulus K,* which is defined as

$$K = \lim_{\Delta p \to 0} \frac{\Delta p}{\Delta \rho/\rho},\qquad(1\text{–}6)$$

where $\Delta\rho/\rho$ is the fractional increase of the density of a fluid element caused by an increase of pressure Δp. The more compressible the material, the lower its value of K. The value of K for a gas depends on the process during this change of ρ. In many cases, the process may be assumed to be isentropic. For a perfect gas in such a process, K can be computed from Eq. 1–5 as follows:

$$K = \rho \frac{dp}{d\rho} = \rho \frac{d}{d\rho}(C\rho^\gamma) = \gamma C\rho^\gamma = \gamma p.\qquad(1\text{–}7)$$

For a liquid, the value of K is practically constant (see Problem 1–5).

PROBLEMS

1–3. What is the relationship between ρ and T of a perfect gas during an isentropic process?

1–4. Find the bulk modulus in terms of p of a perfect gas in an isothermal process (with constant T).

1–5. The equation of state of water is approximately

$$p = A \left(\frac{\rho}{\rho_0}\right)^n - B,$$

where the constants A, B, and n are approximately 3001 atmospheres, 3000 atmospheres, and 7, respectively, and ρ_0 is the density at 0°C. Show that K for water is about 310,000 psi under ordinary pressure.

2 Dimensional Analysis

2-1. Dimensions

While the rest of this book deals mainly with the theoretical analysis of fluid flows, the powerful method of *dimensional analysis* is presented in this chapter. With this method, useful information is obtained by examining the nature of the parameters involved in the case under investigation. This method will be found useful in guiding experimental and theoretical studies of fluid mechanics and many other subjects (see for example Sections 13–2 and 13–5 and Example 11–6). The method will be illustrated first with examples where only mechanical quantities are involved, and compressibility is described with the bulk modulus K. Problems involving thermal quantities explicitly will be discussed in Section 2–7.

In Newton's equation of motion $\mathbf{F} = k\, d(m\mathbf{q})/dt$, four fundamentally different physical quantities are involved, namely, force, mass, length, and time. The nature of all physical quantities defined in mechanics can be expressed in terms of these four fundamental quantities; e.g., velocity is in length per unit time, and the moment of a force is the product of the force and a distance. This combination of the fundamental quantities can be represented by what is called a *dimensional formula*; e.g., the dimensional formula of velocity is L/T or $L^1 T^{-1}$, where L and T represent length and time, respectively. The exponents in the dimensional formula are called the *dimensions* of the quantity. If they are all zero, the quantity is said to be *dimensionless*. For example, being the ratio of the subtending arc to the radius, an angle has the dimensional formula L/L or L^0, and is therefore dimensionless.

The dimensional formula of the constant k in the equation of motion is that of $\mathbf{F}/[d(m\mathbf{q})/dt]$ and is FT^2/ML, where F and M stand for force and mass, respectively. In the study of mechanics, we use for convenience the equation $\mathbf{F} = d(m\mathbf{q})/dt$. In doing so, we must first use proper units so that $k = 1$ numerically. We must also regard F as dimensionally equivalent to ML/T^2 so that k is dimensionless and the equation is *dimensionally homogeneous*, i.e., all terms are similar in dimensions. The number of fundamental quantities in mechanics is thus reduced to three, either M, L, and T with $F = ML/T^2$, or F, L, and T with $M = FT^2/L$. The dimensional formula of moment of a force, for example, is FL in the F-L-T system, and $(ML/T^2)L$ or ML^2/T^2 in the M-L-T system.

8

2–1. Write out the dimensions of displacement, velocity, acceleration, angular displacement, angular velocity, angular acceleration, force, moment of force, work, energy, momentum, impulse, moment of inertia, and angular momentum in the F-L-T system or the M-L-T system.

2–2. Write out the dimensions of the following mechanical properties of fluids: ρ, μ, σ, and K.

2–2. Method of dimensional analysis

In dimensional analysis, we consider problems in which the nature of the quantities Q_1, Q_2, ... involved can be expressed by dimensional formulas in terms of some fundamental quantities, and these quantities Q_1, Q_2, ... are related by a dimensionally homogeneous equation:

$$F(Q_1, Q_2, \ldots) = 0. \tag{2–1}$$

In mechanics and many other subjects, we are dealing with such problems.

Since Eq. 2–1 is dimensionally homogeneous, the relationship can be expressed in terms of dimensionless quantities. For example, the displacement s in linear motion during interval t with constant acceleration a and initial speed u is given by

$$F(s, u, a, t) = s - ut - \tfrac{1}{2}at^2 = 0.$$

This equation is dimensionally homogeneous, and can be reduced to involve only dimensionless terms simply by dividing all the terms by one of them; e.g., dividing by ut,

$$f\left(\frac{s}{ut}, \frac{at}{u}\right) = \frac{s}{ut} - 1 - \frac{1}{2}\frac{at}{u} = 0.$$

Note that a relationship among four quantities has been expressed in terms of two dimensionless products, namely, $s/(ut)$ and at/u. In general, Eq. 2–1 can be expressed in terms of dimensionless products Π_1, Π_2, ... formed with Q_1, Q_2, ...; that is,

$$f(\Pi_1, \Pi_2, \ldots) = 0. \tag{2–2}$$

The number of dimensionless Π-terms involved in this equation is in general less than the number of quantities Q_1, Q_2, ... in Eq. 2–1. The usefulness of dimensional analysis comes from the fact that the number of parameters required to describe a relationship can be reduced. This is a great step forward in the simplification of a problem.

The question now is how many and what the dimensionless products are in a particular case. The first part of the question is answered by the *pi-theorem* which will be further discussed in Section 2–4. This theorem may be stated as follows.

When m quantities $Q_1, Q_2, \ldots, Q_m$, which can be expressed dimensionally in terms of n ($<m$) fundamental quantities, are related by a dimensionally homogeneous equation, the relationship among m the quantities can always be expressed in terms of exactly ($m - r$) dimensionless products, where r ($\leqq n$) is the rank of the dimensional matrix of the m quantities:

$$f(\Pi_1, \Pi_2, \ldots, \Pi_{m-r}) = 0. \tag{2-3}$$

The definition of the dimensional matrix and its rank will be given in Section 2–4. Actually, r is the number of independent fundamental quantities in a particular problem. For example, in problems of statics, there are only two independent fundamental quantities, namely, F and L. Thus $n = r = 2$. If one chooses to use the M-L-T system, then $n = 3$ while r remains 2. In mechanics, we have $r = n$ if the F-L-T system is used.

In Eq. 2–3, any one of the dimensionless products can be written as a function of the others; e.g.,

$$\Pi_1 = \phi(\Pi_2, \Pi_3, \ldots, \Pi_{m-r}). \tag{2-4}$$

In particular, when there is only one independent dimensionless product (i.e., when $m - r = 1$), this product Π_1 is not a function of any other quantity; i.e.,

$$\Pi_1 = \text{constant.} \tag{2-5}$$

The functions f in Eq. 2–3 and ϕ in Eq. 2–4, and the constant in Eq. 2–5 cannot be determined by dimensional consideration, but must be found by other means, i.e., by theoretical analysis or by experimental observations. By virtue of dimensional analysis, however, the amount of theoretical or experimental work can be greatly reduced. For example, in the case where $m - r = 1$, only one observation is necessary to determine the constant in Eq. 2–5.

The determination of a complete set of dimensionless products in a given case will be discussed in the next section. To emphasize the general validity of dimensional analysis in many subjects, problems and examples are taken from solid mechanics in this section and the next. Applications in fluid mechanics will be presented in Section 2–5.

Example 2–1. Assume that the maximum deflection d of a beam simply supported at its ends depends on its length l, the concentrated load P at midspan, and the product (EI) (where E is the Young's modulus of the material, and I is the second moment of the uniform cross-sectional area about the neutral axis). Find by the pi-theorem the number of dimensionless products involved in the description of d. (For simplicity, it is assumed that the yield stress of the material is not reached.)

The problem is to reduce the number of parameters in the relationship $d = F(l, P, EI)$. With the F-L system for this static problem, we have $r = n = 2$. Since $m = 4$ in this case, there are $m - r = 2$ independent dimensionless products involved.

While a systematic method of determining these dimensionless products will be presented in the next section, they can be found by inspection in this simple case. The dimensional formulas for the four quantities are $[d] = L^1$, $[l] = L^1$, $[P] = F^1$, and $[EI] = F^1L^2$. By inspection, we get, among other possibilities, $\Pi_1 = d/l$ and $\Pi_2 = Pl^2/EI$. According to Eq. 2–4,

$$\frac{d}{l} = \phi\left(\frac{Pl^2}{EI}\right).$$

This is the desired result of dimensional analysis.

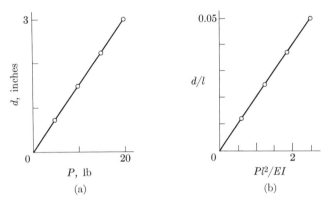

FIGURE 2–1

This functional relationship can be determined by one series of experiments. For example, a beam with $l = 5$ ft and $EI = 30{,}000$ lb-in^2 is tested under various loads. The observed data show that d is proportional to P. By virtue of dimensional analysis, we can conclude that d/l is proportional to $Pl^2/(EI)$. When the data are plotted with d/l versus $Pl^2/(EI)$ as shown in Fig. 2-1(b), the graph can be used for all similar beams if the yield stress is not reached as assumed.

PROBLEMS

2–3. The displacement s in the interval t of a particle in linear motion depends on the initial speed u and the constant acceleration a. Using the pi-theorem, determine the number of dimensionless products involved in the description of s.

2–4. A circular surface of diameter D is submerged in a liquid of unit weight (ρg). The distance from the center of the surface to the free surface is H. The liquid exerts a force P on the surface. (a) How many dimensionless products are involved in the description of P? (b) Find by inspection the necessary dimensionless products. (c) It is observed that P is proportional to H. What must be the relationship among these dimensionless products?

2–3. Determination of a complete set of dimensionless products

The pi-theorem is merely an algebraic theorem and does not deal with the physics of the problem. In order to apply this theorem, one must first know or guess correctly the significant quantities involved. When the m quantities involved have been decided upon, the pi-theorem states that they can form exactly $(m - r)$ independent dimensionless products. Any set of $(m - r)$ independent dimensionless products is called a *complete set*.

It should be realized that with the m quantities, an infinite number of dimensionless products can be formed. For example, in the case of linear displacement s, interval t, initial speed u, and constant acceleration a, it can be shown that $(m - r) = 2$. Examples of complete sets are $(s/ut,\ at/u)$ and $(as/u^2,\ at/u)$. Any one of these sets can be used to express the well-known relationship $s = ut + \frac{1}{2}at^2$; for example,

$$f_1\left(\frac{s}{ut},\ \frac{at}{u}\right) = \frac{s}{ut} - 1 - \frac{1}{2}\frac{at}{u} = 0$$

or

$$f_2\left(\frac{as}{u^2},\ \frac{at}{u}\right) = \frac{as}{u^2} - \frac{at}{u} - \frac{1}{2}\left(\frac{at}{u}\right)^2 = 0.$$

Any complete set, obtained in any manner, can be used in Eq. 2–3. The following is a recommended method for systematically obtaining a complete set in a more complicated problem.

From the m quantities, choose r of them which are not dimensionless and do not have identical or multiple dimensions but, together, contain all the n fundamental quantities. Combine these r chosen quantities with each of the other $(m - r)$ quantities to form a dimensionless product. This can often be done simply by inspection or by the systematic method used in Example 2–2. In this manner, $(m - r)$ dimensionless products are obtained which are independent, since each of them contains at least one different quantity. They therefore form a complete set.

The choice of a complete set is arbitrary and depends on the purpose of the analysis. Often it is desirable to change a dimensionless product in a complete set. This can be done by replacing this dimensionless product by a combination of itself and some of the others in the complete set. For example, Π_2 may be replaced by a combination of Π_2 itself and Π_3, such as $\Pi_{2'} = \Pi_2\Pi_3$.

Example 2–2. A circular bar is under a twisting torque. The angle of twist per unit length (θ/l) is believed to depend on the modulus of rigidity G of the material, the diameter D of the bar, and the twisting torque τ. (a) Find a general expression for θ/l by dimensional analysis. (b) An experiment is performed with a bar under different torques. It is found that θ of this bar is proportional to τ. Simplify the expression obtained in (a) accordingly.

With the F-L system for this problem of statics, the dimensional formulas of these quantities are $[\theta/l] = L^{-1}$, $[G] = FL^{-2}$, $[D] = L$, and $[\tau] = FL$. Here $m = 4$, $n = 2$, and $r = n = 2$. To find a complete set, choose 2 ($=r$) of the quantities, say G and D. These two quantities are different in their dimensions, and together they contain all the fundamental quantities, namely, F and L.

The first dimensionless product is obtained by combining θ/l with the chosen G and D. Let

$$\Pi_1 = \left(\frac{\theta}{l}\right)^{k_1} G^{k_2} D^{k_3},$$

where the exponents k_1, k_2, and k_3 are to be determined. Since, dimensionally, $[\Pi_1] = F^0 L^0$, we have the dimensional relationship

$$F^0 L^0 = (L^{-1})^{k_1}(F^1 L^{-2})^{k_2}(L^1)^{k_3} = F^{k_2} L^{-k_1 - 2k_2 + k_3}.$$

Equating the powers of F and L separately, we have

$$k_2 = 0 \quad \text{and} \quad -k_1 - 2k_2 + k_3 = 0.$$

Here we have three unknowns to satisfy two equations. With $k_2 = 0$, one of the other exponents can be assigned an arbitrary value, say, $k_3 = 1$. Then $k_1 = 1$, and

$$\Pi_1 = \left(\frac{\theta}{l}\right)^1 G^0 D^1 = \frac{\theta D}{l}.$$

The second dimensionless product is to be formed by τ, G, and D. Let

$$\Pi_2 = \tau^{k_4} G^{k_5} D^{k_6}.$$

Dimensionally,

$$F^0 L^0 = (F^1 L^1)^{k_4}(F^1 L^{-2})^{k_5}(L^1)^{k_6},$$

from which we have

$$k_4 + k_5 = 0 \quad \text{and} \quad k_4 - 2k_5 + k_6 = 0.$$

One of the exponents can be assigned an arbitrary value, say, $k_4 = 1$. Then $k_5 = -1$, $k_6 = -3$, and

$$\Pi_2 = \tau^1 G^{-1} D^{-3} = \frac{\tau}{GD^3}.$$

According to Eq. 2–4, we have

$$\frac{\theta D}{l} = \phi\left(\frac{\tau}{GD^3}\right). \tag{2–6}$$

This is a general expression for θ by dimensional analysis. Since it is observed that θ is proportional to τ, the function ϕ must be

$$\frac{\theta D}{l} = C \frac{\tau}{GD^3},$$

where C is a constant which can be computed from the experimental data. Therefore

$$\theta = C \, \frac{\tau l}{GD^4}.$$

If for one reason or another, it is desired to change Π_1, this can be done by replacing it with any combination of Π_1 itself and Π_2; for example, $\Pi_{1'} = \Pi_1/\Pi_2 = \theta GD^4/(\tau l)$. Then Eq. 2–4 gives

$$\frac{\theta GD^4}{\tau l} = \phi_1 \left(\frac{\tau}{GD^3} \right).$$

This replaces Eq. 2–6 as the result of dimensional analysis. Since it is observed that θ is proportional to τ, the function ϕ_1 must be a constant, so that

$$\theta = C \, \frac{\tau l}{GD^4}.$$

PROBLEMS

2–5. Assuming that the force P acting on a mass moving in a circle depends on its mass m, its speed q, and the radius R of the circle, show by dimensional analysis that P is proportional to mq^2/R.

2–6. A simple pendulum consists of a small mass m attached to a weightless string of length l. Show by dimensional analysis that if the pendulum is slightly disturbed and oscillates under gravitational acceleration g, the period τ of oscillations is proportional to $\sqrt{l/g}$ and is independent of m.

2–7. A small particle of mass m starts from the top of a stationary smooth sphere of radius R. After traveling under gravitational acceleration g through an arc of the sphere, the particle leaves the sphere. Show by dimensional analysis that the angle α subtended at the center by this arc is the same for any combination of m, R, and g.

2–8. Assuming that the load P required to buckle a column depends on its length l and its minimum (EI), find an expression for P by dimensional analysis.

2–9. When a satellite of mass m_2 is at the perigee of its orbit, its distance from the earth is a and its speed is U, as shown in Fig. 2–2. At any position, the only force acting on m_2 is due to gravitation, and is equal to $(km_1)m_2$ divided by the square of the distance between the two masses. Show that the distance b of the apogee is given by

$$\frac{b}{a} = f\left(\frac{aU^2}{km_1}\right).$$

[*Hint:* Consider km_1 as one of the quantities involved.]

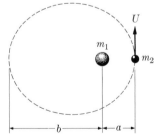

FIGURE 2–2

2–10. Assume that the frequency f (cycles per unit time) of a plucked string depends on its length l, its tension force F, its mass per unit length m, and (AE) (where A is its cross-sectional area). (a) Find an expression for f. (b) If it is observed experimentally that (AE) has no effect on f, simplify your answer in (a).

2–4. The pi-theorem

The basis of the pi-theorem is to be presented in this section. Let m quantities $Q_1, Q_2, \ldots, Q_m$ be involved in a phenomenon, and their dimensional formulas be expressed in terms of $n(<m)$ fundamental quantities $L_1, L_2, \ldots, L_n$ (e.g., F, L, T in mechanics):

$$[Q_1] = L_1^{b_{11}} L_2^{b_{21}} \cdots L_n^{b_{n1}},$$

$$[Q_2] = L_1^{b_{12}} L_2^{b_{22}} \cdots L_n^{b_{n2}},$$

$$\vdots$$

$$[Q_m] = L_1^{b_{1m}} L_2^{b_{2m}} \cdots L_n^{b_{nm}},$$

where the exponents are the known dimensions of the m quantities (e.g., for speed, $[u] = L^1 T^{-1}$). Let Π be a dimensionless product formed by these quantities:

$$\Pi = Q_1^{k_1} Q_2^{k_2} \cdots Q_m^{k_m}, \tag{2-7}$$

where the exponents k_1, etc., are to be determined. From this equation, we have the dimensional relationship

$$L_1^0 L_2^0 \cdots L_n^0 = (L_1^{b_{11}} L_2^{b_{21}} \cdots L_n^{b_{n1}})^{k_1} (L_1^{b_{12}} L_2^{b_{22}} \cdots L_n^{b_{n2}})^{k_2} \cdots (L_1^{b_{1m}} L_2^{b_{2m}} \cdots L_n^{b_{nm}})^{k_m}.$$

Equating the powers of $L_1, L_2, \ldots,$ and L_n separately, we obtain n linear algebraic equations:

$$\text{for} \quad L_1: \quad b_{11}k_1 + b_{12}k_2 + \cdots + b_{1m}k_m = 0,$$

$$\text{for} \quad L_2: \quad b_{21}k_1 + b_{22}k_2 + \cdots + b_{2m}k_m = 0, \tag{2-8}$$

$$\vdots$$

$$\text{for} \quad L_n: \quad b_{n1}k_1 + b_{n2}k_2 + \cdots + b_{nm}k_m = 0.$$

The m exponents $k_1, k_2, \ldots, k_m$ in Eq. 2–7 must satisfy these n equations. For $m > n$, there are more unknowns than equations. Arbitrary values can be assigned for $m - n$ of these exponents, and the other n exponents can then be determined from these n equations. Thus, there is an infinite number of solutions of $(k_1, k_2, \ldots, k_m)$ for forming dimensionless products. In other words, the m quantities can form an infinite number of dimensionless products.

For example, take the displacement s in linear motion during interval t with initial speed u and constant acceleration a. All these quantities can be expressed in terms of L and T. Thus $m = 4$ and $n = 2$. Let a dimensionless product be

$$\Pi = s^{k_1} t^{k_2} u^{k_3} a^{k_4}.$$

Dimensionally,

$$L^0 T^0 = (L^1)^{k_1} (T^1)^{k_2} (L^1 T^{-1})^{k_3} (L^1 T^{-2})^{k_4}.$$

Equating powers of L and T separately, we have, corresponding to Eqs. 2–8,

$$\text{for} \quad L: \quad k_1 \quad\quad + k_3 + k_4 = 0,$$

$$\text{for} \quad T: \quad\quad k_2 - k_3 - 2k_4 = 0.$$

To satisfy these n, or two, equations, arbitrary values can be assigned to $m - n$, or two, of the unknowns. For example, let $k_3 = -1$ and $k_4 = 0$. Then $k_1 = 1$, $k_2 = -1$, and $\Pi = s^1 t^{-1} u^{-1} a^0 = s/(ut)$. This and two other sets of solutions are shown below:

k_1	k_2	k_3	k_4	Π
1	−1	−1	0	$s/(ut)$
0	−1	1	−1	$u/(at)$
−1	0	2	−1	$u^2/(as)$

In a similar manner, an infinite number of Π's can be obtained. However, these dimensionless products are not independent; for example, $u^2/as = (u/at)/(s/ut)$. The last set of (k_1, k_2, k_3, k_4) is equal to the second set minus the first set.

In Eq. 2–2, dimensionless products which can be expressed in terms of the others in the equation need not be included. The important question is therefore how many *independent* dimensionless products can be formed by the m quantities; i.e., how many linearly independent sets of $(k_1, k_2, \ldots, k_m)$ can be obtained from Eq. 2–8. By linearly independent solutions, we mean sets of solutions which cannot be obtained by a linear combination of some of the other sets. In the example above, the three sets of (k_1, k_2, k_3, k_4) are not linear independent, since the second set is equal to the sum of the other two. From the theory of linear algebra, Eq. 2–8 has, other than the trivial solution $k_1 = k_2 = \cdots = k_m = 0$, $m - r$ linearly independent solutions, where r is the rank of the matrix of the coefficients of the equations

$$\begin{bmatrix} b_{11} & b_{12} & \cdots & b_{1m} \\ b_{21} & b_{22} & \cdots & b_{2m} \\ \vdots & & & \\ b_{n1} & b_{n2} & \cdots & b_{nm} \end{bmatrix}.$$

It can be seen readily that this matrix contains the dimensions of the m quantities $Q_1, Q_2, \ldots, Q_m$ involved and is therefore called the *dimensional matrix*.

The matrix has m columns and n rows:

Fundamental quantities	Dimensions of			
	Q_1	Q_2	$\cdots$	Q_m
L_1	b_{11}	b_{12}	$\cdots$	b_{1m}
L_2	b_{21}	b_{22}	$\cdots$	b_{2m}
$\vdots$	$\vdots$			
L_n	b_{n1}	b_{n2}	$\cdots$	b_{nm}

Hence we have the pi-theorem as stated in connection with Eq. 2–3.

The *rank* r of a matrix is equal to the highest order of a nonzero determinant contained by the matrix. The rank is also equal to the number of rows $n(<m)$ unless one or more rows of the matrix can be obtained by a linear combination of some of the other rows. In determining the rank of a matrix, it is well to remember that the rank is not altered if a row which can be obtained by a linear combination of the other rows is removed.

Example 2–3. Repeat Example 2–1 by determining the rank of the dimensional matrix of the quantities involved.

With the F-L-T system, the dimensional matrix of the quantities d, l, P, and EI is

Fundamental quantities	Dimensions of			
	d	l	P	EI
F	0	0	1	1
L	1	1	0	2

The rank of this matrix is 2, since there is a nonzero second-order determinant contained in this matrix; e.g.,

$$\begin{vmatrix} 0 & 1 \\ 1 & 0 \end{vmatrix} = -1.$$

In this case, we have $r = n = 2$. The number $m - r$ of independent dimensionless products that can be formed by the four quantities is therefore 2.

If for one reason or another, the M-L-T system is used, we have $m = 4$ and $n = 3$, and the dimensional matrix

Fundamental quantities	Dimensions of			
	d	l	P	EI
M	0	0	1	1
L	1	1	1	3
T	0	0	-2	-2

It can be verified that the third-order determinants in this matrix are all equal to zero; e.g.,

$$\begin{vmatrix} 0 & 0 & 1 \\ 1 & 1 & 1 \\ 0 & 0 & -2 \end{vmatrix} = 0.$$

However, there are nonzero second-order determinants; e.g.,

$$\begin{vmatrix} 0 & 1 \\ 1 & 1 \end{vmatrix} = -1.$$

The rank r of this matrix is therefore 2. (This can also be seen from the fact that the rows of the matrix are not linearly independent. The last row is -2 times the first row. One of these rows can be removed without altering the rank of the matrix. The resultant matrix has two rows and is of rank 2.) In this case, $r \neq n$ but $m - r$ remains 2.

PROBLEMS

2–11. Repeat Problem 2–4(a) by determining the rank of the dimensional matrix of the quantities involved. First use the F-L-T system, and then repeat by using the M-L-T system.

2–12. If P in Problem 2–4 is believed to depend on D, H, the density ρ of the liquid, and the gravitational acceleration g, thus making $m = 5$, what is the number of independent dimensionless products that can be formed?

2–5. Dimensional analysis in fluid mechanics

In problems of fluid mechanics, the quantities involved may consist of the following:

(a) The dependent variable (e.g., the drag resistance of a body, the speed of flow, or the pressure at a point).

(b) The geometrical description of the boundary of the flow system, in terms of linear dimensions L, a, b, etc.

(c) The mechanical properties of the fluid, such as density ρ, unit weight ρg, viscosity μ, surface tension σ, and the bulk modulus K.

(d) Kinematic description of the flow (e.g., the speed U of an approaching stream).

In problems of fluid dynamics, we have usually $r = 3$. Usually L, ρ, and U (one from each of the last three groups) are chosen to form dimensionless products with the other quantities involved. The dimensionless product obtained with the dependent variable is sometimes called a coefficient. With the other linear dimensions of the boundary, the dimensionless products are a/L,

b/L, etc., which describe the geometrical shape of the system. Combining ρ, L, and U with the mechanical properties of the fluid, we have the following dimensionless products. Because of their importance, they are given special names:

$$\text{Froude number} \quad F = \frac{U}{\sqrt{gL}},$$

$$\text{Reynolds number} \quad R = \frac{UL\rho}{\mu},$$

$$\text{Weber number} \quad W = U\sqrt{\frac{L\rho}{\sigma}},$$

$$\text{Mach number} \quad M = \frac{U}{\sqrt{K/\rho}}.$$

(2–9)

They represent the effects of gravity, viscosity, surface tension, and compressibility, respectively.

In a particular flow, however, not all of these mechanical properties are important. For example, the Weber number is not involved in a flow system without an interface of fluids. On the other hand, there may be other important physical properties involved. Of particular interest is the fact that vapor cavities will form in a flowing liquid where the pressure drops to its vapor pressure p_v (for water, p_v is 0.339 psi abs. at 68°F). These vapor bubbles travel with the liquid, and when they collapse in regions of higher pressure, very high pressure is generated, which has been observed to be very damaging to nearby wall surfaces. In the study of cavitation, p_v of the liquid is therefore an important property to be considered. For a flow with a representative pressure P (e.g., the pressure in an approaching stream), it is the margin $P - p_v$ that is significant in determining the occurrence of cavitation. For this quantity, the dimensionless product usually used is the *cavitation number*

$$\text{Cavitation number } C = \frac{P - p_v}{\frac{1}{2}\rho U^2}.$$

(2–9a)

Through the use of dimensional analysis, we can therefore conclude, in accordance with Eq. 2–4, that any dimensionless product involving a flow variable may be expressed in terms of the geometrical shape of the boundary and the dimensionless numbers representing the significant mchanical properties of the fluid:

$$\text{Dimensionless product} = \phi\left(\frac{a}{L}, \frac{b}{L}, \cdots, F, R, W, M\right).$$

(2–10)

The influence of the four dimensionless numbers on a fluid flow can be seen as follows. Other things being equal, the smaller the value of μ, the higher the value of R. Since the principle of dimensional analysis indicates that it is not the value of μ alone but the value of R which determines the viscous effects, one can conclude that the influence of viscosity is slight when the Reynolds number becomes large. By similar reasoning, the effects of gravity and of surface tension decrease with increase of F and W, respectively. Since the more compressible the material, the lower its value of K and the higher the Mach number, a higher value of M indicates a greater influence of compressibility on the flow. The following examples will demonstrate the influence of these numbers in several flow systems.

Example 2–4. A liquid flows steadily through a long straight horizontal pipe. Find an expression for the drop of pressure per unit distance along the pipe.

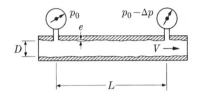

FIGURE 2–3

In Fig. 2–3, $\Delta p/L$ is the pressure drop per unit distance. It is expected that this quantity depends on the diameter D of the pipe, the size e of the wall roughness, ρ and μ of the liquid, and the mean speed V of the flow (V will be defined precisely in Section 5–7). Since there is no free surface, there is no effect of surface tension. The effect of compressibility of the liquid is assumed to be unimportant. In other words,

$$\frac{\Delta p}{L} = F(D, e, \rho, \mu, V).$$

It can be shown that $r = 3$, and these six quantities can form three independent dimensionless products. Combining D, ρ, and V with each of the other three quantites, we obtain the following complete set:

$$\Pi_1 = \frac{\Delta p D}{L\rho V^2}, \qquad \Pi_2 = \frac{e}{D}, \qquad \text{and} \qquad \Pi_3 = \frac{VD\rho}{\mu}.$$

It is the convention to call $2\Pi_1 = f$ the *resistance coefficient* of pipe flow. Thus, according to Eq. 2–4,

$$f = \frac{2\,\Delta p D}{L\rho V^2} = \phi\left(\frac{e}{D}, \frac{VD\rho}{\mu}\right). \tag{2–11}$$

This is the desired expression for Δp obtained by dimensional analysis.

The function ϕ can be determined experimentally by observing the simultaneous values of Δp and V in a pipe. In each test, $2\,\Delta p D/(L\rho V^2)$ and $VD\rho/\mu$ can be computed. From tests with one pipe and one fluid at various speeds, one curve in Fig. 2–4 can be obtained. With pipes of different values of *relative*

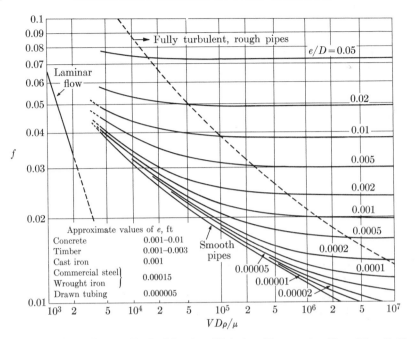

FIG. 2–4. Adapted from L. F. Moody, "Friction Factors for Pipe Flow," *Trans. ASME*, **66,** 8, November 1944.

roughness e/D, Fig. 2–4 can be completed for wall roughness of a particular shape. By virtue of dimensional analysis, this graph is valid for pipes and fluids other than those used in the tests.

Note that there is a sudden change of behavior when $VD\rho/\mu$ is about 2000. Below this value, the flow has been observed to be laminar, i.e., layered, as shown by the path of dye in Fig. 2–5(a). Above this value, laminar flow has seldom been observed, and the flow is more likely to be turbulent, i.e., agitated, as shown in Fig. 2–5(b). The value 2000 has been called the *critical Reynolds number* of pipe flow.

(a)   (b)

FIGURE 2–5

Note also that at low values of $VD\rho/\mu$, the effect of wall roughness is unimportant. At high values of $VD\rho/\mu$, on the other hand, f is independent of the Reynolds number and Δp is independent of μ. The influence of viscosity can therefore be seen to be unimportant when the Reynolds number is large.

Fxample 2-5. The *Venturi meter* shown in Fig. 2-6 is used for the measurement of the *discharge* (volume per unit time) through a pipe. Find an expression for the discharge Q in terms of the pressure drop Δp between the two gages and the diameters D_1 and D_2 of the pipe. Consider only cases where the effects of compressibility are unimportant.

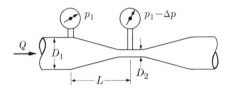

FIGURE 2-6

The pressure drop Δp may be assumed to vary with Q, the ρ and μ of the fluid, size e of the wall roughness, and D_1, D_2, and L of the meter. It is assumed that the pipe upstream is long (more than ten diameters, say) so that the geometrical details upstream have little effect on the flow in the meter. Here $m = 8$ and $r = 3$. Combining ρ, Q, and D_1 with each of the others, we obtain the following complete set of five dimensionless products:

$$\Pi_1 = \frac{Q}{D_1^2}\sqrt{\frac{\rho}{\Delta p}}, \qquad \Pi_2 = \frac{D_2}{D_1}, \qquad \Pi_3 = \frac{L}{D_1}, \qquad \Pi_4 = \frac{e}{D_1}, \qquad \Pi_5 = \frac{Q\rho}{D_1\mu}.$$

Therefore

$$\frac{Q}{D_1^2}\sqrt{\frac{\rho}{\Delta p}} = \phi_1\left(\frac{D_2}{D_1}, \frac{L}{D_1}, \frac{e}{D_1}, \frac{Q\rho}{D_1\mu}\right).$$

This is the desired expression obtained by dimensional analysis.

In engineering practice, the discharge is usually expressed as (see Example 5-5)

$$Q = C_v \frac{A_2}{\sqrt{1 - (D_2/D_1)^4}}\sqrt{\frac{2\,\Delta p}{\rho}}, \qquad (2\text{-}12)$$

where C_v is called the *velocity coefficient*, and $A_2 = \pi D_2^2/4$. To conform with this convention, we replace Π_1 with

$$\Pi_1' = C_v = \frac{4}{\pi\sqrt{2}}\,\Pi_1\sqrt{\frac{1}{\Pi_2^4} - 1}\,.$$

Π_5 can also be replaced by a Reynolds number Π_5' in a more familiar form: with the mean speed $V_1 = Q/A_1 = 4Q/(\pi D_1^2)$,

$$\Pi_5' = \frac{4}{\pi}\,\Pi_5 = \frac{V_1 D_1\rho}{\mu}\,.$$

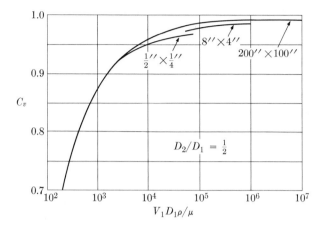

FIG. 2–7. Adapted from *Fluid Meters, ASME*, 1937.

Then, we have for Eq. 2–12,

$$C_v = \phi\left(\frac{D_2}{D_1}, \frac{L}{D_1}, \frac{e}{D_1}, \frac{V_1 D_1 \rho}{\mu}\right).$$

This functional relationship can be determined experimentally. For standard Venturi meters (see *Fluid Meters, ASME*, 1937) with $D_2/D_1 = \frac{1}{2}$, the observed data are shown in Fig. 2–7. The small variation of C_v with the size of the meter is apparently due to the difference in the relative roughness e/D_1. When the Reynolds number is large, the influence of viscosity is unimportant, since C_v is constant and Q in Eq. 2–12 is independent of μ.

Example 2–6. A rectangular plate of sides L and a moves in a direction normal to its plane with a constant speed U in a large body of fluid. Find the factors influencing its drag resistance. Also study the drag of other moving bodies.

Since it is assumed that the plate is far away from any free surface, there is no effect due to surface tension. The effect, if any, of gravity (i.e., weight) is expected to be small. We therefore assume that the drag D depends on L, a, ρ, μ, K, and U. Choosing L, ρ, and U to combine with each of the others, we obtain the following complete set of four independent dimensionless products:

$$\Pi_1 = \frac{D}{\rho U^2 L^2}, \quad \Pi_2 = \frac{a}{L}, \quad \Pi_3 = \frac{U L \rho}{\mu}, \quad \Pi_4 = \frac{U}{\sqrt{K/\rho}}.$$

Since the *drag coefficient* C_D is conventionally defined by

$$D = C_D \tfrac{1}{2}\rho U^2 A, \tag{2–13}$$

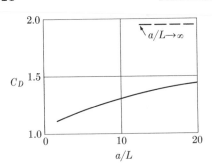

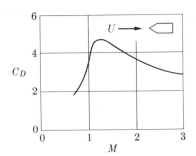

FIGURE 2–8

FIG. 2–9. Adapted from L. Prandtl, *Abriss der Strömungslehre*, Friedrich Vieweg und Sohn, Brunswick, 1935.

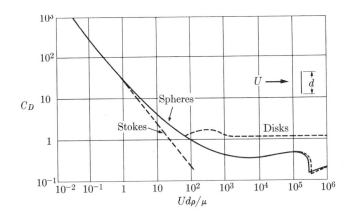

FIG. 2–10. Adapted from F. Eisner, "Das Widerstands-problem," *Proc. Third Int. Congress App. Mech.*, 1931.

where A is a projected area of the body, we replace Π_1 with $\Pi_1' = 2\Pi_1/\Pi_2 = C_D$. Thus

$$C_D = \phi\left(\frac{a}{L}, \frac{UL\rho}{\mu}, \frac{U}{\sqrt{K/\rho}}\right). \qquad (2\text{–}14)$$

This functional relationship can be determined experimentally. It is observed with many plates at different speeds in various fluids that as long as a/L and $UL\rho/\mu$ are the same, the same C_D is obtained irrespective of the value of the Mach number if this number is lower than about $\frac{1}{3}$. This means that the effect of compressibility is felt only at higher values of the Mach number. It is also observed at low Mach number that C_D is constant for a plate as long as $UL\rho/\mu$ is greater than about 10^4, indicating that at a high Reynolds number, the effect of viscosity is secondary. In the range with the Mach number less than $\frac{1}{3}$ and the Reynolds number larger than 10^4, the experimental data indicate that C_D depends on the aspect ratio a/L only, as shown in Fig. 2–8.

Similar analyses can be performed to determine the drag coefficient of bodies of other shapes. In Fig. 2–10, where the observed drag coefficients of spheres and circular disks at low Mach number are shown, the influence of the Reynolds number can be seen. Figure 2–9 shows the influence of compressibility on the drag coefficient of a body of revolution at high Mach number (the Reynolds number being necessarily high in these cases).

Example 2–7. If the effects of viscosity are unimportant, it is believed that sinusoidal waves of very small amplitude on the surface of a liquid of great depth can propagate with a constant speed. Study the wave speed by dimensional analysis under the following three assumed conditions: (a) surface tension is predominant, (b) gravity is predominant, and (c) both surface tension and gravity are effective.

(a) The linear dimensions involved are the wavelength λ and the amplitude a. Surface tension being predominant, the wave speed c is assumed to depend on ρ, σ, λ, and a. The density of the gas above is neglected. For these five quantities, r can be shown to be 3. There are therefore two independent dimensionless products, say

$$\Pi_1 = c\sqrt{\rho\lambda/\sigma} \qquad \text{and} \qquad \Pi_2 = a/\lambda.$$

Note that Π_1 is a Weber number. Thus

$$W = c\sqrt{\frac{\rho\lambda}{\sigma}} = f_1\left(\frac{a}{\lambda}\right).$$

As a/λ approaches zero, f_1 takes a certain value. Then

$$W = c\sqrt{\frac{\rho\lambda}{\sigma}} = \text{constant.} \qquad (2\text{–}15)$$

In this case, c varies inversely with λ.

(b) Gravity being predominant, c is assumed to depend on ρ, g, λ, and a. Again there are two independent dimensionless products, say, the Froude number $c/\sqrt{g\lambda}$ and a/λ. As a/λ approaches zero,

$$F = \frac{c}{\sqrt{g\lambda}} = \text{constant.} \qquad (2\text{–}16)$$

In this case, c varies directly with λ.

(c) When both surface tension and gravity are effective, assume that c depends on ρ, σ, g, λ, and a. With $m = 6$ and $r = 3$, there are three independent dimensionless products. To study the dependence of c on λ, choose ρ, σ, and g to combine with c and λ separately to obtain

$$\Pi_1 = c\sqrt[4]{\rho/g\sigma} \qquad \text{and} \qquad \Pi_2 = \lambda\sqrt{\rho g/\sigma}.$$

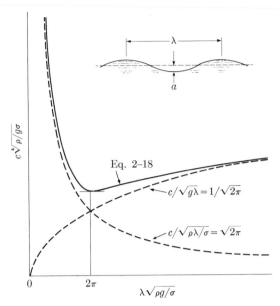

FIGURE 2–11

Use also $\Pi_3 = a/\lambda$. Thus, as a/λ approaches zero,

$$c \sqrt[4]{\frac{\rho}{g\sigma}} = \phi\left(\lambda \sqrt{\frac{\rho g}{\sigma}}\right). \tag{2–17}$$

This functional relationship may be determined by experimental or theoretical study.

It has been found theoretically that

$$\left(c \sqrt[4]{\frac{\rho}{g\sigma}}\right)^2 = \frac{1}{2\pi}\left(\lambda \sqrt{\frac{\rho g}{\sigma}}\right) + 2\pi\left(\lambda \sqrt{\frac{\rho g}{\sigma}}\right)^{-1}, \tag{2–18}$$

which can be rewritten as

$$c = \sqrt{\frac{g\lambda}{2\pi} + 2\pi \frac{\sigma}{\rho\lambda}}.$$

According to this equation, the constant in Eq. 2–15 can be seen to be $\sqrt{2\pi}$ when g can be ignored, and the constant in Eq. 2–16 is $1/\sqrt{2\pi}$ when σ can be neglected.

Note that $\lambda\sqrt{\rho g/\sigma} = W/F$. When Eq. 2–18 is plotted as in Fig. 2–11, it can be seen that for small W ($<2\pi F$), surface tension is predominant, and Eq. 2–15 gives a good approximation for c. For large $W(>2\pi F)$, the effect of surface tension is unimportant, and Eq. 2–16 is a good approximate solution of the wave speed.

PROBLEMS

2–13. A fluid flows through a horizontal tube of cross-sectional area A filled with fine sand. It is believed that the drop of pressure per unit length $(\Delta p/l)$ depends on the discharge per unit cross-sectional area (Q/A), ρ, and μ of the fluid, the sand diameter D, and the (dimensionless) void ratio R which depends on the manner of packing of the sand. (a) Find an expression for $\Delta p/l$. (b) When the flow is extremely slow, it is expected that inertia is relatively unimportant and $\Delta p/l$ should be independent of ρ. Show that

$$\frac{\Delta p}{l} = \frac{Q\mu}{AD^2} f(R),$$

which is *Darcy's law* for slow flow through porous media.

2–14. An infinite plate oscillates in its own plane with maximum speed U and frequency f (dimensionally, T^{-1}) in an infinite body of fluid of density ρ and viscosity μ. (a) Find an expression for the maximum speed u created in the fluid at a distance y from the plate. (b) It is expected that u is proportional to U. Show that

$$u/U = \phi(y\sqrt{\rho f/\mu}).$$

2–15. An infinite plate, originally at rest in a body of viscous fluid, is moved suddenly at constant speed U in its own plane. (a) Find an expression for the speed u created in time t at a distance y from the plate. (b) It is expected that u is proportional to U. Show that

$$u/U = \phi(y\sqrt{\rho/\mu t}).$$

2–16. The discharge of a liquid in a pipe can be measured with an *orifice meter* as shown in Fig. 2–12. (a) The discharge Q may be assumed to depend on $\Delta p(=p_1 - p_2)$, ρ, μ, D, and D_0. Find an expression for Δp. (b) In practice, Q is usually expressed as $Q = CA_0\sqrt{2\,\Delta p/\rho}$, where A_0 is the area of the orifice opening. Do you expect that the coefficient C will become constant as the discharge is increased? Why?

2–17. A liquid of density ρ and viscosity μ flows down an *open channel* of width B as shown in Fig. 2–13. The mean speed V is believed to depend on, among other things, the depth of flow Y, the component of gravitational acceleration $(g \sin \theta)$ along the flow, and the size e of the roughness of the walls. Find the dimensionless parameters that may affect the coefficient k in the formula $V = k\sqrt{Yg \sin \theta}$. When will you expect that the effect of μ is unimportant?

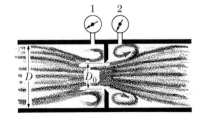

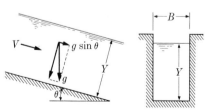

FIGURE 2–12   FIGURE 2–13

2-18. A *weir* is a notch in a barrier across an open channel for measuring the discharge by observing the height H (see Fig. 2–14). For a *triangular notch* in a channel with great depth, show that when the effects of viscosity and surface tension are not important, the discharge can be expected to be $Q = k\sqrt{g}\,H^{5/2}$, where k is a constant depending only on the angle θ of the notch. (k has been experimentally found to be about $0.44 \tan \frac{1}{2}\theta$.)

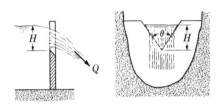

FIGURE 2–14   FIGURE 2–15

2-19. (a) For the *rectangular weir* shown in Fig. 2–15, in an open channel of great depth and width, find by dimensional analysis an expression for Q when the effects of μ and σ are unimportant. (b) It has been found experimentally that

$$Q = 0.586\sqrt{g}\left(1 - 0.2\,\frac{H}{L}\right)LH^{3/2}.$$

(This is known as the *Francis formula*.) Does this formula appear to be reasonable in the light of your analysis?

2-20. If surface tension σ is included in Problem 2–18, what are the factors affecting k? Under what condition would you expect that the effect of σ is unimportant?

2-21. A centrifugal pump is driven by a motor at a certain angular speed ω. The discharge Q through the pump is regulated by a valve in the pipeline. Call the energy gained per unit mass of the fluid gH (dimensionally L^2T^{-2}). Show that when the effects of compressibility and viscosity are unimportant, the quantity $gH/(\omega^2 D^2)$ depends only on the quantity $Q/(\omega D^3)$ for geometrically similar pumps, where D is the size of the pump. In other words, the curve in Fig. 2–16 obtained experimentally with one pump at one speed is valid for these pumps irrespective of their sizes and speeds.

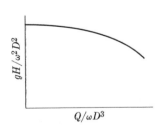

FIGURE 2–16

2-22. A sphere of diameter D oscillates in a large body of fluid with properties ρ, μ, and K. The amplitude and the frequency of the oscillations are L and f, respectively. Find by dimensional analysis an expression for the maximum dynamic pressure p created in the fluid. When would you expect that the effect of compressibility is unimportant?

2–6. Scale models of fluid systems

As will be seen in subsequent chapters, theoretical studies can lead to the understanding of the mechanism of many flow phenomena. However, many practical problems are of such complexity as to defy a theoretical approach. One then has to resort to experimental methods. Since a test on a full-size prototype is often impossible or too expensive, scale models are used. The data obtained from scale models are meaningful and can be related to the prototype because, according to Eq. 2–10,

$$\text{Dimensionless product} = \phi\left(\frac{a}{L}, \frac{b}{L}, \cdots, F, R, W, M\right),$$

the actual size of the flow system does not come into consideration. The same value of a dimensionless product will be produced in the model if it is similar in shape to the prototype (same a/L, etc.) and has the same values of the dimensionless numbers (such as F, R, etc.) involving the significant mechanical properties.

Let the subscripts p and m designate quantities in the prototype and the model, respectively. In designing a model, we have to determine the following three items: the length scale L_m/L_p, the speed scale U_m/U_p, and the fluid for the model. First consider the case where none of the dimensionless numbers in the equation above need to be considered. For example, when gravity is unimportant, both R and W are high, and M is low in both the model and the prototype, the exact values of F, R, W, and M are not important. Therefore, within these limits, any fluid can be used with any length scale and speed scale. For any dimensionless product, the same value will be observed in a geometrically similar model. For example, the time t_p for an event to occur in the prototype can be computed from the chosen L_m/L_p and U_m/U_p, and the observed t_m in the model, since for the dimensionless product Ut/L

$$\frac{U_p t_p}{L_p} = \frac{U_m t_m}{L_m}. \tag{2–19}$$

The force F_p and the discharge Q_p in the prototype can be computed from

$$\frac{F_p}{\rho_p U_p^2 L_p^2} = \frac{F_m}{\rho_m U_m^2 L_m^2} \tag{2–20}$$

and

$$\frac{Q_p}{U_p L_p^2} = \frac{Q_m}{U_m L_m^2}. \tag{2–21}$$

Consider next the case where one type of force is predominant in both the model and the prototype. If viscosity is predominant, both systems must have equal Reynolds numbers:

$$\frac{U_m L_m \rho_m}{\mu_m} = \frac{U_p L_p \rho_p}{\mu_p}. \tag{2–22}$$

If, instead of viscosity, gravitational force is predominant, we must have the same Froude number in both systems:

$$\frac{U_m}{\sqrt{g_m L_m}} = \frac{U_p}{\sqrt{g_p L_p}} \cdot \tag{2–23}$$

(Usually $g_m = g_p$.)   If surface tension is the predominant force, the Weber number must be the same:

$$U_m \sqrt{\frac{L_m \rho_m}{\sigma_m}} = U_p \sqrt{\frac{L_p \rho_p}{\sigma_p}} \cdot \tag{2–24}$$

When only compressibility has to be considered, we must have the same Mach number:

$$U_m \sqrt{\frac{\rho_m}{K_m}} = U_p \sqrt{\frac{\rho_p}{K_p}} \cdot \tag{2–25}$$

In any case, the fluid for the model and the length scale can be arbitrarily chosen, and the speed scale is then determined from one of Eqs. 2–22 to 2–25. With the model operated with U_m thus determined, the same value for any dimensionless product will be observed in the geometrically similar model. Equations such as Eqs. 2–19 to 2–21 can be used to predict the events in the prototype. However, the fluid and the model size must be so chosen that this U_m is practical, and the influence of the other forces remains small in the model as assumed (see Example 2–8a).

If two types of forces must be considered, only one of the three items in the design can be arbitrarily chosen. For example, if both viscosity and gravity must be considered, both Eqs. 2–22 and 2–23 must be satisfied. If the fluid of the model has been chosen, the length and speed scales must be, according to these equations,

$$\frac{U_m}{U_p} = \left(\frac{\rho_p \mu_m g_m}{\rho_m \mu_p g_p} \right)^{1/3}$$

and

$$\frac{L_m}{L_p} = \left(\frac{\rho_p^2 \mu_m^2 g_p}{\rho_m^2 \mu_p^2 g_m} \right)^{1/3}$$

This means that in order to use a scale model with $L_m \neq L_p$, the fluid for the model must be different from that in the prototype (unless $g_m \neq g_p$). To do this is not usually practical. For example, in a geometrically similar model of a river designed in this manner, mercury (with μ/ρ about 0.1 that of water) would have to be used with a length scale of about $\frac{1}{4}$.

If three (or four) types of forces must be considered, three (or four) of Eqs. 2–22 to 2–25 must be satisfied. With any length scale, these equations specify several properties of the fluid for the model. Such a fluid usually does not exist.

In other words, a dynamically similar scale model is practically impossible in such cases.

Fortunately, in many practical problems there is usually only one force that is predominant. The model is designed to produce the same value for the dimensionless number representing the predominant force. The difference between the model and the prototype in the other dimensionless numbers will affect the accuracy of the data observed in the model. The errors thus introduced are called *scale effects*. Sometimes these scale effects can be estimated and a correction can be applied to the observed data (see Example 2–8b).

In problems where cavitation is involved, the cavitation number in Eq. 2–9a must also be made the same. This can usually be accomplished by controlling the fluid pressure in the model (see Example 2–9).

Example 2–8. The resistance to a sailing ship is believed to come from two sources, namely, the creation of surface waves against gravity, and skin friction. It is also believed that gravity is the more important force in this case. (a) Design a model on the assumption that gravity is predominant for a ship with a submerged surface of 4000 ft^2 sailing at 15 ft/sec. (b) Apply a correction if possible for the scale effect due to viscosity.

(a) To design a model for one predominant force, namely gravity in this case, we can choose arbitrarily the fluid for the model and the length scale. With water in the model, we have $\rho_m/\rho_p = 1$. A $\frac{1}{25}$ scale model is considered convenient, and the effect of surface tension is believed to be negligible for the size of the waves in the model. To produce the same Froude number, the speed scale should be, according to Eq. 2–23,

$$\frac{U_m}{U_p} = \sqrt{\frac{g_m L_m}{g_p L_p}} = \sqrt{\frac{1}{25}} = \frac{1}{5}.$$

The speed at which the ship model is to be towed should therefore be $\frac{15}{5} = 3$ ft/sec.

With a geometrically similar ship model running at this speed, the tension in the tow line indicates that the resistance is 1 lb. Ignoring the scale effect due to skin friction, we have, from Eq. 2–20, the resistance of the prototype:

$$F_p = F_m \frac{\rho_p U_p^2 L_p^2}{\rho_m U_m^2 L_m^2} = 1 \times 1 \times 5^2 \times 25^2 = 15{,}600 \text{ lb.}$$

(b) In this case, a correction for the scale effect due to skin friction can be made by estimating the skin friction. This can be done, for example, by using the *skin friction formula* for a flat plate,

$$F = c_f \tfrac{1}{2} \rho U^2 A,$$

where the coefficient c_f depends on the Reynolds number. At the Reynolds number of the model, published experimental data indicate that $c_f = 0.004$. At the Reynolds number of the prototype, $c_f = 0.003$. The total resistance of the prototype can be computed as follows:

Submerged surface of prototype $= 4000 \text{ ft}^2$.

Submerged surface of model $= 4000/25^2 = 6.4 \text{ ft}^2$.

Skin friction of model $= 0.004 \times \frac{1}{2} \times 1.93 \times 3^2 \times 6.4 = 0.22 \text{ lb}$.

Wave resistance in model $= 1 - 0.22 = 0.78 \text{ lb}$.

Wave resistance in prototype (from Eq. 2–20) $= 0.78 \times 1 \times 5^2 \times 25^2$
$$= 12{,}200 \text{ lb}.$$

Skin friction of prototype $= 0.003 \times \frac{1}{2} \times 1.93 \times 15^2 \times 4000 = 2600 \text{ lb}$.

Total resistance of prototype $= 12{,}200 + 2600 = 14{,}800 \text{ lb}$.

This result is believed to be more accurate than that obtained in (a) without correction for the scale effect due to skin friction.

Example 2–9. A ship propeller, L_p in diameter, is to be used in water with a vapor pressure of 100 lb/ft² abs. (at about 90°F). Design a scale model to find the thrust at angular speed ω_p while traveling at speed U_p, and to see if cavitation would occur in the prototype.

The model is to be so designed that the Reynolds number is high enough in the model that it is not necessary to keep this number the same. The size of the surface waves in the model will be large enough that the Weber number in the model can be different from that of the prototype. The effects of compressibility need not be considered since the Mach number is low. The model is to be designed to have the same Froude number.

A length scale of $\frac{1}{4}$ is chosen, and the water at 68°F is to be used in the model. The speed scale should be, according to Eq. 2–23,

$$\frac{U_m}{U_p} = \sqrt{\frac{g_m L_m}{g_p L_p}} = \sqrt{\frac{1}{4}} = \frac{1}{2}.$$

The speed of the ship model should be $\frac{1}{2}$ of that of the prototype. Since all linear speeds in the model should be reduced to the same scale,

$$\frac{\omega_m L_m}{\omega_p L_p} = \frac{U_m}{U_p} = \frac{1}{2}, \qquad \text{or} \qquad \frac{\omega_m}{\omega_p} = \frac{U_m L_p}{U_p L_m} = 2.$$

The angular speed of the model should be twice that of the prototype. When the thrust F_m is observed, the thrust F_p in the prototype can be computed from Eq. 2–20.

To study cavitation, it is necessary to have the same cavitation number $C = (P - p_v)/(\frac{1}{2}\rho U^2)$, where the reference pressure P in this case may be taken to be the pressure above the water surface. This can be accomplished by controlling the pressure P_m in the model. In the prototype, we have atmospheric

$P_p = 14.7$ psi abs. or 2120 lb/ft^2 abs., and $(p_v)_p = 100$ lb/ft^2 abs. In the model, we have $(p_v)_m = 48.8$ lb/ft^2 abs. (at 68°F). For the same value of C in both systems,

$$\frac{P_m - 48.8}{\frac{1}{2}\rho_m U_m^2} = \frac{2120 - 100}{\frac{1}{2}\rho_p U_p^2},$$

$$P_m = 2020 \frac{\rho_m}{\rho_p}\left(\frac{U_m}{U_p}\right)^2 + 48.8 = 554 \text{ lb/ft}^2 \text{ abs.} = 3.85 \text{ psi abs.}$$

The tests should therefore be conducted in a water tunnel under a partial vacuum. (It should be mentioned that cavitation inception is influenced also by other factors, such as the number and size of air bubbles in the water. The determination of the scale effect due to these factors remains a difficult task.)

PROBLEMS

2–23. The increase of pressure Δp in a gradual expansion of a water pipe (see Fig. 2–17) is to be observed with a $\frac{1}{5}$ scale model. It is believed that the effect of viscosity should be considered. If water is used in the model, find the scales for speed, discharge, and pressure change.

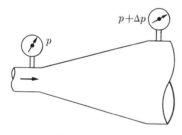

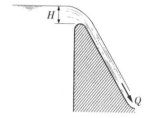

FIGURE 2–17   FIGURE 2–18

2–24. The spillway of a proposed dam is to pass a discharge of 5000 ft^3/sec. A scale model is used to determine the surface elevation H above the spillway crest (see Fig. 2–18). The water supply in the laboratory is limited to 1 ft^3/sec. Choose a length scale for the model. Gravity is the predominant force in this case.

2–25. The drag of an airplane of $L = 10$ ft at a speed $U = 300$ ft/sec is to be predicted with a $\frac{1}{20}$ model in a wind tunnel under atmospheric pressure (μ/ρ of air $= 1.60 \times 10^{-4}$ ft^2/sec). It is believed that the drag coefficient is constant if $UL\rho/\mu > 10^5$. (a) Choose a speed scale for the model so that even the effect of viscosity is unimportant. (b) What is the drag of the prototype when the observed drag of the model is 10 lb at your chosen speed?

2–26. To study the flow around the structural members of an airplane, it is found desirable to have the same Reynolds number in a $\frac{1}{20}$ model. To avoid the scale effect due to compressibility, the speed in the model is to be kept below 450 ft/sec. The speed of the prototype is 300 ft/sec. (a) What is the required value of μ/ρ of the fluid for the model? (b) With the wind tunnel pressurized, μ of air remains practically the same, while ρ increases directly with the absolute pressure. What is the required pressure in the wind tunnel for this purpose?

2–27. Mercury (specific gravity = 13.60, σ = 0.036 lb/ft in air at 68°F) in a vessel is drained by a pipe at the bottom at the rate of 10^{-3} ft^3/sec. A model with water (σ = 0.0055 lb/ft) is to be used to determine the minimum depth of mercury to prevent air from being sucked into the pipe. It is believed that the shape of the free surface is influenced by gravity and surface tension. Find the length scale and the discharge in the model.

2–28. The pressure at the entrance to the impeller of a large water pump is expected to be 10 psi below the atmospheric. A $\frac{1}{25}$ scale model is used to investigate whether cavitation would occur inside the pump. It is believed that the influence of viscosity is negligible with water in the model and a speed scale of $\frac{1}{5}$. Determine the required pressure at the entrance of the pump model.

2–7. Dimensional analysis involving heat transfer

In the previous sections of this chapter, the discussion has been limited to problems where heat transfer is not important. In fluid systems where a significant temperature gradient exists, however, heat transfer must be considered.

In problems involving heat transfer, new physical quantities are encountered which do not appear in purely mechanical problems. The question is: How many new fundamental quantities are involved, in addition to F-L-T or M-L-T? In the equation of state of perfect gases, $p = \rho RT$, and the equation relating C_v and specific internal energy e, $e = C_v T$, there are two new physical quantities, namely, temperature and heat energy. However, heat energy can be expressed in units of mechanical energy and can be expressed dimensionally as FL. Thus for e, internal energy per unit mass, the dimensional formula is $(FL)M^{-1}$ or $L^2 T^{-2}$, and the unit may be ft^2/sec^2, etc. If temperature is regarded as a new fundamental quantity,* to be identified by θ, the dimensional formula for C_v

---

* Actually, it is not necessary to consider temperature as a new fundamental quantity, since temperature T never appears singly but is always associated with such coefficients as R, C_v, and k. Thus, for example, (RT) can be considered as a quantity with the dimensions of p/ρ or $L^2 T^{-2}$ (where T represents time). Equation 2-26 in Example 2-10 can be written as

$$\frac{QL}{A(kT_1 - kT_0)} = f\left[\frac{b}{L}, \frac{\rho_0 UL}{\mu_0}, \frac{U}{\sqrt{gL}}, \frac{U}{\sqrt{\gamma(RT_0)}}, \frac{\mu_0 \gamma(C_v T_0)}{(kT_0)}, \frac{(kT_1)}{(kT_0)}, \gamma\right].$$

In technical applications, however, it is more convenient to treat temperature as a separate entity, and θ as a new fundamental quantity. The number $m - r$ of independent dimensionless products involved in a problem is not changed, since both m and r are increased by unity.

will be $L^2T^{-2}\theta^{-1}$. In fluid flow problems with heat transfer, the number of fundamental quantities involved is then four ($n = 4$), namely, F-L-T-θ or M-L-T-θ.

To describe the thermal properties of a perfect gas with constant heat capacity, two constants are needed; e.g., R and C_v. Other thermal properties of the gas can be expressed in terms of these two constants. For examples, we have $\gamma = 1 + (R/C_v)$, and the bulk modulus K in an isentropic process is γp (see Eq. 1–7). In addition, there is the *thermal conductivity* k, which is defined as the heat energy conducted through a unit area per unit time under a temperature gradient of one degree per unit distance. The dimensional formula for k is $FT^{-1}\theta^{-1}$. The dimensional analysis of problems of heat transfer is demonstrated in the following example.

Example 2–10. A flat plate of length L and width b is placed in a stream of a perfect gas of speed U and temperature T_0, as shown in Fig. 2–19. The temperature of the plate is maintained at a constant temperature T_1. Find the dimensionless products that may affect the rate of flow of heat from (or into) the plate.

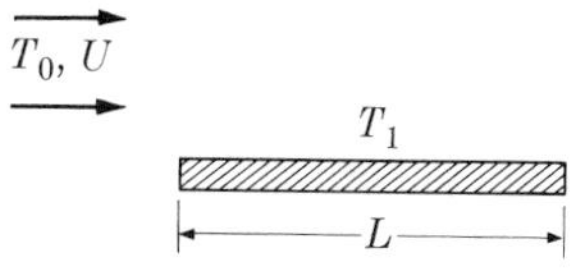

FIGURE 2–19

Let Q be the flow of heat from (or into) the plate per unit time. Q must depend on the speed U and temperature T_0 of the stream, the temperature T_1, and the linear dimensions L and b of the plate. The gravitational acceleration g is involved in creating natural convection. The mechanical properties involved are density ρ_0 and viscosity μ_0 of the stream. (For simplicity, the variation of μ will not be considered here.) The thermal properties involved are described with C_v, R, and k. The dimensional formulas of these twelve quantities are:

$$[Q] = FLT^{-1}, \qquad [U] = LT^{-1}, \qquad [T_0] = [T_1] = \theta,$$

$$[g] = LT^{-2}, \qquad [L] = [b] = L, \qquad [\rho_0] = FL^{-4}T^2,$$

$$[\mu_0] = FL^{-2}T, \qquad [C_v] = [R] = L^2T^{-2}\theta^{-1}, \qquad \text{and} \qquad [k] = FT^{-1}\theta^{-1}.$$

With $m = 12$ and $r = 4$, there are eight independent dimensionless products; for example,

$$\frac{QL}{Ak(T_1 - T_0)} = f\left(\frac{b}{L}, \frac{\rho_0 UL}{\mu_0}, \frac{U}{\sqrt{gL}}, \frac{U}{\sqrt{\gamma RT_0}}, \frac{\mu_0 \gamma C_v}{k}, \frac{T_1}{T_0}, \gamma\right), \qquad (2\text{–}26)$$

where $A = bL$ is the area of the plate, and $\gamma = 1 + (R/C_v)$. The product $QL/Ak(T_1 - T_0)$ is named after Nusselt. The products $\rho_0 UL/\mu_0$ and $U/\sqrt{gL}$ are of course the Reynolds number and the Froude number, respectively. Since $\gamma RT_0 = \gamma p_0/\rho_0 = K_0/\rho_0$ (see Eqs. 1–4 and 1–7), the product $U/\sqrt{\gamma RT_0}$ is the Mach number M of the stream. The dimensionless product $\mu_0 \gamma C_v/k$ is very important in heat transfer phenomena, and is called the *Prandtl number:*

$$\text{Prandtl number } P = \frac{\mu \gamma C_v}{k}. \tag{2-27}$$

This number depends on the physical properties of the fluid and is independent of the flow. The value of P for gases is approximately equal to unity.

It is the convention to replace the product T_1/T_0 with the *Eckert number*

$$\text{Eckert number } E = \frac{U^2}{\gamma C_v(T_1 - T_0)}, \tag{2-28}$$

which is equal to $(\gamma - 1)M^2/[(T_1/T_0) - 1]$. The Froude number F, which represents the effect of gravity, is usually replaced by the *Grashof number*

$$\text{Grashof number } G = \frac{gL^3\rho_0^2(T_1 - T_0)}{\mu_0^2 T_0}, \tag{2-29}$$

which is equal to $[(T_1/T_0) - 1] \cdot (R/F)^2$, where R is the Reynolds number. Thus, Eq. 2-26 can be rewritten as

$$\frac{QL}{Ak(T_1 - T_0)} = \phi\left(\frac{b}{L}, \frac{\rho_0 UL}{\mu_0}, G, M, P, E, \gamma\right). \tag{2-30}$$

Experiments show that when the Mach number is low, the effects of M and γ are not important. At low Mach numbers, the effect of the Eckert number is very small except when T_1/T_0 is nearly equal to unity. The Grashof number is important only in the case of natural convection. Usually, in the case of forced convection, as shown in Fig. 2-19, the quantity $QL/Ak(T_1 - T_0)$ can be said to depend only on the shape of the plate, the Reynolds number, and the Prandtl number.

3 Fluid Statics

3-1. Fluid statics

In Chapter 2, the method of dimensional analysis has been found to be a powerful tool in the study of fluid mechanics. However, to obtain quantitative results, experimental or theoretical analyses must be performed. The rest of this book deals mainly with the theoretical study of fluid mechanics and experimental verification of the results. The theoretical analysis will be based on Newton's laws of motion, the laws of thermodynamics, the principle of conservation of mass, and the properties of the fluid. In this chapter, problems of fluid statics are to be considered. In *fluid statics*, the fluid may be at rest or in constant linear motion like a rigid body.

In considering fluids as continuous materials, the momentum $m\mathbf{q}$ of a small fluid element is actually the vector sum of those of its molecules. When this vector sum is zero, the element is said to be at rest with $m\mathbf{q} = 0$, although the individual molecules are in motion. Thus, a "static" body of fluid will exhibit effects of molecular diffusion, e.g., the slow dispersion of a drop of ink in a cup of still water.

Since the fluid elements are not being deformed in fluid statics, the shearing stresses in the fluid must be zero everywhere, in accordance with the definition of a fluid. According to Eq. 1-2, the total force acting on each fluid element is zero. It is to be shown that as a consequence, the normal stress at a point in the fluid is independent of the orientation of the surface on which it acts. Take a small volume around the point P in question, as shown in Fig. 3-1. Let the mean

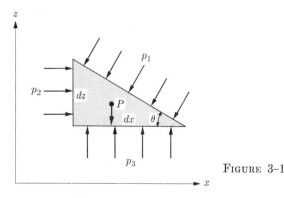

FIGURE 3-1

37

normal stress on the three surfaces be p_1, p_2, and p_3. It is to be shown that $p_1 = p_2 = p_3$ as dx and dz become infinitesimal. From Eq. 1–2, we have $\mathbf{F} = 0$; therefore $F_x = 0$ and $F_z = 0$:

$$F_x = p_2 \, dy \, dz - \left(p_1 \, dy \, \frac{dz}{\sin \theta} \right) \sin \theta = 0,$$

giving $p_1 = p_2$, and

$$F_z = p_3 \, dy \, dx - \left(p_1 \, dy \, \frac{dx}{\cos \theta} \right) \cos \theta - \rho g \, \frac{dx \, dy \, dz}{2} = 0,$$

giving $p_3 = p_1 + (\rho g \, dz/2)$. As dx and dz become infinitesimal, the last term, which is a higher-order term, vanishes in comparison with the other terms. Thus we have $p_1 = p_2 = p_3$ at point P for any value of θ. Let this scalar quantity be called the *pressure p* at this point in a static body of fluid.

3–2. Equation of fluid statics

An equation governing the variation of pressure p from point to point in fluid statics can be derived according to Newton's equation of motion. Consider an infinitesimal fluid element at an arbitrary elevation z_0, as shown in Fig. 3–2. The forces acting on this element are the forces due to fluid pressure and gravity. Let ρ_0 and p_0 be the density and the mean pressure, respectively, at the bottom of the element. By Taylor's series, the pressure at the top can be expressed as $p_0 + (\partial p/\partial z)_0 \, dz + \cdots$, and the mean density of the element can be expressed as $\rho_0 + \cdots$. Here the dots represent higher-order terms, i.e., terms with higher powers of dz. The partial derivative $(\partial p/\partial z)_0$ is evaluated at z_0 and is the rate of change of p in the z-direction. For equilibrium, we have $F_z = 0$:

$$p_0 \, dx \, dy - \left[p_0 + \left(\frac{\partial p}{\partial z} \right)_0 dz + \cdots \right] dx \, dy$$

$$- (\rho_0 + \cdots)g \, dx \, dy \, dz = 0,$$

$$\left(\frac{\partial p}{\partial z} \right)_0 + \rho_0 g + \cdots = 0.$$

Since dz is infinitesimal, the higher-order terms vanish in comparison with the first two finite terms. Therefore

$$\left(\frac{\partial p}{\partial z} \right)_0 + \rho_0 g = 0.$$

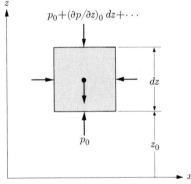

FIGURE 3–2

(Note that in deriving this equation by using an infinitesimal element, the value ρ_0 can be used in effect for the mean value of ρ in dz, and $(\partial p/\partial z)_0 \, dz$ can be used as the change of p in a distance dz.) Since this position z_0 is arbitrary, we have the following equation for every fluid element:

$$\frac{\partial p}{\partial z} = -\rho g,$$

which states that pressure drops with elevation at a rate equal to the local value of the unit weight. In a similar manner, we have from $F_x = 0$ and $F_y = 0$,

$$\frac{\partial p}{\partial x} = 0 \qquad \text{and} \qquad \frac{\partial p}{\partial y} = 0.$$

These equations state that p does not change in a horizontal direction. If p does not change with time, it is then a function of z only. It follows that dp/dz and therefore ρg are functions of z only. Thus density does not vary horizontally in a static body of fluid.

In this book, we shall use h to denote the elevation of a point. With $dh = dz$, we have the *equation of fluid statics*

$$\frac{dp}{dh} = -\rho g. \tag{3–1}$$

There are two dependent variables in this equation, namely, p and ρ. In order to obtain the pressure distribution $p(h)$ by integration, additional information is needed to specify the density as a function of either p or h. Such information may come from observation or other studies.

In many cases, the percentage variation of ρ is small (e.g., when the domain of interest is limited vertically). The coefficient ρ in Eq. 3–1 may then be approximated by a constant in integrating for p. With ρ as a constant, we obtain

$$\frac{p}{\rho} + gh = \text{constant}. \tag{3–2}$$

This equation is called the *equation of hydrostatics*. The constant of integration is to be determined with a known pressure at a certain point in the fluid. Let p_1 be the known pressure at a point at elevation h_1. Then for any other point in the same body of static fluid of constant ρ,

$$\frac{p}{\rho} + gh = \frac{p_1}{\rho} + gh_1,$$

or

$$p - p_1 = -\rho g(h - h_1). \tag{3–3}$$

Example 3–1. A *manometer* with a light fluid is used to measure the pressure difference between two vessels, as shown in Fig. 3–3. Given the density ρ of the liquid in the vessels and ρ_m of the immiscible manometer fluid, find the pressure difference for a manometer reading R.

In this case, the variation of density in each fluid can be neglected. Using Eq. 3–3 repeatedly, we have

$$p_1 - p_2 = -\rho g(h_1 - h_2),$$

$$p_2 - p_3 = -\rho_m g(h_2 - h_3),$$

$$p_3 - p_4 = -\rho g(h_3 - h_4).$$

To find $p_1 - p_4$, we add the three equations together:

$$p_1 - p_4 = \rho g(h_4 - h_1) + (\rho - \rho_m)g(h_2 - h_3)$$

$$= \rho g a + (\rho - \rho_m)g R.$$

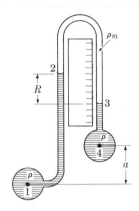

FIGURE 3–3

Other things being the same, the smaller the density difference $\rho - \rho_m$, the larger the reading R. Thus in measuring small pressure differences, a small density difference is desirable. If a gas is used as the manometer fluid, ρ_m will be negligible compared with ρ. Then $p_1 - p_4 = \rho g(a + R)$.

PROBLEMS

3–1. (a) A manometer with a heavy liquid, such as mercury, is commonly used. Given the density ρ of the fluid in the vessels and ρ_m of the manometer liquid, find the pressure difference between the vessels shown in Fig. 3–4. (b) Given $a = 0$ and the pressure difference is 10 psi. To facilitate ease of observation, what must be ρ_m such that R is not more than 3 ft?

3–2. The pressure gage in Fig. 3–5 reads 3.5 psi. Is this gage accurate? The specific gravity of mercury is 13.6.

3–3. From the manometer reading R and the depth L, compute the depth of water in the diving bell shown in Fig. 3–6.

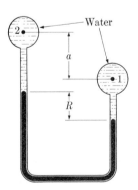

FIGURE 3–4

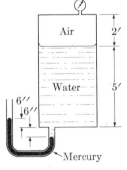

FIGURE 3–5

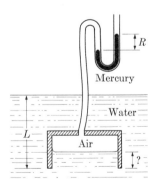

FIGURE 3–6

3–4. An inclined manometer is used to magnify the reading. In Fig. 3–7, the scale reads zero when there is no difference of pressure between the two vessels of gas. The cross-sectional area of the reservoir is 50 times that of the inclined tube, and the angle of inclination is 30°. The manometer liquid weighs 50 lb/ft³. Find the pressure difference between the two vessels when the reading R is 2 ft.

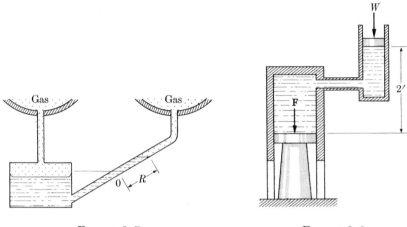

FIGURE 3–7   FIGURE 3–8

3–5. A force **F** of 20,000 lb is to be developed in a hydraulic press as shown in Fig. 3–8. Find the required weight W. The area of the large piston is 100 ft² and that of the small one is 1 ft². The fluid is an oil of specific gravity 0.80.

3–3. Hydrostatic force on a plane surface

When a surface is submerged in a liquid, the pressure acting on it varies from one point to another (see Fig. 3–9). On a plane surface, the hydrostatic forces form a system of parallel forces. It is often of practical interest to find the magnitude, the direction, and the line of action of the resultant of these forces. Usually one side of a submerged body is exposed to atmospheric pressure (e.g., the door of a pressure vessel). For these cases, it is the force in excess of that due to the atmospheric pressure that is of practical interest. In this and the next sections, p is used to denote the gage pressure, and the computed forces are those due to pressure in excess of the atmospheric.

Consider a free surface under atmospheric pressure as shown in Fig. 3–9. By calling the surface elevation h_1, we have then $p_1 = 0$. Equation 3–3 gives for any other point in the fluid

$$p = \rho g(h_1 - h) = \rho g D,$$

where D is the depth of a point measured from the free surface. The gage pressure at a point is therefore proportional to its depth below the free surface.

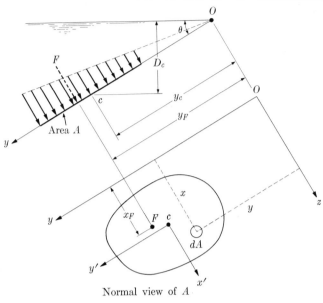

Normal view of A

FIGURE 3–9

The resultant of the hydrostatic forces on a plane is normal to the plane. To find the magnitude F of the resultant, it is convenient to place the xy-plane in the plane of the surface under consideration, and the x-axis in the plane of the free surface (extended if necessary), as shown in Fig. 3–9. For a differential area dA located at coordinates (x, y), we have

$$p = \rho g D = \rho g y \sin \theta,$$

and

$$dF = p \, dA = \rho g y \sin \theta \, dA.$$

The magnitude F is therefore

$$F = \int dF = \rho g \sin \theta \iint y \, dA. \tag{3–4}$$

The double integral may be evaluated simply as shown in Example 3–2. Since $\iint y \, dA = y_c A$, where y_c is the ordinate of the centroid c of the plane area A, we can also write

$$F = \rho g \sin \theta \, y_c A = \rho g D_c A = p_c A, \tag{3–5}$$

where D_c is the depth of the centroid c below the free surface. The average pressure F/A has been found to be equal to p_c at the centroid.

Next, find the line of action of this resultant. Let x_F and y_F locate its point of action on the plane. By the definition of a resultant, its moment about any point or axis must be the same as the distributed hydrostatic forces. By taking the moment about the x-axis, we have

$$Fy_F = \int y \, dF = \rho g \sin \theta \iint y^2 \, dA; \qquad (3\text{–}6)$$

y_F can be determined by evaluating the double integral. This double integral is the second moment I_x of the area with respect to the x-axis. By the theorem of parallel axes, we have $I_x = I_{x'} + Ay_c^2$, where $I_{x'}$ is the second moment of the area with respect to the x'-axis, which passes through the centroid c and is parallel to the x-axis. With F from Eq. 3–5,

$$y_F = \frac{\rho g \sin \theta I_x}{F} = \frac{I_x}{y_c A} = \frac{I_{x'}}{y_c A} + y_c. \qquad (3\text{–}7)$$

Thus the point of action always lies at a lower elevation than the centroid. However, for a deeply submerged surface with large y_c, the difference $y_F - y_c$ is small.

By taking the moment about the y-axis, we have

$$Fx_F = \int x \, dF = \rho g \sin \theta \iint xy \, dA, \qquad (3\text{–}8)$$

from which x_F can be determined by performing the double integration. The double integral $\iint xy \, dA$ is usually called the product of "inertia" P_{xy} of the area with respect to the x- and y-axes. By the theorem of parallel axes, $P_{xy} = P_{x'y'} + Ax_c y_c$, where $P_{x'y'}$ is the product of "inertia" of the area with respect to the x'- and y'-axes through the centroid. Thus

$$x_F = \frac{\rho g \sin \theta P_{xy}}{F} = \frac{P_{xy}}{y_c A} = \frac{P_{x'y'}}{y_c A} + x_c. \qquad (3\text{–}9)$$

For an area symmetrical about the y'-axis, $P_{x'y'} = 0$ and $x_F = x_c$.

Example 3–2. Find the resultant of the hydrostatic forces acting on the triangular area shown in Fig. 3–10.

If the values of y_c, $I_{x'}$, and $P_{x'y'}$ for the area are available (e.g., from handbooks), this problem is immediately solved by using Eqs. 3–5, 3–7, and 3–9. We shall solve the problem by evaluating the integrals in Eqs. 3–4, 3–6, and 3–8 by integrating with respect to one variable at a time. For example,

$$\iint xy \, dA = \iint xy \, dx \, dy = \int_{y=0}^{y=12} \left[\int_{x=-y/2}^{x=y} xy \, dx \right] dy$$

$$= \int_{y=0}^{y=12} \left[y \frac{x^2}{2} \Big|_{x=-y/2}^{x=y} \right] dy = \int_0^{12} \frac{y}{2} \left(y^2 - \frac{y^2}{4} \right) dy = 1944 \text{ ft}^4.$$

Here integration with respect to x is first performed, with y kept constant. The limits of x are given by the equations of the two boundaries. The next integration

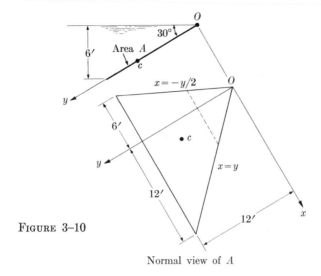

FIGURE 3–10

Normal view of A

is performed with respect to y, with the maximum and minimum y as limits. Similarly,

$$\iint y \, dA = \iint y \, dx \, dy = \int_{y=0}^{y=12} \left[\int_{x=-y/2}^{x=y} y \, dx \right] dy = \int_0^{12} y \left(y + \frac{y}{2} \right) dy$$

$$= 864 \text{ ft}^3,$$

$$\iint y^2 \, dA = \iint y^2 \, dx \, dy = 7776 \text{ ft}^4.$$

Thus

$$F = \rho g \sin \theta \iint y \, dA = 62.4 \times \tfrac{1}{2} \times 864 = 27{,}000 \text{ lb},$$

$$y_F = \frac{1}{F} \rho g \sin \theta \iint y^2 \, dA = \frac{\iint y^2 \, dA}{\iint y \, dA} = \frac{7776}{864} = 9.00 \text{ ft},$$

$$x_F = \frac{1}{F} \rho g \sin \theta \iint xy \, dA = \frac{\iint xy \, dA}{\iint y \, dA} = \frac{1944}{864} = 2.25 \text{ ft}.$$

PROBLEMS

3–6. The upstream face of a dam is vertical and the depth of the water is 18 ft. Find the hydrostatic force per foot of dam, and show that the line of action of the resultant is located at $\frac{2}{3}$ of the depth from the water surface.

3–7. Find the net force and its line of action acting on the circular gate under water shown in Fig. 3–11.

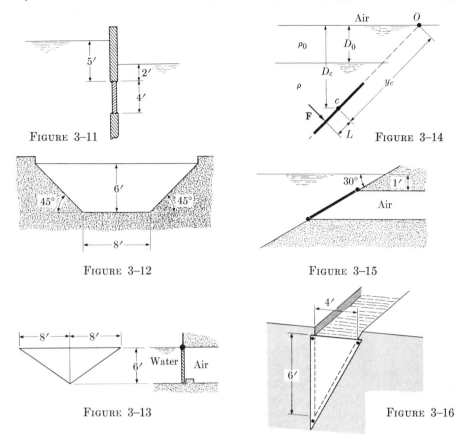

FIGURE 3–11

FIGURE 3–14

FIGURE 3–12

FIGURE 3–15

FIGURE 3–13

FIGURE 3–16

3–8. A vertical trapezoidal barrier is erected at the end of a flume, as shown in Fig. 3–12. Find the hydrostatic force acting on the barrier when the flume is filled.

3–9. A vertical triangular gate is held in place by a hinge along the top and a key at the bottom, as shown in Fig. 3–13. Find the reaction at the key.

3–10. Show that in Fig. 3–14, (a) the magnitude F of the resultant hydrostatic force is equal to Ap_c, where A is the area of the plane, and p_c is the pressure at its centroid, and (b) the distance L locating the resultant is

$$ L = \frac{I_{x'}}{y_c A} \bigg/ \left[1 - \left(1 - \frac{\rho_0}{\rho} \right) \frac{D_0}{D_c} \right]. $$

3–11. A circular gate 4 ft in diameter covers the entrance to a tunnel under water, as shown in Fig. 3–15. The gate is supported at the top and at the bottom by hinges. Find the reactions at the hinges.

3–12. A triangular plate is bolted to the end of a triangular channel, as shown in Fig. 3–16. The channel is filled with water to the top of the plate. Find the tensile forces in the three bolts at the corners of the plate.

3–4. Hydrostatic forces on a general surface

The hydrostatic forces acting on a submerged curved surface vary in direction from point to point (see Fig. 3–17). Unlike the special case considered in the previous section, a general system of forces must be represented by a force and a couple. In this case, it can also be represented by the resultant of each of the three components of the hydrostatic forces along one vertical and two orthogonal horizontal axes. Only when these three orthogonal resultants intersect at one point can the system be represented by a single force as its resultant.

Consider first the resultant of the components along a horizontal x-axis. On any area dA whose normal is at an angle θ with the horizontal axis (see Fig. 3–17), the x-component of the hydrostatic force is

$$dF_x = dF \cos \theta = p \, dA \cos \theta.$$

But $dA \cos \theta$ is the projection of dA on a plane normal to the x-axis. Let this projection be denoted by dA_x. Then

$$F_x = \int dF_x = \iint p \, dA_x = \rho g \iint D \, dA_x. \qquad (3\text{--}10)$$

By comparing this equation with Eq. 3–4, we see that F_x is the same in magnitude and in location as the hydrostatic force on the vertical projection of the surface. Thus the resultant F_x of the horizontal x-component of the hydrostatic forces on surface ABC in Fig. 3–18 is equal to the sum of the forces on the planes $A'B'$ and $B''C''$. Since the forces acting on planes $B'C'$ and $B''C''$ are equal and opposite, this sum is equal to the force on the plane $A'C'$.

Next find the resultant of the vertical components of the hydrostatic forces. In Fig. 3–19, the vertical component of the force on dA is

$$dF_v = dF \sin \theta = p \, dA \sin \theta = \rho g D \, dA \sin \theta.$$

But $dA \sin \theta$ is the horizontal projection of dA, and $D \, dA \sin \theta$ is a volume dV

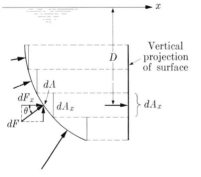

FIGURE 3–17

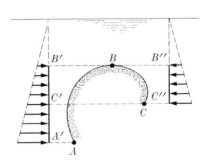

FIGURE 3–18

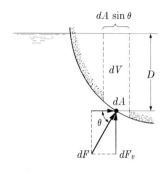

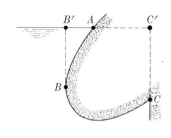

FIGURE 3–19   FIGURE 3–20

equal to that lying between dA and the plane of the free surface. Thus

$$F_v = \int dF_v = \rho g \iiint dV. \tag{3–11}$$

Therefore, F_v is equal to the weight of liquid in a volume equal to that lying between the surface and the plane of the free surface. This force acts through the
centroid of this volume. Thus, the vertical force on surface AB in Fig. 3–20 is
the weight of volume ABB', acting downward, and that on surface BC is equal
to the weight of volume $B'BCC'$, acting upward. (Note that this volume of liquid
may not actually exist.) The total vertical hydrostatic forces on surface ABC
is therefore equal to the weight of the volume $ABCC'$, acting upward through
the centroid of this volume.

From this, *Archimedes' principle* follows immediately: the buoyant force on a
solid is equal to the weight of liquid it has displaced, and acts through the
centroid of the displaced volume.

Example 3–3. In Fig. 3–21, the three surfaces *adcd*, *bcf*, and *cdef* of a triangular
block are exposed to a body of liquid. Find the hydrostatic forces acting on
them.

To demonstrate the application of the results obtained in this section, the
three planes are treated together as a continuous surface. The vertical force on
this continuous surface is equal to the weight of liquid in a volume equal to

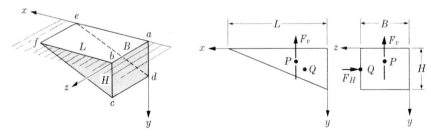

FIGURE 3–21

$\frac{1}{2}HLB$ between this surface and the plane of the free surface, acting through the centroid P of this volume at $x = \frac{1}{3}L$ and $z = \frac{1}{2}B$.

Since the vertical projection normal to the x-axis of area $cdef$ is equal to the area $abcd$, there is no net horizontal force in the x-direction. The horizontal force in the z-direction is equal to the force on the vertical plane bcf. Using Eqs. 3–4 to 3–9, we have $F_H = \frac{1}{6}\rho g L H^2$, acting through point Q at $x = \frac{1}{4}L$ and $y = \frac{1}{2}H$. Note that the lines of action of F_v and F_H do not intersect in this case. The hydrostatic forces on this continuous surface therefore cannot be represented by a single force.

PROBLEMS

3–13. The vessel in Fig. 3–22 is filled with a liquid. Find the hydrostatic force acting on each of its two horizontal surfaces. What is the relationship among these two forces and the weight of the liquid inside?

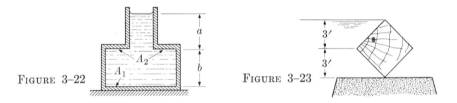

FIGURE 3–22   FIGURE 3–23

3–14. A long bar with a square cross section is fixed along the top of a dam in Fig. 3–23. Find the magnitude and the line of action of the resultant hydrostatic force acting on a foot of the bar.

3–15. A solid cylinder is used as an automatic gate in Fig. 3–24. The cylinder is hinged at A. The gate opens by turning about A when the depth of water H is 10 ft. What must be the weight per foot of the cylinder?

FIGURE 3–24

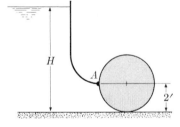

3–5. Stability of submerged and floating bodies

A position of equilibrium of a body is said to be *stable* if the departure of the body from this position remains small after being slightly disturbed. Otherwise, the position is said to be *unstable*. Thus a pendulum hanging with the bob down is in a stable position of equilibrium. If the bob is directly above the support, the pendulum is in equilibrium, but the position is unstable as it will overturn if slightly disturbed. An unstable equilibrium position is therefore a possible but not a probable position of a body.

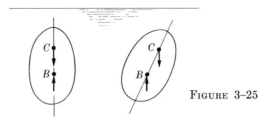

FIGURE 3–25

When a completely submerged body is in equilibrium, its weight acting through its center of mass C is colinear with an equal and opposite buoyant force acting through the *center of buoyancy B*. Point B is the centroid of the submerged volume and may not coincide with point C if the body is not homogeneous. If C is located above B, these two forces will form an overturning couple when the body is slightly disturbed (see Fig. 3–25). This couple will turn the body further from its original position of equilibrium. Thus, for completely submerged bodies, top heaviness causes instability. If C is located below B, these forces will form a righting couple when the body is slightly disturbed. The position is therefore stable.

When a floating body is in equilibrium, the volume of displacement V is such that the weight $\rho g V$ of the displaced liquid is equal to the weight W of the body:

$$\rho g V = W.$$

The *buoyant force* $\rho g V$ acts through the *center of buoyancy B*, which is the centroid of volume V. When the body is in equilibrium, B lies on the vertical through the center of mass C of the body, as shown in Fig. 3–26(a). It is to be shown that this position can sometimes be stable even when C is located above B. The reason for this is that when the body is in a disturbed position, the volume of displacement V assumes a different shape and, as a result, the center of buoyancy takes a new position B'. It is possible that the resultant couple is a righting couple, as shown in Fig. 3–26(b).

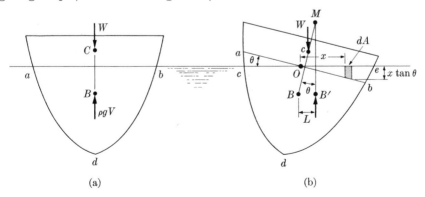

(a)   (b)

FIGURE 3–26

Let the body, in being disturbed, turn a small angle θ about an axis as shown. To locate B', take the moment about B of the weight of the new displaced volume $cdbe$, which is equal to volume $acdb$ plus the wedge obe minus the wedge aco. Since B is the centroid of volume $acdb$, its moment about B is zero. Since the volume of displacement remains the same, the weights of the wedges obe and aco are equal, and their moment is a couple. Since the moment of a couple is independent of the moment center, we can therefore write for a small disturbance

$$\rho g V \cdot L = \text{moment of } cdbe \text{ about axis through } B$$

$$= \text{moment of wedges about axis through } O$$

$$= \iint x(\rho g x \tan \theta \, dA)$$

$$= \rho g \tan \theta I \doteq \rho g \theta I,$$

where dA is a differential area of the plane ce bounded by the waterline, $x \tan \theta \, dA$ is a differential volume of the wedges as shown in Fig. 3–26(b), and $I = \iint x^2 \, dA$ is the second moment of plane ce with respect to the axis of turning. For a small disturbance, the second moment of plane ab can be used for I. It can be shown that this axis of turning passes through the centroid of plane ab (see Problem 3–18). Thus $L = \theta I / V$. The position is stable if L is larger than $\overline{CB} \sin \theta$, that is, if $I/V > \overline{CB}$.

In the discussion of the stability of floating bodies, it is convenient to refer to the intersection M of the vertical through B' and the line BC. The point M is located by

$$\overline{BM} = \frac{L}{\sin \theta} \doteq \frac{L}{\theta} = \frac{I}{V}. \tag{3–12}$$

If point M is found to be above C, the position is stable. Point M is called the *metacenter*, and the distance $\overline{MC}$ is called the *metacentric height*. Since a body may turn about different axes, there is a metacenter for each axis of turning. The body is stable if the center of mass C is below the lowest of the metacenters corresponding to the axis about which I of plane ab is minimum.

It should be mentioned that a body in a stable position of equilibrium may overturn if the disturbance is sufficiently large; e.g., the sinking of a ship in a storm. On the other hand, a body departing from an unstable position does not necessarily overturn, but may reach another position of equilibrium. (It can be verified that the intersection M in Fig. 3–26 is higher for a larger angle θ. Although the lowest position of M for small θ is below C, it is possible that when θ is larger than a certain value θ_0, M becomes higher than C. Then θ_0 is a stable position of equilibrium.) A more detailed analysis is necessary to reveal all the possibilities. However, a study of stability as defined above for small disturbances gives a significant indication of the situation.

PROBLEMS

3–16. A solid cube of specific gravity 0.8 floats on water. Will it probably float with one surface horizontal? Will it if the specific gravity is 0.7 instead?

3–17. Two barges are used to transport a piece of machinery, as shown in Fig. 3–27. Is the system stable? The center of gravity of the whole system is 1 ft above the water surface.

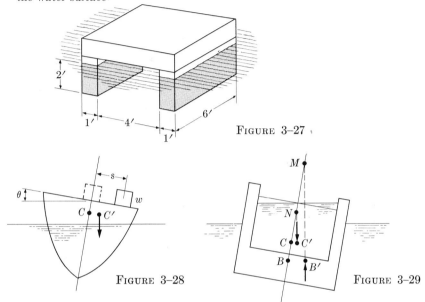

FIGURE 3–27

FIGURE 3–28

FIGURE 3–29

3–18. Show that in Fig. 3–26, any axis of turning due to a small disturbance must pass through the centroid of the plane ab bounded by the waterline.

3–19. The metacentric height of a ship can be determined experimentally by shifting a known weight w through a distance s on the ship, as shown in Fig. 3–28, and measuring the small angle of tilt θ thus created. Show that the metacentric height is $ws/W\theta$, where W is the weight of the ship and contents.

3–20. When a floating body contains a liquid, the center of mass of the body will shift to C' when disturbed, as shown in Fig. 3–29. Derive an expression for the distance $\overline{NC}$. This distance is of interest because the metacenter M must be above N to give stable equilibrium.

3–6. Equilibrium and stability of stratified fluids

Any horizontal stratification of density $\rho(h)$ can produce equilibrium in a body of fluid. The equation of fluid statics, $dp/dh = -\rho g$, will give the corresponding pressure distribution $p(h)$. Let us consider a perfect gas with its temperature T decreasing with elevation at a constant rate λ:

$$\frac{dT}{dh} = -\lambda. \tag{3–13}$$

The density of a perfect gas varies with the temperature and pressure according to the equation of state:

$$\rho = \frac{p}{RT},\tag{3-14}$$

where p and T are the absolute pressure and absolute temperature, respectively. This case serves as an approximation of the atmosphere within a few miles of the earth's surface, with air currents neglected. The value λ is called the *lapse rate* of the atmosphere. With the equation of fluid statics

$$\frac{dp}{dh} = -\rho g,\tag{3-15}$$

we have three equations for the determination of the three functions p, ρ, and T. Integrating Eq. 3-13 and using the condition that $T = T_1$ at $h = h_1$, we obtain

$$T = T_1 + \lambda(h_1 - h).\tag{3-16}$$

To relate p to T, we eliminate ρ from Eqs. 3-14 and 3-15:

$$dp = -\rho g\, dh = -\frac{pg}{RT}\left(-\frac{dT}{\lambda}\right) = \frac{pg}{R\lambda}\frac{dT}{T}.$$

Separating the variables p and T, and integrating, we have

$$\frac{p^{\lambda R/g}}{T} = \frac{p_1^{\lambda R/g}}{T_1},$$

where p_1 is the pressure at h_1. Substituting in T from Eq. 3-16, we have p in terms of h:

$$h - h_1 = \frac{T_1}{\lambda}\left[1 - \left(\frac{p}{p_1}\right)^{\lambda R/g}\right].\tag{3-17}$$

Thus, from known values of T_1 and p_1 at h_1 and the lapse rate, the elevation h of another location can be computed from the pressure p there.

While any horizontal stratification of density can produce equilibrium, the condition may not be stable. For example, a layer of heavier liquid can be in equilibrium on top of a layer of lighter liquid, but the condition will not remain if the system is slightly disturbed. If a drop of the heavier liquid is displaced into the lighter liquid, being heavier, it will drop farther away from its original position. For stratified liquids, bottom heaviness generally produces stability (see Problem 3-24). In gases, however, a particle upon being disturbed will experience the pressure of its new environment and will therefore change in density. Its new density may be larger or smaller than that of the surrounding gas. As a result, bottom heaviness does not necessarily mean stability, as will be shown presently.

Let a small parcel of a perfect gas be displaced through dh upward from its position of equilibrium at h_0. It will acquire the pressure of its new environment, but its density and temperature will be different from those of the new surroundings. It will rise farther if its new density is less than that of the surrounding gas. The condition is then unstable. Let p_0 and ρ_0 be the undisturbed values associated with h_0. During the disturbance, the effects of heat conduction and internal friction are probably unimportant. For a perfect gas in this isentropic process, its new pressure p_1 and density ρ_1 are related by the isentropic relation

$$\frac{p_1}{\rho_1^\gamma} = \frac{p_0}{\rho_0^\gamma} \qquad (\gamma > 1). \tag{3–18}$$

The pressure p_1 is that of its new surroundings:

$$p_1 = p_0 + \left(\frac{dp}{dh}\right)_0 dh.$$

Its new density is therefore

$$\rho_1 = \rho_0 \left(\frac{p_1}{p_0}\right)^{1/\gamma} = \rho_0 \left[1 + \frac{1}{p_0}\left(\frac{dp}{dh}\right)_0 dh\right]^{1/\gamma} = \rho_0 \left[1 + \frac{1}{\gamma p_0}\left(\frac{dp}{dh}\right)_0 dh + \cdots\right].$$

Here we have used the binomial series.

$$(1 \pm e)^n = 1 \pm ne + \cdots \qquad (e < 1).$$

Since the density of the surrounding gas is $\rho_0 + (d\rho/dh)_0\, dh$, the condition is unstable if

$$\rho_0 \left[1 + \frac{1}{\gamma p_0}\left(\frac{dp}{dh}\right)_0 dh\right] < \rho_0 + \left(\frac{d\rho}{dh}\right)_0 dh.$$

Since the elevation h_0 is arbitrary, we have for instability

$$\frac{\rho}{\gamma p}\frac{dp}{dh} < \frac{d\rho}{dh}.$$

Since dp/dh is negative, we can have instability even with negative $d\rho/dh$, i.e., with bottom heaviness.

It is instructive to express in terms of dT/dh the critical condition for stability when

$$\frac{\rho}{\gamma p}\frac{dp}{dh} = \frac{d\rho}{dh}. \tag{3–19}$$

Using Eqs. 3–14, 3–19, and 3–15 successively, we have

$$\frac{dT}{dh} = \frac{d}{dh}\left(\frac{p}{R\rho}\right) = \frac{1}{R\rho}\left(\frac{dp}{dh} - \frac{p}{\rho}\frac{d\rho}{dh}\right) = \frac{1}{R\rho}\left(1 - \frac{1}{\gamma}\right)\frac{dp}{dh} = -\frac{g}{R}\left(1 - \frac{1}{\gamma}\right).$$

$$\tag{3–20}$$

This critical rate of decrease of temperature with height is called *"adiabatic"* *lapse rate*. For air, $R/g = 53.35$ ft/°F or 29.27 m/°C, and $\gamma = 1.40$. This rate is 0.54°F per 100 ft or 0.98°C per 100 m. Any local lapse rate greater than this means instability at this level. For example, when the ground is heated during a summer day, the lower strata of air become heated and the lapse rate increases. The resultant instability causes large-scale convective rise of the hot air, with possible thunderstorms as a result.

PROBLEMS

3–21. Given p_0 and ρ_0 at $h = 0$, show that at another h in a perfect gas with a uniform temperature, $p = p_0 e^{-\rho_0 g h / p_0}$.

3–22. At sea level, the temperature is 70°F and the pressure is 14.7 psi abs. At the same time, the temperature at Station A is 50°F and the pressure is 11.76 psi abs. Assuming a constant lapse rate, find the elevation of Station A.

3–23. Can the atmosphere described in the previous problem be stable?

3–24. Assuming that the bulk modulus K ($= \rho \, dp/d\rho$) for ocean water is constant, show that the critical condition for stability is

$$\frac{\rho}{K} \frac{dp}{dh} = \frac{d\rho}{dh}.$$

Also find the density distribution $\rho(h)$ under this condition. [*Hint:* By Taylor's expansion, $e^x = 1 + x + \frac{1}{2}x^2 + \cdots .$]

3–7. Surface tension

The interface between two immiscible fluids behaves as if there were a film under tension. The magnitude of this tension depends on the fluids involved and the temperature. We now consider two cases with three materials in contact.

When three immiscible fluids are in equilibrium as shown in Fig. 3–30, the angles A and B must be such that

$$\sigma_{13} = \sigma_{12} \cos A + \sigma_{23} \cos B, \qquad \sigma_{12} \sin A = \sigma_{23} \sin B.$$

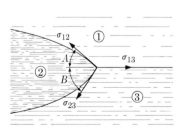

FIGURE 3–30

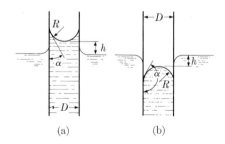

(a)   (b)

FIGURE 3–31

If $\sigma_{13} > \sigma_{12} + \sigma_{23}$, equilibrium is impossible and fluid 2 will spread, as in the case of a lubricating oil dropped on a water-air interface. When this is not the case, equilibrium is possible and fluid 2 will form a lense, as in the case of drops of melted fat on the surface of a bowl of soup.

A problem of practical interest is capillary rise or depression in tubes. When two immiscible fluids come into contact with a solid surface, the contact angle α, as shown in Fig. 3–31, depends on the three materials and the temperature. When the contact angle is known, the capillary rise or depression in small tubes can be computed. If the diameter of the tube is less than a fraction of an inch, the meniscus is spherical in shape. As can be seen in Fig. 3–31, the radius of the meniscus in $R = \frac{1}{2}D/|\cos \alpha|$. The gage pressure just below the meniscus is therefore

$$p = \mp \sigma \left(\frac{1}{R} + \frac{1}{R} \right) = \mp \frac{4\sigma |\cos \alpha|}{D},$$

with the upper sign and the lower sign for Fig. 3–31(a) and 3–31(b), respectively. Therefore

$$h = \frac{4\sigma |\cos \alpha|}{\rho g D}.$$

At 68°F, $\sigma = 0.0050$ lb/ft for water-air, and 0.035 lb/ft for mercury-air; and $\alpha = 25.5°$ for water and 129° for mercury in clean glass tubes.

PROBLEM

3–25. A drop of water at 68°F forms a circular spot of radius R between two plates at a small distance d apart, as shown in Fig. 3–32. Compute the force required to pull the plates further apart.

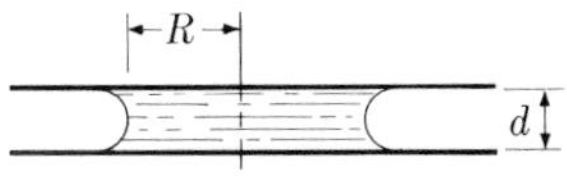

FIGURE 3–32

4 Kinematics of Fluids

4–1. Pictorial methods of description

In the study of fluid flow, it is necessary to describe the motion of an infinite number of fluid particles. This task is similar to that of describing the movement of numerous vehicles in a busy city street. Pictorially, this can be done by two methods. One is to trace the movement of the individual vehicles. This can be done by taking long-exposure photographs. Each vehicle will trace a continuous line in the picture. Another method is to describe the instantaneous flow pattern. This can be done by taking successive photographs of short exposure. In each of these pictures, each vehicle will appear as a short dash which indicates the direction and the speed of the vehicle movement. This picture shows the instantaneous flow pattern formed by the numerous dashes.

These methods are used in the study of fluid motion. The motion can be made visible, for example, by distributing on the surface and throughout the fluid small solid particles which follow the fluid in motion. However, we are not interested so much in the technique of taking the photographs as in the methods of describing the motion. In the first (long-exposure) method, attention is concentrated on the life history of the particles. In the second (short-exposure) method, interest is centered on the instantaneous flow pattern throughout the whole field (see Fig. 4–1). In fluid mechanics, the second method is usually more advantageous.

A line traced out by a particle through a period of time is called a *pathline*. The lines in the long-exposure photographs are pathlines. A line which is tangential to the instantaneous velocity vectors of the particles on it is called a *streamline* (see Fig. 6–10). Streamlines can be drawn on the short-exposure photographs by sketching lines tangential to the short dashes. These streamlines form the instantaneous *flow pattern*. The flow pattern may change with time. If the flow pattern and the speed at each point remain unchanged, the flow is said to be *steady*. In a steady flow, the pathline and the streamline through any point coincide.

PROBLEM

4–1. (a) Soldiers are marching in a column with four in a row. At a certain instant, all turn left and continue marching. Draw several pathlines. Also draw the flow patterns before and after the left turn. (b) A *streak line* for a given point is a line joining the particles which have passed this point. Sketch the streak line for the starting point of a soldier in the front row. (c) Under what condition will the streamline and the streak line through a point coincide?

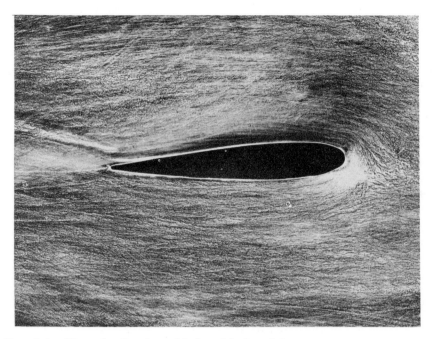

FIG. 4–1. Flow visualization with lampblack and kerosene vapor. (By Aerolab Supply Co., Hyattsville, Md.)

4–2. The Lagrangian and Eulerian systems

Corresponding to the two pictorial methods presented in the previous section, there are two mathematical schemes of description of fluid motion. One describes directly the flow characteristics of fluid particles, and the other describes directly the flow pattern throughout the field. The former system is named *Lagrangian*. In this system, the independent variables are a, b, c, and t, where t is time, and a, b, and c are the space coordinates of the fluid elements at a certain time t_0. Thus each element is identified by a set of three values of these coordinates according to where it was at t_0. All dependent variables, such as the subsequent location, density, etc., of the elements, are expressed in terms of these four independent variables. For example, in a flow with uniform speed U in the x-direction, the position of each and every element is given by

$$x = a + Ut,$$

where a is the x-coordinate of the particle at $t = 0$. Here the position x is a dependent variable, expressed in terms of the independent variables a and t. This system describes directly what happens to each element, identified by its original position (a, b, c). This system is used occasionally in fluid mechanics.

The other system of description is called the *Eulerian* system. In this system, the independent variables are three space coordinates and time, such as x, y, z, and t with rectangular coordinates, and r, θ, z, and t with cylindrical coordinates. All dependent variables, such as temperature, velocity, etc., are expressed in terms of these independent variables. For example, the temperature distribution in an atmosphere may be described as

$$T = T_0 - \lambda z + kt.$$

Here the temperature decreases at a lapse rate λ with elevation z, but is increasing at a rate k degrees per unit time. No reference is made as to the fluid elements. Any element that happens to be at z at time t will have the indicated T. Instead, this system describes primarily what is happening throughout the whole field. This method is usually found to be more convenient mathematically in the study of fluid motion, and will be used exclusively in this book.

PROBLEMS

4–2. In Fig. 4–2, the flow is in the x-direction with speed $u = U[1 - (1 - y/H)^2]$, where U is a constant. Compute the speed at the following points:

$x = H, y = 0$; $x = H, y = \frac{1}{4}H$;

$x = H, y = \frac{1}{2}H$; $x = H, y = \frac{3}{4}H$;

$x = H, y = H$;   and $x = 0, y = \frac{1}{2}H$.

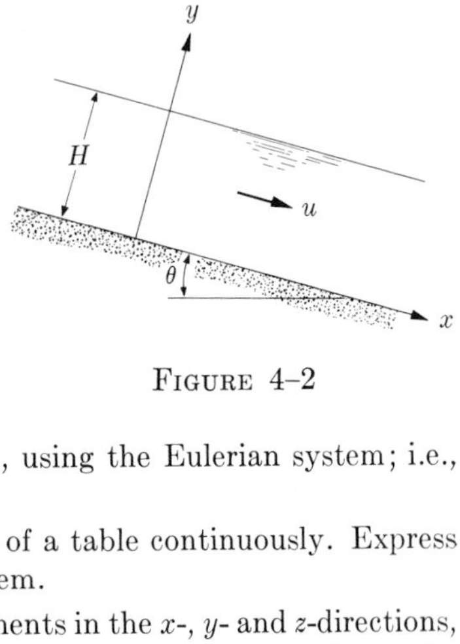

FIGURE 4–2

Sketch several streamlines.

4–3. Express the velocity distribution in Fig. 1–1, using the Eulerian system; i.e., express u as a function of the coordinates.

4–4. Particles are dropping from rest off the edge of a table continuously. Express the velocity distribution in the Eulerian system.

4–5. With u, v, and w denoting the velocity components in the x-, y- and z-directions, respectively, a flow is described as $u = -\frac{1}{2}x$, $v = \frac{1}{2}y$, and $w = 0$. Draw the velocity vectors at the points $(2, \frac{1}{2}, 0)$, $(2, 1, 0)$, $(1, 1, 0)$, $(\frac{1}{2}, 2, 0)$, $(1, 2, 0)$, and $(2, 2, 0)$. Sketch a streamline (which does not necessarily pass through these points).

4–3. Particle derivative

In Newton's law of motion as stated in Eq. 1–2, the quantity $d(m\mathbf{q})/dt$ is the rate of change of the momentum $m\mathbf{q}$ of a particle, and is called the *particle derivative* of $m\mathbf{q}$. We shall first discuss the particle derivative of a scalar property of a particle, such as its temperature T. In the Eulerian system with rectangular coordinates, the temperature distribution in the flow is described with T expressed as a function of x, y, z, and t. For changes dx, dy, dz, and dt, the total change of $T(x, y, z, t)$ is

$$dT = \frac{\partial T}{\partial x}\, dx + \frac{\partial T}{\partial y}\, dy + \frac{\partial T}{\partial z}\, dz + \frac{\partial T}{\partial t}\, dt.$$

In following a particle in this flow, there will be displacements dx, dy, and dz of the particle during dt. The expression above gives the total change of T experienced by this particle during dt. Thus the particle derivative is

$$\frac{DT}{Dt} = \frac{\partial T}{\partial x}\frac{dx}{dt} + \frac{\partial T}{\partial y}\frac{dy}{dt} + \frac{\partial T}{\partial z}\frac{dz}{dt} + \frac{\partial T}{\partial t}.$$

[The symbol D/Dt is used to indicate a particle derivative of a function of several variables, such as $T(x, y, z, t)$; while the symbol d/dt indicates the time derivative of a function of t only. d/dt can also indicate a particle derivative; e.g., $d(mq)/dt$ in Eq. 1–2.] Since dx, dy, and dz are the particle displacements in dt, dx/dt, dy/dt, and dz/dt are the velocity components u, v, and w of the particle in the x-, y-, and z-directions, respectively. Thus

$$\frac{DT}{Dt} = u\frac{\partial T}{\partial x} + v\frac{\partial T}{\partial y} + w\frac{\partial T}{\partial z} + \frac{\partial T}{\partial t}. \tag{4–1}$$

The term $\partial T/\partial t$, the partial derivative obtained with x, y, and z held constant, is the *local rate of change*. It is the rate of change of T at a point as different particles pass by. The other three terms on the right of Eq. 4–1 are called the *convective terms*, which represent the rate of change due to the fact that the particle moves into a new position with a different T. The total change experienced by a moving particle is the sum of the local and the convective changes.

For ease of instruction, Eq. 4–1 has been derived for the particle derivative of a scalar quantity. The same derivation is completely valid for a vector quantity. For example, take the velocity distribution $\mathbf{q}(x, y, z, t)$ of a fluid flow. The rate of change of velocity of a particle moving in this field is

$$\frac{D\mathbf{q}}{Dt} = u\frac{\partial \mathbf{q}}{\partial x} + v\frac{\partial \mathbf{q}}{\partial y} + w\frac{\partial \mathbf{q}}{\partial z} + \frac{\partial \mathbf{q}}{\partial t}. \tag{4–2}$$

For a discussion of the derivative of a vector quantity, see the next section.

Example 4–1. The temperature T in a long tunnel is known to be

$$T = T_0 - \alpha e^{-x/L} \sin\frac{2\pi t}{\tau},$$

where T_0, α, L, and τ are constants, and x is measured from the entrance. A particle moves into the tunnel with a constant speed U. Find the rate of change of temperature it experiences.

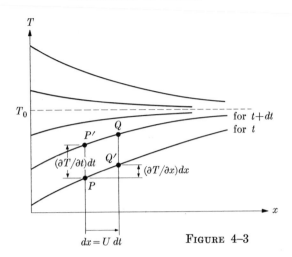

FIGURE 4-3

The temperature in the tunnel is expressed in the Eulerian system and is shown graphically for several values of t in Fig. 4-3. As the particle moves in the tunnel, its temperature is indicated successively by dots such as P and Q. If it did not move, it would experience only the local change from P to P' during dt. If the temperature distribution did not vary with time (i.e., if there were only one curve for all t in Fig. 4-3), the moving particle would experience only a convective change between P and Q'. It can be seen from Fig. 4-3 that for an infinitesimal dt, the change of T from P to Q is the sum of the local change from P to P' and the convective change from P to Q'; that is, $dT = (\partial T/\partial t)\, dt + (\partial T/\partial x) U\, dt$.

From the given temperature distribution $T(x, t)$, we have

$$\frac{DT}{Dt} = \frac{\partial T}{\partial t} + U\frac{\partial T}{\partial x} = -\frac{2\pi \alpha e^{-x/L}}{\tau}\cos\frac{2\pi t}{\tau} + \frac{U\alpha e^{-x/L}}{L}\sin\frac{2\pi t}{\tau}.$$

This gives DT/Dt at any x the particle happens to be at t.

PROBLEMS

4-6. At $t = 0$, particle A passes the origin with $T = 0°C$. At $t = 1$ sec, it passes the point $(1, 0, 0)$ one foot away with $T = 3°C$, while particle B passes the origin with $T = 2°C$. Estimate $\partial T/\partial t$ and $\partial T/\partial x$ at the origin, and DT/Dt of particle A during this interval. Can you see any relationship among these three values?

4-7. The market price P in dollars of used cars of a certain model is found to be $P = 1000 + 0.02x - 2.00t$, where x is the distance in miles west of Detroit and t is time in days. If a car of this model is driven from Detroit at $t = 0$ towards the west at a rate of 400 miles per day, find by using a particle derivative whether its value is increasing or decreasing. How much of this change is due to depreciation, and how much is due to moving into a better market?

4-8. When T is expressed in terms of cylindrical coordinates as $T(r, \theta, z, t)$ in the Eulerian system,
show that

$$\frac{DT}{Dt} = u \frac{\partial T}{\partial r} + \frac{v}{r} \frac{\partial T}{\partial \theta} + w \frac{\partial T}{\partial z} + \frac{\partial T}{\partial t}, \qquad (4\text{-}3)$$

where u, v, and w are the velocity components
in the r-, θ-, and z-directions, respectively, as
shown in Fig. 4-4.

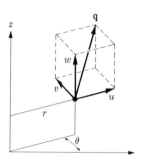

FIGURE 4-4

4-4. Derivative of a vector quantity

To study the term $d(m\mathbf{q})/dt$ in the equation of motion, first define the addition
of vectors and the derivative of a vector. The *sum of vectors* $\mathbf{A}$ and $\mathbf{B}$ is a vector
with its magnitude and direction determined as shown in Fig. 4-5. A vector
quantity may be a function of a scalar; e.g.,
velocity $\mathbf{q}$ of a particle may vary with time t.
The *derivative* of $\mathbf{q}(t)$ with respect to t is defined similarly to that of a scalar function:

$$\frac{d}{dt} \mathbf{q}(t) \equiv \lim_{\Delta t \to 0} \frac{\mathbf{q}(t + \Delta t) - \mathbf{q}(t)}{\Delta t} = \lim_{\Delta t \to 0} \frac{\Delta \mathbf{q}}{\Delta t},$$

$$(4\text{-}4)$$

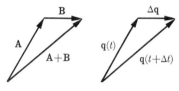

FIGURE 4-5   FIGURE 4-6

where $\Delta \mathbf{q}$ is the vector difference between $\mathbf{q}$ at $(t + \Delta t)$ and $\mathbf{q}$ at t (see Fig. 4-6).
The direction of the derivative which is a vector is in the direction of $\Delta \mathbf{q}$ (as
$\Delta t \to 0$). It should be noted that the derivative is zero only when the vector $\mathbf{q}$
is constant in magnitude as well as in direction. If a vector is constant in magnitude but varies in direction, its derivative, according to a construction like
Fig. 4-6, is perpendicular to it. From the definition above, it is not difficult to
prove the following equalities, which will be found useful subsequently:

$$\frac{d}{dt} (\mathbf{A} + \mathbf{B}) = \frac{d\mathbf{A}}{dt} + \frac{d\mathbf{B}}{dt} \qquad (4\text{-}5)$$

and

$$\frac{d}{dt} (m\mathbf{A}) = m \frac{d\mathbf{A}}{dt} + \mathbf{A} \frac{dm}{dt}, \qquad (4\text{-}6)$$

where m is a scalar. For example, neglecting higher-order terms, we have

$$\frac{d}{dt} (m\mathbf{A}) = \lim_{\Delta t \to 0} \frac{(m + \Delta m)(\mathbf{A} + \Delta \mathbf{A}) - m\mathbf{A}}{\Delta t} = \lim_{\Delta t \to 0} \frac{m \, \Delta \mathbf{A} + \mathbf{A} \, \Delta m}{\Delta t}$$

$$= m \frac{d\mathbf{A}}{dt} + \mathbf{A} \frac{dm}{dt}.$$

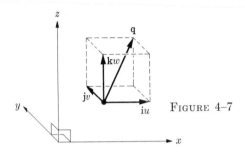

FIGURE 4–7

In analysis, it is often convenient to express a vector as the sum of three orthogonal components. When fixed rectangular coordinate axes are used, we have

$$\mathbf{q} = \mathbf{i}u + \mathbf{j}v + \mathbf{k}w, \qquad (4\text{–}7)$$

where u, v, and w are the (scalar) components of vector $\mathbf{q}$ in the directions of the x-, y-, and z-axes, respectively, and $\mathbf{i}$, $\mathbf{j}$, and $\mathbf{k}$ are vectors with magnitudes equal to unity pointing in the directions of the x-, y-, and z-axes, respectively (see Fig. 4–7). They are called *unit vectors*. Because they are vectors with constant magnitude and direction, we have $d\mathbf{i}/dt = d\mathbf{j}/dt = d\mathbf{k}/dt = 0$. From Eqs. 4–5 and 4–6, we have

$$\frac{d\mathbf{q}}{dt} = \frac{d}{dt}\,(\mathbf{i}u + \mathbf{j}v + \mathbf{k}w) = \mathbf{i}\,\frac{du}{dt} + \mathbf{j}\,\frac{dv}{dt} + \mathbf{k}\,\frac{dw}{dt}. \qquad (4\text{–}8)$$

Thus, with rectangular coordinate axes, the x-component of the derivative of $\mathbf{q}$, for example, is equal to the derivative of the x-component of $\mathbf{q}$.

When cylindrical coordinates are used, $\mathbf{q}$ may be expressed in terms of its components u, v, and w in the radial, circumferential, and axial directions, respectively, as shown in Fig. 4–4:

$$\mathbf{q} = \mathbf{e}_r u + \mathbf{e}_\theta v + \mathbf{e}_z w, \qquad (4\text{–}9)$$

where $\mathbf{e}_r$, $\mathbf{e}_\theta$, and $\mathbf{e}_z$ are unit vectors in the respective directions. The unit vectors $\mathbf{e}_r$ and $\mathbf{e}_\theta$ are not constant vectors but vary in direction depending on the coordinate θ. To differentiate $\mathbf{q}$, one must first find how these unit vectors vary with θ. For a small change of coordinate $d\theta$, as shown in Fig. 4–8, the direction of $d\mathbf{e}_\theta$ is seen to be opposite to $\mathbf{e}_r$. Since the triangles abc and lmn are similar, the magnitude of $d\mathbf{e}_\theta$ is $d\theta$. Thus

$$d\mathbf{e}_\theta = -\mathbf{e}_r\, d\theta. \qquad (4\text{–}10)$$

It can be shown in a similar manner that

$$d\mathbf{e}_r = \mathbf{e}_\theta\, d\theta. \qquad (4\text{–}11)$$

We shall use these equations in finding the components of $d\mathbf{q}/dt$ in the cylindrical coordinate system.

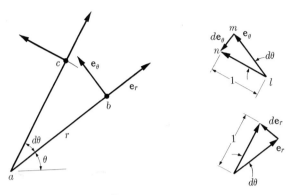

FIGURE 4–8

Throughout this book, the letters u, v, and w are used to denote three orthogonal components of the velocity vector $\mathbf{q}$. In the case of rectangular coordinates, they are in the x-, y-, and z-directions, respectively. When cylindrical coordinates are used, they are in the r-, θ-, and z-directions, respectively. In any case, the (scalar) *speed* q can be computed from

$$q^2 = u^2 + v^2 + w^2. \tag{4–12}$$

PROBLEM

4–9. For the cylindrical coordinate system, show that

$$\mathbf{e}_r = \mathbf{i}\cos\theta + \mathbf{j}\sin\theta, \qquad \mathbf{e}_\theta = -\mathbf{i}\sin\theta + \mathbf{j}\cos\theta.$$

From these, derive Eqs. 4–10 and 4–11.

4–5. Particle acceleration

We can now express the particle derivative $d(m\mathbf{q})/dt$ in the Eulerian system. We shall consider only cases where the mass m of a particle does not change with time. Then

$$\frac{d(m\mathbf{q})}{dt} = m\frac{d\mathbf{q}}{dt} = m\mathbf{a}, \tag{4–13}$$

where the vector $\mathbf{a}$ is the particle derivative of velocity $\mathbf{q}$, and is called the *acceleration* of the particle.

First consider an Eulerian system with fixed rectangular coordinate axes. The vector $\mathbf{q}$ and its components u, v, and w are then functions of x, y, z, and t. According to Eq. 4–2, the particle derivative of $\mathbf{q}$ is

$$\mathbf{a} = \frac{D\mathbf{q}}{Dt} = \frac{\partial \mathbf{q}}{\partial t} + u\frac{\partial \mathbf{q}}{\partial x} + v\frac{\partial \mathbf{q}}{\partial y} + w\frac{\partial \mathbf{q}}{\partial z}. \tag{4–14}$$

With $\mathbf{q} = \mathbf{i}u + \mathbf{j}v + \mathbf{k}w$, where $\mathbf{i}$, $\mathbf{j}$, and $\mathbf{k}$ are constant vectors, we have, for example,

$$\frac{\partial \mathbf{q}}{\partial t} = \frac{\partial}{\partial t}(\mathbf{i}u + \mathbf{j}v + \mathbf{k}w) = \mathbf{i}\frac{\partial u}{\partial t} + \mathbf{j}\frac{\partial v}{\partial t} + \mathbf{k}\frac{\partial w}{\partial t}.$$

Equation 4–14 can therefore be shown to be

$$\mathbf{a} = \mathbf{i}\left(\frac{\partial u}{\partial t} + u\frac{\partial u}{\partial x} + v\frac{\partial u}{\partial y} + w\frac{\partial u}{\partial z}\right) + \mathbf{j}\left(\frac{\partial v}{\partial t} + u\frac{\partial v}{\partial x} + v\frac{\partial v}{\partial y} + w\frac{\partial v}{\partial z}\right)$$

$$+ \mathbf{k}\left(\frac{\partial w}{\partial t} + u\frac{\partial w}{\partial x} + v\frac{\partial w}{\partial y} + w\frac{\partial w}{\partial z}\right).$$

The components of the particle acceleration are therefore

$$a_x = \frac{\partial u}{\partial t} + u\frac{\partial u}{\partial x} + v\frac{\partial u}{\partial y} + w\frac{\partial u}{\partial z},$$

$$a_y = \frac{\partial v}{\partial t} + u\frac{\partial v}{\partial x} + v\frac{\partial v}{\partial y} + w\frac{\partial v}{\partial z}, \tag{4–15}$$

$$a_z = \frac{\partial w}{\partial t} + u\frac{\partial w}{\partial x} + v\frac{\partial w}{\partial y} + w\frac{\partial w}{\partial z}.$$

The physical meaning of each term in these equations is not difficult to understand. For example, in the expression for a_x, the term $\partial u/\partial t$ is that part of a_x due to the local change of the x-component of u of the velocity, and the other terms are the convective change of u due to the change of the particle position.

When cylindrical coordinates are used, the particle derivative is obtained with Eq. 4–3:

$$\mathbf{a} = \frac{D\mathbf{q}}{Dt} = \frac{\partial \mathbf{q}}{\partial t} + u\frac{\partial \mathbf{q}}{\partial r} + \frac{v}{r}\frac{\partial \mathbf{q}}{\partial \theta} + w\frac{\partial \mathbf{q}}{\partial z},$$

where $\mathbf{q} = \mathbf{e}_r u + \mathbf{e}_\theta v + \mathbf{e}_z w$. Since $\mathbf{e}_r$ and $\mathbf{e}_\theta$ vary with θ, we have

$$\frac{\partial \mathbf{q}}{\partial \theta} = \frac{\partial}{\partial \theta}(\mathbf{e}_r u + \mathbf{e}_\theta v + \mathbf{e}_z w) = \mathbf{e}_r\frac{\partial u}{\partial \theta} + u\frac{\partial \mathbf{e}_r}{\partial \theta} + \mathbf{e}_\theta\frac{\partial v}{\partial \theta} + v\frac{\partial \mathbf{e}_\theta}{\partial \theta} + \mathbf{e}_z\frac{\partial w}{\partial \theta}.$$

From Eqs. 4–10 and 4–11, we have

$$\frac{\partial \mathbf{e}_r}{\partial \theta} = \mathbf{e}_\theta \quad \text{and} \quad \frac{\partial \mathbf{e}_\theta}{\partial \theta} = -\mathbf{e}_r.$$

Therefore

$$\frac{v}{r}\frac{\partial \mathbf{q}}{\partial \theta} = \mathbf{e}_r\left(\frac{v}{r}\frac{\partial u}{\partial \theta} - \frac{v^2}{r}\right) + \mathbf{e}_\theta\left(\frac{v}{r}\frac{\partial v}{\partial \theta} + \frac{uv}{r}\right) + \mathbf{e}_z\frac{v}{r}\frac{\partial w}{\partial \theta}.$$

Since the unit vectors are independent of r, z, and t, we have, for example,

$$\frac{\partial \mathbf{q}}{\partial t} = \frac{\partial}{\partial t}(\mathbf{e}_r u + \mathbf{e}_\theta v + \mathbf{e}_z w) = \mathbf{e}_r \frac{\partial u}{\partial t} + \mathbf{e}_\theta \frac{\partial v}{\partial t} + \mathbf{e}_z \frac{\partial w}{\partial t}.$$

Therefore

$$\mathbf{a} = \mathbf{e}_r \left(\frac{\partial u}{\partial t} + u \frac{\partial u}{\partial r} + \frac{v}{r}\frac{\partial u}{\partial \theta} - \frac{v^2}{r} + w \frac{\partial u}{\partial z}\right)$$

$$+ \mathbf{e}_\theta \left(\frac{\partial v}{\partial t} + u \frac{\partial v}{\partial r} + \frac{v}{r}\frac{\partial v}{\partial \theta} + \frac{uv}{r} + w \frac{\partial v}{\partial z}\right)$$

$$+ \mathbf{e}_z \left(\frac{\partial w}{\partial t} + u \frac{\partial w}{\partial r} + \frac{v}{r}\frac{\partial w}{\partial \theta} + w \frac{\partial w}{\partial z}\right),$$

that is,

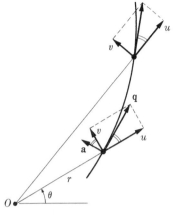

FIGURE 4–9

$$a_r = \frac{\partial u}{\partial t} + u \frac{\partial u}{\partial r} + \frac{v}{r}\frac{\partial u}{\partial \theta} + w \frac{\partial u}{\partial z} - \frac{v^2}{r},$$

$$a_\theta = \frac{\partial v}{\partial t} + u \frac{\partial v}{\partial r} + \frac{v}{r}\frac{\partial v}{\partial \theta} + w \frac{\partial v}{\partial z} + \frac{uv}{r}, \qquad (4\text{–}16)$$

$$a_z = \frac{\partial w}{\partial t} + u \frac{\partial w}{\partial r} + \frac{v}{r}\frac{\partial w}{\partial \theta} + w \frac{\partial w}{\partial z}.$$

The meaning of the first four terms on the right of these equations is clear. It is well known that due to the change of direction of $\mathbf{e}_\theta$, we have $a_r = -v^2/r$ for a particle moving along a circular path ($u = 0$). Due to the change of direction of $\mathbf{e}_r$, we have the term uv/r in a_θ. For example, consider a particle in a flow with constant u and v, as shown in Fig. 4–9. Since $\mathbf{q}$ is constant in magnitude, the particle acceleration must be perpendicular to $\mathbf{q}$. It can be seen that $a_r/a_\theta = -v/u$ or $(-v^2/r)/(uv/r)$.

PROBLEMS

4–10. In a steady flow in the xy-plane, the following velocity components at four points are given (all quantities in ft-sec units): $u = 1$, $v = 1$ at the origin; $u = 1.1$, $v = 1$ at $(1, 0, 0)$; $u = 1.3$, $v = 1$ at $(0, 1, 0)$; and $u = 1.4$, $v = 1$ at point A at $(1, 1, 0)$. A particle passing the origin will be close to A in one second later, with a change of u of about 0.4. Compare this value with a_x computed from Eq. 4–15.

4–11. (a) If the amount of liquid flowing down the plane in Fig. 4–2 is increasing slowly, the velocity components can be approximated as $u = kt[1 - (1 - y/mt^2)^2]$, $v = w = 0$, where k and m are constants. Compute the particle acceleration component a_x. Is this due to local change or convective change or both? (b) The flow between two converging walls which, if extended, intersect at the z-axis is approximately radial with $u = -k/r$, where k is a constant, and r is the radial coordinate. Compute the particle acceleration at $r = L$. Is this due to local change or convective change or both?

4–6. Steady flow

A flow is said to be *steady* if, at any point, the velocity does not change with time. In the Eulerian system, the velocity q (and therefore its components u, v, and w) is then a function of the spatial coordinates only, and therefore $\partial q / \partial t = 0$ (with $\partial u / \partial t = \partial v / \partial t = \partial w / \partial t = 0$). The flow pattern of a steady flow remains the same, and the streamlines and pathlines coincide.

In nature, steady flow is rather an exception than the usual case. If we examine closely the velocity at a point in a supposedly steady stream, we will find that it varies rapidly with time. If a velocity component at a point is recorded continuously, the record will appear similar to Fig. 4–10, even when the stream is discharging at a constant rate. The fluctuations are observed to be random and of high frequency, and the flow is referred to as being *turbulent* (see Fig. 2–5b). Turbulent flows are unsteady by definition. However, one can define a *temporal mean* velocity component $\bar{u}$ at a point:

$$\bar{u} = \frac{1}{\tau} \int_{t-\tau/2}^{t+\tau/2} u\, dt,$$

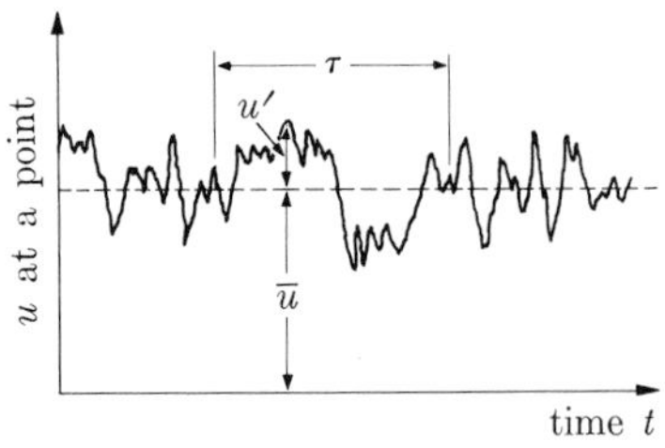

FIGURE 4–10

where τ is a period long enough to include a sufficient number of fluctuations for computing a stable mean value $\bar{u}$. If $\bar{u}$ and similarly defined $\bar{v}$ and $\bar{w}$ are independent of time at every point, as indicated, for example, by the horizontal dashed line in Fig. 4–10, the turbulent flow is said to be *quasi-steady*.

A flow in which rapid random fluctuations are absent is called a *laminar* flow; e.g., a sluggish flow of syrup (see also Fig. 2–5a). A laminar flow can be steady or unsteady.

Since velocity is the rate of particle displacement relative to a frame of reference, whether the flow is steady or not may depend also on the choice of the frame of reference. The earth is usually used for reference, and coordinate axes are fixed relative to the earth. However, a flow which is unsteady relative to the earth may sometimes be steady relative to axes moving relative to the earth. For example, consider the flow of air caused by a body moving through it at a constant velocity. When velocities relative to the earth are considered, the velocity at a point in space is zero before the body gets near, acquires a certain value when the body passes by, and finally returns to zero. Thus locally the velocity changes with time, and therefore the flow is unsteady relative to the earth. However, if coordinate axes are attached to the moving body, the velocity relative to the moving axes and body is steady. The flow pattern observed in the moving frame is identical with that of a steady flow (relative to the earth) passing a fixed body, as shown in Fig. 4–11(b). Since steady flows are easier to study, moving axes are often used when possible to reduce the flow to a steady one.

PROBLEM

4–12. Does steadiness of a flow imply that there is no particle acceleration? Classify the flows in Problem 4–11 as to their steadiness.

4–7. Moving axes

In the equation of motion $\mathbf{F} = d(m\mathbf{q})/dt$, the velocity $\mathbf{q}$ of the fluid particle is referred to an inertial system such as the earth unless the large-scale motion in the atmosphere or the oceans is being studied. When the frame of reference is moving at a constant velocity $\mathbf{U}$ relative to the earth, the velocity $\mathbf{q}_r$ at any point relative to the moving axes is related to the absolute velocity $\mathbf{q}$ relative to the earth by the principle of relative motion

$$\mathbf{q} = \mathbf{U} + \mathbf{q}_r, \qquad (4\text{–}17)$$

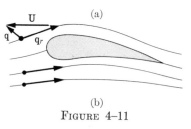

as indicated in Fig. 4–11. The particle acceleration $d\mathbf{q}/dt$ is therefore

$$\mathbf{a} = \frac{d\mathbf{q}}{dt} = \frac{d}{dt}(\mathbf{U} + \mathbf{q}_r) = \frac{d\mathbf{q}_r}{dt}.$$

In other words, the particle acceleration can be computed from the relative velocity in this case. The velocity components in Eqs. 4–15 and 4–16 can be those relative to a coordinate system moving at a constant velocity.

FIGURE 4–11

If the origin of the coordinate system is accelerating or the axes are rotating, the relationship between the relative and absolute velocities is more complicated. The particle acceleration to be used in the equation of motion cannot be computed from Eqs. 4–15 and 4–16 with the velocity components relative to the axes. Such moving axes will not be used in this book.

PROBLEMS

4–13. A circular cylinder of radius R moves with a constant velocity in a fluid. Given the radial and circumferential components of the velocity relative to the moving body

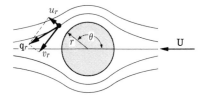

$$u_r = -U \cos \theta (1 - R^2/r^2),$$

$$v_r = U \sin \theta (1 + R^2/r^2)$$

FIGURE 4–12

(see Fig. 4–12), find the velocity relative to the earth at (a) $r = R$, $\theta = 0$; (b) $r = R$, $\theta = \pi/2$; and (c) $r = R$, $\theta = \pi/4$.

4–14. In the previous problem, find the particle acceleration at the three points.

4–8 Equation of continuity

As stated in Chapter 1, fluid mechanics is based on Newton's laws of motion, the laws of thermodynamics, the principle of conservation of mass, and the physical properties of fluids. In this section, the *principle of conservation of mass* is to be stated mathematically as an equation, called an *equation of continuity*. In deriving such an equation, imagine a volume in space. There is fluid flowing into and out of this volume. If the outflow of mass is greater than the inflow, the principle of conservation of mass requires that there is an equal decrease of mass stored in the volume.

In the case of rectangular coordinates, imagine a small frame in space dx, dy, and dz in size, as shown in Fig. 4–13. Let u_0 and ρ_0 be the values of u and ρ at the center at time t. At the same instant, the mean values of u and ρ at face $abcd$ are $u_0 + (\partial u/\partial x)_0(-dx/2)$ and $\rho_0 + (\partial\rho/\partial x)_0(-dx/2)$, respectively; and at face $efgh$, $u_0 + (\partial u/\partial x)_0(dx/2)$ and $\rho_0 + (\partial\rho/\partial x)_0(dx/2)$, respectively. Therefore, during a short interval dt,

Mass inflow through $abcd$

$$= \rho u \, dA \, dt = \left[\rho_0 - \left(\frac{\partial\rho}{\partial x}\right)_0 \frac{dx}{2}\right]\left[u_0 - \left(\frac{\partial u}{\partial x}\right)_0 \frac{dx}{2}\right] dy \, dz \, dt$$

$$= \rho_0 u_0 \, dy \, dz \, dt - \frac{1}{2}\left[u_0\left(\frac{\partial\rho}{\partial x}\right)_0 + \rho_0\left(\frac{\partial u}{\partial x}\right)_0\right] dx \, dy \, dz \, dt,$$

Mass outflow through $efgh$

$$= \left[\rho_0 + \left(\frac{\partial\rho}{\partial x}\right)_0 \frac{dx}{2}\right]\left[u_0 + \left(\frac{\partial u}{\partial x}\right)_0 \frac{dx}{2}\right] dy \, dz \, dt$$

$$= \rho_0 u_0 \, dy \, dz \, dt + \frac{1}{2}\left[u_0\left(\frac{\partial\rho}{\partial x}\right)_0 + \rho_0\left(\frac{\partial u}{\partial x}\right)_0\right] dx \, dy \, dz \, dt.$$

The dimension of $\rho u \, dA \, dt$ can be readily shown to be that of mass. The velocity components v and w at these two surfaces do not contribute to the inflow or outflow. The net outflow of mass across these two surfaces during dt is therefore

Net outflow through $abcd$ and $efgh$

$$= \left[u_0\left(\frac{\partial\rho}{\partial x}\right)_0 + \rho_0\left(\frac{\partial u}{\partial x}\right)_0\right] dx \, dy \, dz \, dt$$

$$= \left[\frac{\partial(\rho u)}{\partial x}\right]_0 dx \, dy \, dz \, dt.$$

Note that this net outflow is caused by the difference of ρu at these two surfaces. Should ρu be the same, that is, $\partial(\rho u)/\partial x = 0$, this net outflow would be zero.

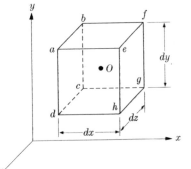

FIGURE 4–13

Similarly, we have

$$\text{Net outflow through } dcgh \text{ and } abfe = \left[\frac{\partial(\rho v)}{\partial y}\right]_0 dx\, dy\, dz\, dt,$$

$$\text{Net outflow through } bcgf \text{ and } adhe = \left[\frac{\partial(\rho w)}{\partial z}\right]_0 dx\, dy\, dz\, dt.$$

The sum of these is the total net outflow of mass from the frame during dt. By the principle of conservation of mass, this net outflow is equal to the decrease of mass during dt inside the frame. This decrease can be expressed as $-[(\partial\rho/\partial t)_0\, dt]\, dx\, dy\, dz$. Thus,

$$\left[\frac{\partial(\rho u)}{\partial x}\right]_0 + \left[\frac{\partial(\rho v)}{\partial y}\right]_0 + \left[\frac{\partial(\rho w)}{\partial z}\right]_0 = -\left(\frac{\partial\rho}{\partial t}\right)_0.$$

Since the point O is arbitrary, we have

$$\frac{\partial(\rho u)}{\partial x} + \frac{\partial(\rho v)}{\partial y} + \frac{\partial(\rho w)}{\partial z} = -\frac{\partial\rho}{\partial t}. \tag{4–18}$$

This is the equation of continuity which must be satisfied at every point in a fluid flow (except at a source or sink. See Example 6–2).

This equation can also be written as

$$\frac{\partial u}{\partial x} + \frac{\partial v}{\partial y} + \frac{\partial w}{\partial z} = -\frac{1}{\rho}\frac{D\rho}{Dt}$$

(see Problem 4–17). When the fluid can be considered to be incompressible (not necessarily homogeneous), the particle derivative of ρ is zero. The equation of continuity then becomes

$$\frac{\partial u}{\partial x} + \frac{\partial v}{\partial y} + \frac{\partial w}{\partial z} = 0. \tag{4–19}$$

In cylindrical coordinates, it can be shown by using a frame as shown in Fig. 4–14 that the equation of continuity is

$$\frac{\partial(\rho ur)}{r\,\partial r} + \frac{\partial(\rho v)}{r\,\partial\theta} + \frac{\partial(\rho w)}{\partial z} = -\frac{\partial\rho}{\partial t}. \tag{4–20}$$

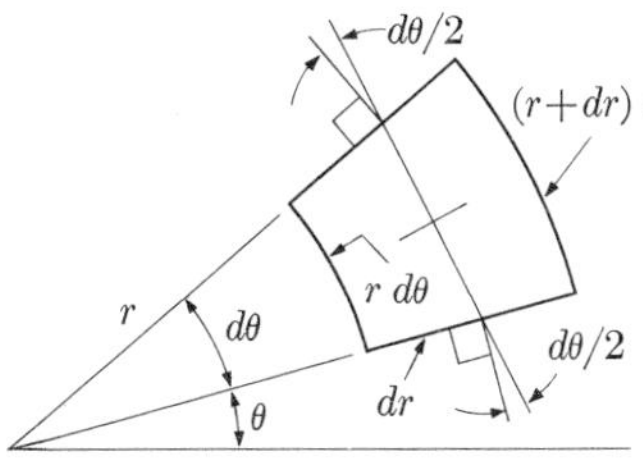

This equation can also be written as

$$\frac{\partial(ur)}{r\,\partial r} + \frac{\partial v}{r\,\partial\theta} + \frac{\partial w}{\partial z} = -\frac{1}{\rho}\frac{D\rho}{Dt}.$$

When the fluid is considered incompressible, we have

FIGURE 14–14

$$\frac{\partial(ur)}{r\,\partial r} + \frac{\partial v}{r\,\partial\theta} + \frac{\partial w}{\partial z} = 0. \tag{4–21}$$

See Eq. 4–27 for the equation of continuity in vector notations.

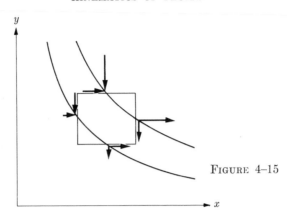

FIGURE 4–15

Example 4–2. Given $u = kx$ for a homogeneous incompressible fluid where all velocity vectors are parallel to the xy-plane, find the other velocity components.

Since the velocity vectors are all parallel to the xy-plane, we have $w = 0$. From Eq. 4–19 for incompressible flow,

$$\frac{\partial v}{\partial y} = -\frac{\partial u}{\partial x} = -k.$$

Integrating, we obtain

$$v = -ky + f(x, z, t).$$

Since $\partial v / \partial y$ is a partial derivative, we have an arbitrary function $f(x, z, t)$ of integration. Continuity demands that the function v must be of this form. The function f is to be determined by other considerations. If $f = 0$, the velocity vectors at various points and the streamlines can be sketched as shown in Fig. 4–15. Note that the net outflow in the x-direction from the frame ($\partial u / \partial x$ being positive) is balanced by the net inflow in the y-direction ($\partial v / \partial y$ being negative).

PROBLEMS

4–15. Using the frame shown in Fig. 4–14, derive directly the equation of continuity for a homogeneous incompressible fluid (ρ = constant) in the cylindrical coordinate system.

4–16. Simplify Eq. 4–20 for each of the following cases: (a) homogeneous incompressible fluid, (b) steady flow, and (c) all velocity vectors perpendicular to the z-axis.

4–17. Show that Eq. 4–18 can be written as

$$\frac{\partial u}{\partial x} + \frac{\partial v}{\partial y} + \frac{\partial w}{\partial z} = -\frac{1}{\rho}\frac{D\rho}{Dt},$$

where $D\rho/Dt$ is a particle derivative.

4–18. Water in a pan rotates steadily about the z-axis, as if it were a rigid body. Write the velocity components in a cylindrical coordinate system, and verify that they satisfy Eq. 4–21. Also write the velocity components in a rectangular coordinate system, and verify that Eq. 4–19 is satisfied.

4–19. Given $u = A_1x + B_1y + C_1z$, $v = A_2x + B_2y + C_2z$, and $w = A_3x + B_3y + C_3z$, what must be the relationship among the constant coefficients so that this is a possible flow of an incompressible fluid?

4–20. Given $u = Cx^2yzt$ and $v = y^2z - Cxy^2zt$, find an expression for w of the incompressible flow.

4–21. In Example 4–2, we have obtained $u = kx$, $v = -ky$, and $w = 0$ for the flow of a homogeneous and incompressible fluid ($\rho =$ constant). Is this velocity distribution also possible for an incompressible but nonhomogeneous fluid?

4–9. Scalar product and divergence of vectors

In the previous section, the principle of conservation of mass has been expressed as the equation of continuity. Although the physical principle involved is the same, the equation appears in different forms depending on the coordinate system used (compare Eqs. 4–18 and 4–20). To make clear the physical meaning of an equation, it is desirable to use notations without reference to coordinate systems. In dealing with vector quantities, this is accomplished by using vector notations. Two operations with vectors are introduced here. They are the scalar product of two vectors and the divergence of a vector field. They will be found useful in subsequent discussions.

The *scalar product* $\mathbf{A} \cdot \mathbf{B}$ of the two vectors is defined as

$$\mathbf{A} \cdot \mathbf{B} = AB \cos (\mathbf{A}, \mathbf{B}), \qquad (4\text{–}22)$$

where A and B are the magnitudes of the respective vectors, and $(\mathbf{A}, \mathbf{B})$ is their included angle. Inasmuch as $\cos (\mathbf{B}, \mathbf{A}) = \cos (\mathbf{A}, \mathbf{B})$, it is evident that the scalar product is commutative; i.e.,

$$\mathbf{A} \cdot \mathbf{B} = \mathbf{B} \cdot \mathbf{A}. \qquad (4\text{–}23)$$

That the scalar product is distributive can be shown as follows. The scalar product $\mathbf{A} \cdot \mathbf{B}$ may be considered as the product AB_A, where B_A is the component of $\mathbf{B}$ in the direction of $\mathbf{A}$. Then $\mathbf{A} \cdot (\mathbf{B} + \mathbf{C})$ is equal to the product of A and the component of the vector $\mathbf{B} + \mathbf{C}$ in the direction of $\mathbf{A}$. But this component of the sum $\mathbf{B} + \mathbf{C}$ is equal to the sum of the components B_A and C_A, in accordance with the definition of vector sum shown in Fig. 4–5. Thus

$$\mathbf{A} \cdot (\mathbf{B} + \mathbf{C}) = \mathbf{A} \cdot \mathbf{B} + \mathbf{A} \cdot \mathbf{C}. \qquad (4\text{–}24)$$

When rectangular coordinates are used, the scalar product can be computed from the components of the two vectors (see Problem 4–23):

$$\mathbf{A} \cdot \mathbf{B} = A_xB_x + A_yB_y + A_zB_z. \qquad (4\text{–}25)$$

To define the divergence at a point P in a vector field, for example, the velocity field $\mathbf{q}$ in a fluid flow, consider an imaginary surface S enclosing a volume V and point P, as shown in Fig. 4–16. The *divergence* of $\mathbf{q}$ at point P is written either as div $\mathbf{q}$ or $\nabla \cdot \mathbf{q}$, and is defined as

$$\text{div } \mathbf{q} = \nabla \cdot \mathbf{q} = \lim_{V \to 0} \frac{1}{V} \int_S \mathbf{n} \cdot \mathbf{q} \, dS, \qquad (4\text{–}26)$$

where $\mathbf{n}$ is the unit vector in the direction of the outward normal of dS. The physical meaning of the divergence can be seen easily. For example, if $\mathbf{q}$ is the velocity at any small surface dS, $\mathbf{n} \cdot \mathbf{q} = q \cos (\mathbf{n}, \mathbf{q})$ is its component along the outward normal $\mathbf{n}$, and $\mathbf{n} \cdot \mathbf{q} \, dS$ is the volume outflow per unit time through dS. The surface integral is the total volume outflow per unit time through S, and the divergence of $\mathbf{q}$ is then the volume outflow per unit time per unit volume at point P.

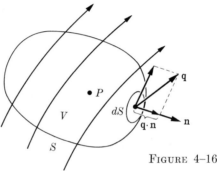

FIGURE 4–16

Similarly, the mass outflow per unit time per unit volume is $\nabla \cdot (\rho\mathbf{q})$. Since for conservation of mass this outflow from an ordinary point must be equal to the rate of decrease of density, the equation of continuity is

$$\nabla \cdot (\rho\mathbf{q}) = -\frac{\partial \rho}{\partial t}. \qquad (4\text{–}27)$$

When rectangular coordinates are used, this equation takes the form of Eq. 4–18:

$$\frac{\partial(\rho u)}{\partial x} + \frac{\partial(\rho v)}{\partial y} + \frac{\partial(\rho w)}{\partial z} = -\frac{\partial \rho}{\partial t}.$$

It is interesting to note that with rectangular coordinates, the vectorial operator ∇ may be treated as

$$\nabla = \mathbf{i}\frac{\partial}{\partial x} + \mathbf{j}\frac{\partial}{\partial y} + \mathbf{k}\frac{\partial}{\partial z}, \qquad (4\text{–}28)$$

such that, according to Eq. 4–25,

$$\nabla \cdot (\rho\mathbf{q}) = \frac{\partial(\rho u)}{\partial x} + \frac{\partial(\rho v)}{\partial y} + \frac{\partial(\rho w)}{\partial z}.$$

PROBLEMS

4–22. Point O is the origin of a rectangular coordinate system. Points P and Q are located at coordinates $(5, 4, 2)$ and $(0, -1, 2)$, respectively. Show that lines OP and OQ are normal to each other.

4–23. Find the values of $\mathbf{i} \cdot \mathbf{i}$ and $\mathbf{i} \cdot \mathbf{j}$, and derive Eq. 4–25.

4–24. The *work* W done by a force $\mathbf{F}$ through a displacement $\mathbf{s}$ is defined as $W = \mathbf{F} \cdot \mathbf{s}$. A train is traveling in a NE direction on a one-percent grade against a horizontal westward wind drag of 200 lb. Find the work done against the wind per mile of travel.

4–25. Using the infinitesimal volume shown in Fig. 4–13, evaluate the surface integral in Eq. 4–26 to show that the divergence of velocity $\mathbf{q}$ is

$$\nabla \cdot \mathbf{q} = \frac{\partial u}{\partial x} + \frac{\partial v}{\partial y} + \frac{\partial w}{\partial z}. \tag{4–29}$$

4–26. Using the infinitesimal volume shown in Fig. 4–14, evaluate the surface integral in Eq. 4–26 to show that the divergence of velocity $\mathbf{q}$ is

$$\nabla \cdot \mathbf{q} = \frac{1}{r} \frac{\partial (ur)}{\partial r} + \frac{1}{r} \frac{\partial v}{\partial \theta} + \frac{\partial w}{\partial z}. \tag{4–30}$$

4–27. Corresponding to the definition of divergence of a vector field in Eq. 4–26, we have that of the *gradient* of a scalar field, such as pressure p:

$$\operatorname{grad} p = \lim_{V \to 0} \frac{1}{V} \int_S p\mathbf{n} \, dS. \tag{4–31}$$

By evaluating the surface integral with the volume shown in Fig. 4–13, show that in rectangular coordinates,

$$\operatorname{grad} p = \nabla p = \mathbf{i}\frac{\partial p}{\partial x} + \mathbf{j}\frac{\partial p}{\partial y} + \mathbf{k}\frac{\partial p}{\partial z}. \tag{4–32}$$

4–10. Other topics of fluid kinematics

Other topics of fluid kinematics, including velocity potential, stream functions, circulation, and vorticity, will be presented in Chapters 6, 7, and 8.

5 Dynamics of Frictionless Incompressible Flow

5-1. Pressure in a frictionless flow

All real fluids exhibit internal friction when the fluid elements continue to deform in a flow. However, when the Reynolds number of a flow is high, the effects of internal friction in most parts of the flow may be so small as to be unimportant. A frictionless flow may then serve as an approximation of the actual flow. We shall first study the consequences if there were no internal friction, and then discuss the validity of the results.

It was shown in Section 3-1 that in fluid statics, the normal stress is independent of directions and is called pressure. It is to be shown presently that this is also the case in a frictionless fluid in motion. For the mass shown in Fig. 3-1, since there are no shearing stresses, we have for $F_x = ma_x$ and $F_z = ma_z$,

$$p_2 \, dy \, dz - \left(p_1 \, dy \, \frac{dz}{\sin \theta} \right) \sin \theta = \rho \, \frac{dx \, dy \, dz}{2} \, a_x$$

and

$$p_3 \, dy \, dx - \left(p_1 \, dy \, \frac{dz}{\cos \theta} \right) \cos \theta - \rho g \, \frac{dx \, dy \, dz}{2} = \rho \, \frac{dx \, dy \, dz}{2} \, a_z.$$

Since the body force and the mass are of the order of $dx \, dy \, dz$, while the surface forces are of the order of $dx \, dy$ and $dy \, dz$, the higher-order terms vanish by comparison, thus making $p_1 = p_2$ and $p_1 = p_3$. In other words, since

$$p_2 - p_1 = \tfrac{1}{2}\rho a_x \, dx, \qquad p_3 - p_1 = \tfrac{1}{2}\rho(a_z + g) \, dz,$$

we have $p_1 = p_2 = p_3$ at a point since dx and dz are infinitesimal. (For a viscous fluid in motion, the normal stresses need not be the same in all directions, as will be discussed in Chapter 11.)

5-2. The momentum equations for frictionless flow

Just as the flow variables u, v, and w must satisfy the equation of continuity in accordance with the principle of conservation of mass, they must also satisfy the momentum equations to be derived from Newton's second law of motion.

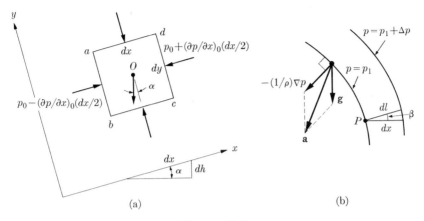

FIGURE 5–1

Consider the fluid element occurring at a certain instant the space in $dx\,dy\,dz$ as shown in Fig. 5–1(a). We shall first write the equation $F_x = ma_x$. Let p_0 be the value of p at the center. The mean value of p on surfaces ab and cd are then $p_0 + (\partial p/\partial x)_0(-dx/2)$ and $p_0 + (\partial p/\partial x)_0(dx/2)$, respectively. The x-component of the body force is $-\rho_0 g\,dx\,dy\,dz(\partial h/\partial x)$, where $h(x, y, z)$ is the elevation of a point and $\partial h/\partial x$ is the rate of increase of elevation in the x-direction. In a given case, $\partial h/\partial x$ is a constant and is equal to $\sin\alpha$ in Fig. 5–1(a). For positive $\partial h/\partial x$, this component of the body force is in the negative x-direction as shown. Since there is no shearing stress, we have

$$F_x = \left[p_0 - \left(\frac{\partial p}{\partial x}\right)_0 \frac{dx}{2}\right] dy\,dz - \left[p_0 + \left(\frac{\partial p}{\partial x}\right)_0 \frac{dx}{2}\right] dy\,dz - \rho_0 g\,dx\,dy\,dz\,\frac{\partial h}{\partial x}$$

$$= -\left(\frac{\partial p}{\partial x}\right)_0 dx\,dy\,dz - \rho_0 g\,\frac{\partial h}{\partial x}\,dx\,dy\,dz.$$

If p is increasing in the x-direction, $\partial p/\partial x$ is positive and there will be a net surface force due to pressure in the negative x-direction. With $m = \rho_0\,dx\,dy\,dz$, and a_x from Eq. 4–15, the equation $a_x = F_x/m$ becomes

$$\frac{\partial u}{\partial t} + u\frac{\partial u}{\partial x} + v\frac{\partial u}{\partial y} + w\frac{\partial u}{\partial z} = a_x = -\frac{1}{\rho}\frac{\partial p}{\partial x} - g\frac{\partial h}{\partial x}. \tag{5–1}$$

Since the point O is arbitrary, the subscript 0 has been omitted. Similarly, for the y- and z-directions, we have

$$\frac{\partial v}{\partial t} + u\frac{\partial v}{\partial x} + v\frac{\partial v}{\partial y} + w\frac{\partial v}{\partial z} = a_y = -\frac{1}{\rho}\frac{\partial p}{\partial y} - g\frac{\partial h}{\partial y} \tag{5–2}$$

and

$$\frac{\partial w}{\partial t} + u\frac{\partial w}{\partial x} + v\frac{\partial w}{\partial y} + w\frac{\partial w}{\partial z} = a_z = -\frac{1}{\rho}\frac{\partial p}{\partial z} - g\frac{\partial h}{\partial z}. \tag{5–3}$$

Equations 5–1 to 5–3 are called the *Euler's equations*, or the *momentum equa-*

tions for frictionless fluids. They must be satisfied at every point in a frictionless flow. In these equations, the constants $\partial h/\partial x$, $\partial h/\partial y$, and $\partial h/\partial z$ are known once the orientation of the axes is chosen.

When cylindrical coordinates are used, the momentum equations for frictionless fluids can be shown to be

$$\frac{\partial u}{\partial t} + u\frac{\partial u}{\partial r} + \frac{v}{r}\frac{\partial u}{\partial \theta} + w\frac{\partial u}{\partial z} - \frac{v^2}{r} = a_r = -\frac{1}{\rho}\frac{\partial p}{\partial r} - g\frac{\partial h}{\partial r}, \qquad (5\text{--}4)$$

$$\frac{\partial v}{\partial t} + u\frac{\partial v}{\partial r} + \frac{v}{r}\frac{\partial v}{\partial \theta} + w\frac{\partial v}{\partial z} + \frac{uv}{r} = a_\theta = -\frac{1}{\rho}\frac{\partial p}{r\,\partial \theta} - g\frac{\partial h}{r\,\partial \theta}, \qquad (5\text{--}5)$$

$$\frac{\partial w}{\partial t} + u\frac{\partial w}{\partial r} + \frac{v}{r}\frac{\partial w}{\partial \theta} + w\frac{\partial w}{\partial z} = a_z = -\frac{1}{\rho}\frac{\partial p}{\partial z} - g\frac{\partial h}{\partial z}. \qquad (5\text{--}6)$$

Note that $\partial h/\partial r$ and $\partial h/\partial \theta$ are not constant but vary with the coordinate θ (see Problem 5–3). Only when the z-axis is chosen to be vertical do they have the constant value of zero.

It can be easily seen from Eq. 4–28 that the momentum equation for frictionless fluids can be written in vector notation as

$$\mathbf{a} = -\frac{1}{\rho}\nabla p - g\nabla h, \qquad (5\text{--}6a)$$

where the vector ∇p is the gradient of p and has the following components:

$$\nabla p = \mathbf{i}\frac{\partial p}{\partial x} + \mathbf{j}\frac{\partial p}{\partial y} + \mathbf{k}\frac{\partial p}{\partial z}. \qquad (5\text{--}6b)$$

It can be shown to be the vector pointing in the direction of the steepest rise of p, and having a magnitude equal to the rate of increase of p in this direction as follows. In Fig. 5–1(b) are shown surfaces joining points of the same pressure. At point P, such a vector has a magnitude of $\partial p/\partial l$. Its x-component is

$$\frac{\partial p}{\partial l}\cos \beta = \frac{\partial p}{\partial l}\cdot\frac{dl}{dx} = \frac{\partial p}{\partial x}.$$

Similarly, its y- and z-components can be shown to be $\partial p/\partial y$ and $\partial p/\partial z$, respectively. Thus, in accordance with Eq. 5–6(b), this vector is ∇p. The vector ∇h is therefore directed upward with a magnitude of unity. According to Eq. 5–6(a), the particle acceleration is equal to the sum of two vectors as shown in Fig. 5–1(b).

PROBLEMS

5–1. Show that Eqs. 5–1 to 5–3 can be reduced to Eq. 3–1 for fluid statics.

5–2. Derive Eqs. 5–4 and 5–5.

5–3. Let the origin be placed at elevation h_0, and the x- and z-axes be horizontal. (a) Find the elevation h of a point at (x, y, z), and from this find $\partial h/\partial x$ and $\partial h/\partial y$. (b) Find h of a point at (r, θ, z), and find $\partial h/\partial r$ and $\partial h/\partial \theta$.

5–3. Incompressible flow

All fluids are compressible to a certain extent, although the compressibility of liquids is much less evident. In many cases, however, the effects of the variation of density of the fluid elements are so small that they can be neglected in analysis. We shall study these cases first, and the effects of compressibility will be studied after Chapter 13. At this time, we shall only comment qualitatively as to the practical validity of the assumption of an incompressible flow.

The density of a fluid varies with temperature and pressure. By considering the density as a constant, we have excluded from our consideration problems in which the variation of density is the primary factor. For example, in systems to which a significant quantity of heat is added, the variation of density due to temperature change is often responsible for the fluid motion; e.g., the convection in a boiling kettle. Another example where density variation must be considered is the propagation of sound waves.

When the change of density is not the primary factor in a phenomenon, compressibility can sometimes be ignored. In Section 2–5, it has been found by dimensional consideration that the effect of compressibility can be expected to be small when the Mach number is low. It has been observed that when the Mach number is less than $\frac{1}{3}$, compressibility may be ignored. Although the pressure and therefore the density are influenced by gravity, this influence is negligible if the domain of interest is limited in the vertical dimension; e.g., in the flow around an airplane.

This discussion applies to both liquids and gases. Although in the absolute sense gases are much more compressible than liquids, both can be assumed to be incompressible as an approximation under suitable conditions.

When the variation of the density of a fluid element can be ignored, we have $D\rho/Dt = 0$. If the fluid is also homogeneous, ρ is a constant. There are then four unknown functions in the equations of motion, namely, u, v, w, and p. We have now an equal number of equations for their solution, namely, three momentum equations and an equation of continuity. The laws of thermodynamics are therefore not needed in the study of incompressible flow. The problem is now reduced to one of mathematical technique of solving these equations and the art of applying the results to flows of real fluids.

5–4 Fluid in rigid-body motion

The use of the momentum equations is to be demonstrated with some simple examples where the fluid elements move without change of shape. For example, a viscous liquid in a rotating pan may finally move with the pan as if it were a rigid body (see Fig. 5–2). We shall assume the possibility of such a motion and study the consequences. Since there is no continuing deformation of the fluid elements in these cases, there is no shearing stress in the fluid, even though the fluid is viscous.

Example 5–1. In Fig. 5–2 is shown a pan rotating at a constant angular speed ω. Assuming that the fluid inside moves with the pan as if the system were a rigid body, find the pressure distribution in the liquid.

For the assumed motion, we have $u = 0$, $v = \omega r$, and $w = 0$ in the coordinate system shown. The density ρ is considered to be constant. It can be readily verified that the equation of continuity, Eq. 4–21, is satisfied.

For the chosen coordinate system, $\partial h/\partial r = \partial h/\partial \theta = 0$, and $\partial h/\partial z = 1$. Substituting these and the velocity components into Eqs. 5–4 to 5–6, we have the following momentum equations:

$$-\omega^2 r = -\frac{1}{\rho}\frac{\partial p}{\partial r},$$

$$0 = -\frac{1}{\rho}\frac{\partial p}{r\,\partial \theta},$$

$$0 = -\frac{1}{\rho}\frac{\partial p}{\partial z} - g.$$

These equations can be integrated separately to yield the following:

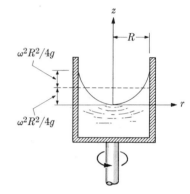

FIGURE 5–2

$$p = \tfrac{1}{2}\rho\omega^2 r^2 + f_1(\theta, z),$$

$$p = f_2(z, r),$$

$$p = -\rho g z + f_3(r, \theta),$$

where f_1, f_2, and f_3 are arbitrary functions of integration. The variable t is not involved because p at any point is assumed to be independent of time. Each of the differential equations demands a certain form for the solution of p. For example, the second equation, $\partial p/\partial \theta = 0$, demands that p must be independent of θ; that is, $p = f_2(z, r)$, meaning that p can at most be a function of the independent variables r and z only. We must seek a solution of p which satisfies all the three required forms. Since from f_2, p is to be independent of θ, f_1 and f_3 must not contain θ. It becomes apparent that $f_1(z) = -\rho g z + C$ and $f_3(r) = \tfrac{1}{2}\rho\omega^2 r^2 + C$, where C is an arbitrary constant. Thus

$$p = \tfrac{1}{2}\rho\omega^2 r^2 - \rho g z + C.$$

The constant C is to be determined with one known condition in the system. For the coordinate system shown in Fig. 5–2, we have $p = 0$ at $r = 0$ and $z = 0$. Therefore C must be zero:

$$p = \tfrac{1}{2}\rho\omega^2 r^2 - \rho g z.$$

This is the solution of p for all points in the fluid.

In obtaining this solution, all the equations of motion have been used. This solution represents therefore all the information that can be obtained from mechanics for the assumed flow. From this solution, however, one can derive other special information. For example, the equation of the surface is obtained by setting $p = 0$:

$$z = \frac{\omega^2}{2g} r^2.$$

The volume of liquid above the horizontal plane containing the origin is

$$V = \int_0^R \left(\frac{\omega^2}{2g} r^2 \right) 2\pi r \, dr = \frac{\pi \omega^2 R^4}{4g}.$$

The original elevation of the surface is

$$z = \frac{V}{\pi R^2} = \frac{\omega^2 R^2}{4g}.$$

Example 5–2. A tank containing a liquid is being acclerated horizontally with constant a. Assuming that the liquid finally moves with the tank without further change of shape, find the final shape of the surface.

Use the fixed rectangular coordinate system shown in Fig. 5–3. With $u(t)$ and $v = w = 0$, the equation of continuity, Eq. 4–19, is satisfied. With $a_x = a$, $a_y = a_z = 0$, $\partial h / \partial x = \partial h / \partial y = 0$, and $\partial h / \partial z = 1$, Eqs. 5–1 to 5–3 become

$$a = -\frac{1}{\rho} \frac{\partial p}{\partial x},$$

$$0 = -\frac{1}{\rho} \frac{\partial p}{\partial y},$$

$$0 = -\frac{1}{\rho} \frac{\partial p}{\partial z} - g.$$

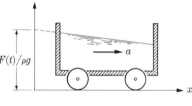

FIGURE 5–3

With ρ considered to be constant, these equations give separately

$$p = -\rho a x + f_1(y, z, t),$$
$$p = f_2(x, z, t),$$
$$p = -\rho g z + f_3(x, y, t).$$

The variable t is involved because the motion is unsteady with u increasing with time. Since the second equation indicates that p is independent of y, it is clear that $f_1(z, t) = -\rho g z + F(t)$ and $f_3(x, t) = -\rho a x + F(t)$. Thus

$$p = -\rho g z - \rho a x + F(t).$$

This is the solution of p for all positions and all time. To find the shape of the surface, it is sufficient to know p at one instant t_0. For t_0, the pressure distribution is

$$p = -\rho g z - \rho a x + C,$$

where $C = F(t_0)$, the value of $F(t)$ at $t = t_0$. This constant is to be determined, if desired, by a known p at a known position at this instant t_0. The equation of the surface at t_0 is obtained by setting $p = 0$:

$$z = \frac{C}{\rho g} - \frac{a}{g} x.$$

The surface is therefore a plane with a slope equal to a/g.

PROBLEMS

5–4. A uniform stream parallel to a horizontal xy-plane flowing at an angle α with the x-axis can be described with $\rho = $ constant, $u = U \cos \alpha$, $v = U \sin \alpha$, and $w = 0$. Find the pressure distribution.

5–5. By using the momentum equations, find the pressure at the bottom of a bucket of water, one foot deep, on an elevator which accelerates downward at the rate of 10 ft/sec².

5–6. A box with an opening at the top, as shown in Fig. 5–4, is filled with a liquid. It is accelerating up an inclined plane with acceleration a. Find the pressure at the corners A and B.

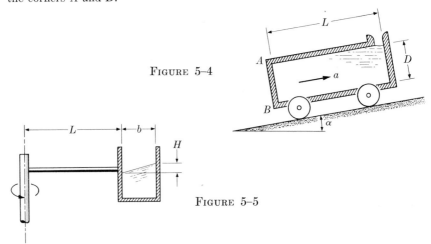

FIGURE 5–4

FIGURE 5–5

5–7. A closed circular cylinder is filled with a fluid and rotates steadily about its horizontal axis. Find an expression for the difference of pressure between the highest and the middle points of the cylinder.

5–8. A bucket of width b is attached to a horizontal arm of length L as shown in Fig. 5–5. The system revolves about a vertical axis at a constant angular speed ω. Find the distance H.

5–5. Bernoulli's equation

With the momentum equations and the equation of continuity, the velocity and pressure at every position in a frictionless incompressible flow can theoretically be determined. However, the solution for a particular flow system is usually difficult mathematically. We shall be content at this moment to find a relationship between the two principal unknowns, p and q. Although this relationship does not describe the flow pattern or the pressure distribution, it can, under suitable conditions, yield useful information.

First rewrite the momentum equations using the l-axis in the direction of the velocity vector $\mathbf{q}$ at a certain point P (see Fig. 5–6). Let R be the radius of curvature of the instantaneous streamline passing this point. Let the principal normal n-axis be directed from the center of curvature. The binormal m-axis is normal to the l- and n-axes as shown. Imagine a cylindrical coordinate system with the origin at the center of curvature. In the neighborhood of point P at this instant, we have $u = 0$ in the radial direction, $v = q(l)$ in the tangential direction, and $w = 0$ in the axial direction. With these values, Eqs. 5–4 to 5–6 become

$$\frac{\partial u}{\partial t} - \frac{q^2}{R} = -\frac{1}{\rho}\frac{\partial p}{\partial n} - g\frac{\partial h}{\partial n}, \qquad (5\text{–}7)$$

$$\frac{\partial q}{\partial t} + q\frac{\partial q}{\partial l} = -\frac{1}{\rho}\frac{\partial p}{\partial l} - g\frac{\partial h}{\partial l}, \qquad (5\text{–}8)$$

$$\frac{\partial w}{\partial t} = -\frac{1}{\rho}\frac{\partial p}{\partial m} - g\frac{\partial h}{\partial m}. \qquad (5\text{–}9)$$

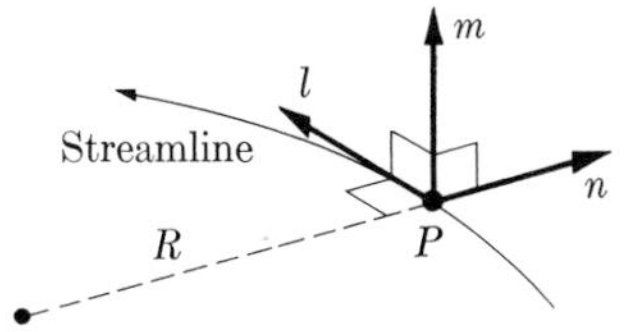

FIGURE 5–6

These equations can be used to study the flow in the neighborhood of point P. In these equations, no assumption regarding the compressibility of the fluid has been made.

With Eqs. 5–7 and 5–9, consider a steady flow of a homogeneous incompressible fluid. With $\partial w/\partial t = 0$ and constant ρ, Eq. 5–9 becomes

$$\frac{\partial}{\partial m}\left(\frac{p}{\rho} + gh\right) = 0.$$

The sum $p/\rho + gh$ is therefore constant in the direction of the binormal axis. For this flow, Eq. 5–7 becomes

$$\frac{\partial}{\partial n}\left(\frac{p}{\rho} + gh\right) = \frac{q^2}{R}.$$

There is an increase of the sum $p/p + gh$ in the direction of the n-axis of a curved streamline. If the streamline is straight ($R \rightarrow \infty$), this sum is then constant in the n-direction. Thus, across straight streamlines in a homogeneous incompressible fluid,

$$\frac{p}{\rho} + gh = \text{constant.} \qquad (5\text{–}10)$$

Next consider Eq. 5–8 for the case of a steady incompressible flow. In this case, the fluid is not necessarily homogeneous (e.g., layers of oil and water). In such a flow, we have $\partial q/\partial t = 0$, and the same ρ along a streamline, that is, $\partial \rho/\partial l = 0$. Then

$$\frac{\partial}{\partial l}\left(\frac{p}{\rho}\right) = \frac{1}{\rho}\frac{\partial p}{\partial l} - \frac{p}{\rho^2}\frac{\partial \rho}{\partial l} = \frac{1}{\rho}\frac{\partial p}{\partial l}.$$

Since $q \cdot \partial q/\partial l = \partial(\frac{1}{2}q^2)/\partial l$, Eq. 5–8 can be rewritten as

$$\frac{\partial}{\partial l}\left(\frac{p}{\rho} + gh + \frac{q^2}{2}\right) = 0.$$

Upon integrating with respect to l, we obtain

$$\frac{p}{\rho} + gh + \frac{q^2}{2} = f(n,\ m,\ t).$$

The variable t is involved because, although the flow is steady, p in the incompressible fluid may change with time, e.g., the change of atmospheric pressure over a pond. However, p can change with time in a steady flow only when the same change is experienced throughout the whole field. Otherwise the terms $\partial p/\partial n$, etc., will change with time, causing unsteady motion. If this overall pressure load does not change with time, we then have in the l-direction (with constant n and m)

$$\frac{p}{\rho} + gh + \frac{q^2}{2} = \text{constant (along streamline).} \tag{5–11}$$

This is *Bernoulli's equation* for incompressible frictionless flows. Although it has been derived for the immediate neighborhood of point P, it can be applied successively at other points of the streamline. Thus it is applicable along the entire streamline in a steady, frictionless, incompressible (though not necessarily homogeneous) flow with constant pressure load. The value of the constant may vary from one streamline to another.

Each term in Bernoulli's equation has the dimensions of energy, or work, per unit mass. The quantity $q^2/2$ is the kinetic energy per unit mass, and gh is the potential energy per unit mass of the fluid. Being the integral of

$$\frac{1}{\rho}\frac{\partial p}{\partial l}\ dl,$$

p/ρ is the work done by unit mass of the incompressible fluid against the pressure difference along the flow. This equation therefore states the conservation of mechanical energy and work along a streamline in a frictionless incompressible flow.

The following are several examples of practical applications of Bernoulli's equation.

Example 5–3. Find the velocity of the jet of a thin liquid flowing from a large tank through an *orifice*.

Since the tank is large, the falling of the liquid surface in the tank is very slow. The flow can therefore be considered as steady for approximation. Neglecting the internal friction in the flow, we may apply Bernoulli's equation along the streamlines. We choose points 1 and 2 along a streamline where the values of the variables are either known or desired (see Fig. 5–7). According to Eq. 5–11,

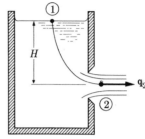

$$\frac{p_1}{\rho} + gh_1 + \frac{q_1^2}{2} = \frac{p_2}{\rho} + gh_2 + \frac{q_2^2}{2},$$

FIGURE 5–7

where $p_1 = 0$ and $q_1 \doteq 0$. The value of p_2 inside the jet is not known exactly. Since $p = 0$ at the surface of the jet, it may be surmised that p_2 is approximately zero. This is possible when the curvature of the jet is such that, in accordance with Eq. 5–7, p may be constant across the streamlines of the jet although they differ in elevation h. Thus

$$q_2 = \sqrt{2g(h_1 - h_2)}.$$

The velocity varies slightly across the jet. If H is the distance from the center of the orifice to the liquid surface, and the size of the orifice is small compared with H, we have

$$q_2 = \sqrt{2gH}.$$

This relationship is called *Torricelli's theorem*.

The discharge of Q of the orifice is usually expressed as

$$Q = C_D A \sqrt{2gH}, \tag{5–12}$$

where A is the area of the orifice opening, and C_D is called the *coefficient of discharge*. This coefficient is necessary in Eq. 5–12 because the area of the jet is less than A, and the actual speed of the jet may be slightly less than $\sqrt{2gH}$ due to internal friction. Theoretically, the value of C_D can be determined from the equations of fluid motion. However, due to mathematical difficulties, this has been done in only a few special cases. Often C_D is determined experimentally. For a sharp-edged orifice, as shown in Fig. 5–7, C_D has been observed to be about 0.6.

Example 5–4. Compute the velocity of flow from the observed pressure difference in a *Pitot-static tube*.

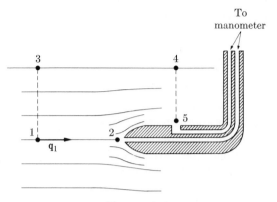

FIGURE 5–8

The Pitot tube shown in Fig. 5–8 is combined with a static-head tube at point 5. This is a device for measuring the velocity of flow at a point, such as point 1. The tube is placed near this point opposing the (presumably known) direction of the velocity. Since the fluid in the tube is stationary, the velocity must be zero at point 2, where the streamline splits. This point is called a *stagnation point*. In Bernoulli's equation,

$$\frac{p_1}{\rho} + gh_1 + \frac{q_1^2}{2} = \frac{p_2}{\rho} + gh_2 + \frac{q_2^2}{2}, \tag{5–13}$$

we have $q_2 = 0$, and p_2 is measured by a pressure connection as indicated in Fig. 5–8. Since p_1 cannot be measured directly, it is to be computed from p_5 which can be measured. Noting that the streamlines passing between points 1 and 3, and those between points 4 and 5, are essentially parallel and straight, we have from Eq. 5–10

$$\frac{p_1}{\rho} + gh_1 = \frac{p_3}{\rho} + gh_3 \qquad \text{and} \qquad \frac{p_4}{\rho} + gh_4 = \frac{p_5}{\rho} + gh_5.$$

As the streamline 3–4 is essentially undisturbed, we have $q_3 = q_4$. Thus from Bernoulli's equation,

$$\frac{p_3}{\rho} + gh_3 = \frac{p_4}{\rho} + gh_4.$$

From these three equations, we obtain

$$\frac{p_1}{\rho} + gh_1 = \frac{p_5}{\rho} + gh_5.$$

Thus, from Eq. 5–13,

$$\frac{q_1^2}{2} = \left(\frac{p_2}{\rho} + gh_2\right) - \left(\frac{p_5}{\rho} + gh_5\right).$$

The value on the right of this equation can be observed with a manometer.

PROBLEMS

5–9. The gage pressure at the top of a water main is p_0. Find the height of a jet through a small hole on the top of the main.

5–10. The water in a closed tank is 16 ft deep. The gage pressure of the air above the water is 10 psi. Find the velocity and discharge through a circular sharp-edged orifice, one inch in diameter, at the bottom of the tank.

5–11. A *Pitot tube* is used for measuring q_1, as shown in Fig. 5–9. Compute q_1 from the observed H.

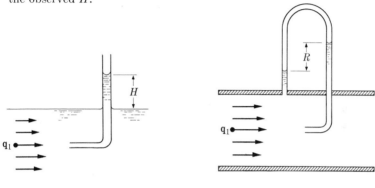

FIGURE 5–9   FIGURE 5–10

5–12. In measuring the velocity at the centerline of a water main, as shown in Fig. 5–10, the manometer reading is 6 in. The specific gravity of the manometer oil is 0.80. Compute the speed q_1.

5–6. Comments on application of Bernoulli's equation

Bernoulli's equation in the form of Eq. 5–11 has been derived for steady, frictionless, incompressible flow. It should be applied to cases where these conditions are nearly realized so that the result is a good approximation of the actual flow. The problem of compressibility has been commented on briefly in Section 5–3. In this section, we shall discuss qualitatively the influence of internal friction and unsteadiness.

Although the effect of viscosity is small in most parts of a flow at high Reynolds number, there are regions of the flow where the influence of viscosity is predominant. The magnitude of shearing stresses depends not only on the viscosity of the fluid, but also on the rate of change of shape of the fluid elements. For example, the shearing stress in Fig. 1–1 is $\mu U/d$, where U/d indicates the rate of change of shape. In order that the shearing stresses are small, the spatial rate of change of velocity, as well as the value of μ, must be small. Thus, while Bernoulli's equation may be applied between points 1 and 2 in Fig. 5–8, it cannot be used between points 2 and 5, although they are on the same streamline. The reason for this is that the spatial rate of change of velocity is very great near a

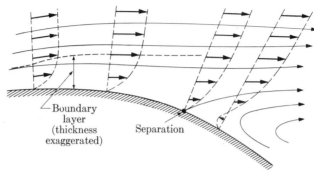

Boundary
layer
(thickness
exaggerated)   Separation

FIGURE 5–11

solid surface, due to the fact that there is no relative velocity between the solid and the contacting fluid (see Fig. 5–11). When the Reynolds number is not too low, the region under the influence of the solid surface may remain extremely thin for some distance, and is called a *boundary layer*. In general, Bernoulli's equation is not applicable to streamlines within a boundary layer.

The boundary layer can influence the flow in another way. If the fluid velocity in general decreases downstream (e.g., in a diverging channel), the low velocity in the boundary layer may become zero and then reverse in direction. As a result, eddies are formed as shown in Fig. 5–11. This phenomenon is called *separation* of the flow from the solid boundary. There is a large spreading zone of eddies downstream. As will be explained below, Bernoulli's equation is not applicable in this region of unsteadiness.

Bernoulli's equation for incompressible fluids states the conservation of mechanical energy and work in a frictionless flow. In a real fluid, mechanical energy may be transferred from one particle to another through the viscous stresses between them, and may be dissipated into heat. This transfer and dissipation may become considerable between points separated by large distances. Thus in Fig. 5–8, the equation is applicable between points 1 and 2 only when they are reasonably close together.

Another difficulty in applying Bernoulli's equation is the fact that fluids usually flow with turbulence (see Figs. 2–5b and 4–10). With the velocity fluctuating at a point, turbulent flow is by definition unsteady. In a turbulent flow, the instantaneous streamlines are tortuous and change rapidly in shape. Application of Bernoulli's equation to the instantaneous flow is therefore not permissible. However, it is to be shown that if the turbulent motion is not strong, and the temporal mean velocity at a point does not change with time (i.e., the flow is quasi-steady), Bernoulli's equation can be applied to the mean flow to yield a good approximation. Let the instantaneous velocity and pressure at a point be expressed as the sum of their respective mean values and fluctuations, that is, $u = \bar{u} + u'$, $v = \bar{v} + v'$, $w = \bar{w} + w'$, and $p = \bar{p} + p'$, where, for example, u is the instantaneous value, $\bar{u}$ is the mean value at the point, and u' is the instantaneous fluctuation (see Fig. 4–10). With these values in Eq. 5–1, it can be reduced for quasi-steady incompressible frictionless flows to the following (see Eq. 13–12):

$$\overline{u}\,\frac{\partial \overline{u}}{\partial x} + \overline{v}\,\frac{\partial \overline{u}}{\partial y} + \overline{w}\,\frac{\partial \overline{u}}{\partial z} + \frac{\partial \overline{u'u'}}{\partial x} + \frac{\partial \overline{u'v'}}{\partial y} + \frac{\partial \overline{u'w'}}{\partial z} = -\,\frac{1}{\rho}\,\frac{\partial \overline{p}}{\partial x} - g\,\frac{\partial h}{\partial x},$$

where, for example, $\overline{u'v'}$ is the temporal mean value of the product of the instantaneous fluctuations u' and v' at a point. Although the mean values $\overline{u'}$ and $\overline{v'}$ are equal to zero, the mean value $\overline{u'v'}$ may not be zero. (If, for example, positive and negative values of u' occur more frequently with positive and negative v', respectively, $u'v'$ will be positive most of the time and $\overline{u'v'}$ will probably be positive.) Except for the terms involving $\overline{u'u'}$, $\overline{u'v'}$, and $\overline{u'w'}$, this equation for the mean flow is similar to Eq. 5–1 for the instantaneous flow. In many cases, these mean values are small compared with $\overline{q}^2$ at the point and can be neglected in an approximate solution for the mean flow. We then have

$$\frac{\overline{p}}{\rho} + gh + \frac{\overline{q}^2}{2} \doteq \text{constant} \tag{5–14}$$

along a streamline constructed according to the mean flow pattern. The error of this equation is of the order of $\overline{u'u'}$. Thus, if the flow in Fig. 5–8 is turbulent, the result obtained in Example 5–4 is the temporal mean speed at point 1. However, if $\overline{u'u'}$ is not small compared with $\overline{q}^2$, Eq. 5–14 is of little practical value.

5–7. One-dimensional analysis

In many practical problems, such as the flow in pipes and channels, one may be interested only in the mean speed and pressure over a cross section. Solutions can be more readily obtained by considering such mean quantities. Such analyses are called *one-dimensional analyses*. In this section, we shall rewrite Bernoulli's equation in terms of these mean quantities for use in such an analysis. Other examples of one-dimensional analyses will be found in later chapters.

Let Q be the *discharge* in volume per unit time passing a cross section:

$$Q = \int_A v_l\, dA, \tag{5–15}$$

where A is the area of the cross section normal to the axis of the channel, and v_l is the velocity component normal to dA of the area A (see Fig. 5–12). The *mean speed* V at this cross section is defined as

$$V = \frac{Q}{A} = \frac{1}{A}\int_A v_l\, dA. \tag{5–16}$$

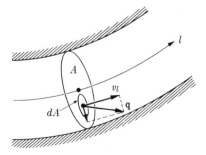

FIGURE 5–12

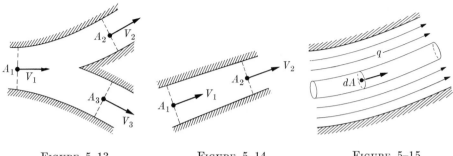

FIGURE 5–13   FIGURE 5–14   FIGURE 5–15

The equation of continuity in one-dimensional analysis is to be expressed in terms of the mean speed. For example, in Fig. 5–13, we have for steady or quasi-steady flow,

$$\rho_1 V_1 A_1 = \rho_2 V_2 A_2 + \rho_3 V_3 A_3.$$

In Fig. 5–14, we have $\rho_1 V_1 A_1 = \rho_2 V_2 A_2$. If ρ is constant,

$$VA = Q = \text{constant.} \tag{5–17}$$

To express Bernoulli's equation in terms of V, consider the *stream tube* bounded by the streamlines passing the edge of a differential area (see Fig. 5–15). Along this tube, while the speed and the cross section dA may vary, $q\,dA$ is constant due to the continuity of the steady incompressible flow. If the fluid can be considered frictionless, we have along this tube

$$\frac{p}{\rho} + gh + \frac{q^2}{2} = \text{constant,}$$

and therefore

$$\left(\frac{p}{\rho} + gh + \frac{q^2}{2}\right) q\,dA = \text{constant.}$$

Summing up all the tubes, we have along the flow

$$\int_A \left(\frac{p}{\rho} + gh\right) q\,dA + \frac{1}{2}\int_A q^3\,dA = \text{constant.}$$

To evaluate these integrals at a cross section, we limit our attention to flows where all the streamlines are practically parallel and normal to A. Then $q = v_l$ and $p/\rho + gh$ can be considered to be the same at every point over the cross section, in accordance with Eq. 5–10. The first integral becomes $(p/\rho + gh)VA$. Dividing the equation by VA, we obtain

$$\frac{p}{\rho} + gh + \alpha\,\frac{V^2}{2} = \text{constant,} \tag{5–18}$$

where
$$\alpha = \frac{1}{A V^3} \int_A q^3 \, dA \tag{5–19}$$

is called the *kinetic energy correction factor* at the cross section. The term $\alpha V^2/2$ is the average kinetic energy per unit mass passing the cross section.

Equation 5–18 is Bernoulli's equation for the one-dimensional analysis of steady, incompressible, frictionless flows. It differs from Eq. 5–11 in that it is expressed in terms of the mean speed V over a cross section. In Eq. 5–18, the value of $p/\rho + gh$ at any point of the cross section may be used to represent the whole section. In the case of pipe flows, the value at the center of the cross section is usually used. In the case of open channels, the value at the free surface is usually used (see Example 5–6). In view of Eq. 5–14, Eq. 5–18 can also be used for approximate analysis of quasi-steady turbulent flows, with an error of the order of $\overline{u'u'}$.

An obvious difficulty in using Eq. 5–18 is that the distribution of q or v_l over a cross section is unknown, and therefore α cannot be computed. In practice, α is usually estimated according to the expected velocity distribution. The minimum value of α is unity when the velocity is uniform over the cross section. In the case of turbulent flow in a straight passage, the temporal mean velocity is nearly the same at all points of a cross section. The value of α is therefore not much greater than unity (about 1.1) and is usually assumed to be unity for convenience. At a cross section where the velocity is very unevenly distributed, e.g., in a diverging channel or downstream of a bend, α may become exceedingly large. At such a cross section, one-dimensional analysis can seldom be applied with confidence.

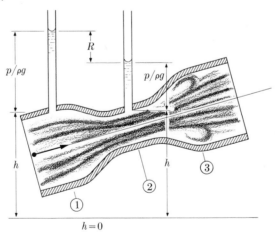

FIGURE 5–16

Example 5–5. In Fig. 5–16 is shown a *venturi meter*, which is a device for measuring the steady discharge Q through a pipe. Find Q for a turbulent flow in terms of the piezometer readings.

In the converging passage between cross sections 1 and 2, the velocity distribution becomes more even, and the degree of turbulence is not expected to increase. The value α_2 at section 2 is expected to be smaller than α_1 at section 1. Since the flow is turbulent, α_1 is not much greater than unity. We assume $\alpha_1 = \alpha_2 = 1$.

From Bernoulli's equation,

$$\frac{p_1}{\rho} + gh_1 + \frac{V_1^2}{2} = \frac{p_2}{\rho} + gh_2 + \frac{V_2^2}{2}.$$

From the equation of continuity, we have $V_1 = Q/A_1$ and $V_2 = Q/A_2$. Therefore

$$Q = \frac{A_1 A_2}{\sqrt{A_1^2 - A_2^2}} \sqrt{2\left[\left(\frac{p_1}{\rho} + gh_1\right) - \left(\frac{p_2}{\rho} + gh_2\right)\right]}.$$

For accurate measurement, a correction is necessary to account for the effects of viscosity and turbulence:

$$Q = \frac{C_v A_1 A_2}{\sqrt{A_1^2 - A_2^2}} \sqrt{2\left[\left(\frac{p_1}{\rho} + gh_1\right) - \left(\frac{p_2}{\rho} + gh_2\right)\right]}, \qquad (5\text{–}20)$$

where C_v is the *velocity coefficient*, whose value depends on the geometry of the meter and the Reynolds number. (For C_v of a standard meter, see Fig. 2–7.) The difference of $p/\rho + gh$ between two sections can be observed with piezometers (see Fig. 5–16), a manometer, etc. In evaluating the sum $p/\rho + gh$ at a cross section, p and h of any point of the section may be used. In Fig. 5–16, p and h at the top of the sections are indicated.

In the diverging passage between sections 2 and 3, there may be separation of the main flow from the wall and an increase of the degree of turbulence. Bernoulli's equation should therefore not be applied between these two sections. If the expansion of the pipe is made very gradual so as to avoid separation, the distance between the two sections will be so long that the effect of internal friction can no longer be safely ignored. The pressure change in the expansion is therefore not used for determining the discharge.

Example 5–6. The thin liquid in an *open channel* with rectangular cross section flows through a gradual contraction, as shown in Fig. 5–17. Given the depth and the mean speed just before the contraction, find the depth of flow at the throat of the contraction.

Since the contraction is gradual, the curvature of the streamlines is probably small in the converging channel, and $p/\rho + gh$ can be expected to be practically the same over a cross section, in accordance with Eq. 5–10. In applying Bernoulli's equation to a flow with a free surface, this quantity can be most conveniently evaluated at the surface where $p = 0$. The value of $p/\rho + gh$ at a section is therefore equal to gh of the surface. For the flow in Fig. 5–17, we can

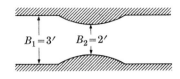

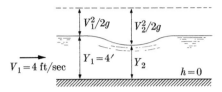

FIGURE 5–17

therefore write Eq. 5–18 as

$$gY_1 + \frac{V_1^2}{2} = gY_2 + \frac{V_2^2}{2},$$

with α_1 and α_2 assumed to be unity. With the equation of continuity, Eq. 5–17,

$$V_1B_1Y_1 = V_2B_2Y_2,$$

we have two equations for the two unknowns V_2 and Y_2. Substituting the given numerical values and eliminating V_2, we obtain

$$Y_2^3 - 4.02\,Y_2^2 + 0.56 = 0.$$

The three solutions of this cubic equation are 3.99, 0.39, and −0.36 ft. The negative value is obviously unrealistic. Requiring that the water surface be continuous, we can show that 3.99 ft is the solution for Y_2 (see Example 10–1).

PROBLEMS

5–13. At a certain cross section of a pipe, it is found that the speed over the upper half is 10 ft/sec, while that at the lower half is practically zero. Compute the mean speed and the kinetic energy correction factor at this cross section.

5–14. The speed in a long circular pipe of radius R is found to be (a) for laminar flow, $q = U[1 - (r/R)^2]$, and (b) for turbulent flow, $\bar{q} = U[1 - (r/R)]^{1/7}$, where U is the maximum speed at the center, and r is the distance from the center. Compute the mean speed and the kinetic energy correction factor in each case.

5–15. A nozzle, one inch in diameter, is connected to a 2-in. hose. Compute the speed and the discharge of the nozzle. The pressure at the centerline of the hose at the connection is observed to be 20 psi gage, and the diameter of the jet is 1 in.

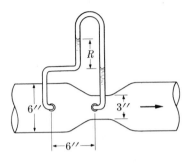

FIGURE 5–18

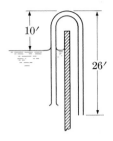

FIGURE 5–19

5–16. In Fig. 5–18 is shown a differential manometer with an oil of specific gravity 0.80. (a) When the reading R is 1 ft, compute the discharge through the water main. (b) Repeat the problem if the pipe is inclined at 30° with the horizontal.

5–17. A U-tube acts as a *siphon*, as shown in Fig. 5–19. Neglecting internal friction, compute the speed of the outflow. Also compute the pressure at the top of the siphon. (If this value is lower than the vapor pressure of the liquid, vapor pockets will form and the flow will stop.)

5–18. In a horizontal open channel with a uniform width, there is a barrier as shown in Fig. 5–20. Given the depth of flow Y_2 over the peak and the speed $V_2 = \sqrt{gY_2}$, find the possible value or values of the upstream depth Y_1.

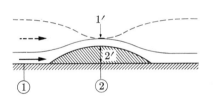

FIGURE 5–20

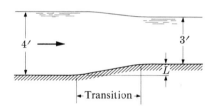

FIGURE 5–21

5–19. An open channel with a rectangular cross section is 8 ft in width and 4 ft in depth. The speed of flow is 3 ft/sec. This channel is connected through a transition to another channel which is 6 ft in width and 3 ft in depth, as shown in Fig. 5–21. Show that for an ideal transition with no loss of energy, the difference L in elevation between the channels should be 0.70 ft.

6 Irrotational Flow

6–1. Rotation of fluid elements

In this chapter, we shall attempt to seek, from the momentum equations and the equation of continuity, solutions of the velocity and pressure distribution in a frictionless fluid. However, due to the presence of the nonlinear terms $u\,\partial u/\partial x$, etc., which involve the product of two unknowns, the solution of these equations is extremely difficult. We shall limit our attention to a special class of flow problems in which all fluid elements do not rotate during motion. Such flows are called *irrotational flows*. It will be shown that for such flows, the problem is reduced to the solution of linear equations.

One may question whether such flows can exist. At this moment, one may be satisfied with the following argument. In a frictionless fluid there are no shearing stresses acting on a fluid element. If its density is uniquely related to the pressure, as in incompressible and isentropic flows, the net force due to pressure acting on the element always passes through its center of mass (see Section 8–5). As a result, there is no turning moment acting on the fluid element. If it is originally not rotating, it will remain not rotating throughout its course. Thus if all the fluid elements are originally not rotating, e.g., starting from rest, the flow will remain irrotational. For real fluids, an irrotational flow can occur with proper boundary conditions such that the net viscous force acting on any fluid element is zero (see Problems 11–21b and 11–24). Usually, the viscous stresses will impart rotational motion to the fluid elements. However, for a fluid starting from an irrotational condition and flowing at a high Reynolds number, the flow remains essentially irrotational except where the change of velocity with distance is very large, such as in the boundary layer at a solid surface. Thus the flow from a reservoir over the top of a spillway or the motion set up in a still atmosphere by a passing airfoil may be approximated by an irrotational flow outside the boundary layer.

To express the rotation of an element in mathematical terms, we follow the fluid which occupies the space $dx\,dy\,dz$ at a certain time. At a time dt later, the same element will have, in general, undergone translation, rotation, and deformation as shown in Fig. 6–1. The z-component of the rotation (vector) of the element is indicated by the difference between the angles $d\alpha$ and $d\beta$. If $d\alpha$ and $d\beta$ are equal, this component of the rotation is zero, as indicated by the

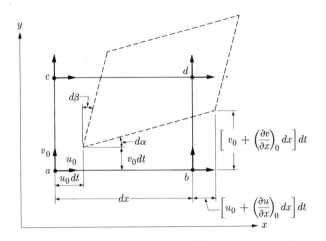

FIGURE 6–1

constant orientation of the diagonal ad. To express $d\alpha$ in terms of the velocity distribution, let u_0 and v_0 be the velocity components at corner a. Then at corner b a distance dx away, we have $u_0 + (\partial u/\partial x)_0\,dx$ and $v_0 + (\partial v/\partial x)_0\,dx$. The displacements of these corners during the time dt have therefore the values shown in Fig. 6–1. Thus

$$\tan(d\alpha) = \frac{[v_0 + (\partial v/\partial x)_0\,dx]\,dt - v_0\,dt}{dx - u_0\,dt + [u_0 + (\partial u/\partial x)_0\,dx]\,dt} = \left(\frac{\partial v}{\partial x}\right)_0 dt,$$

where higher-order terms have been dropped. Similarly,

$$\tan(d\beta) = \left(\frac{\partial u}{\partial y}\right)_0 dt.$$

The subscript can now be omitted. The angle of rotation of the diagonal of the element about the z-axis is $\frac{1}{2}(d\alpha - d\beta)$. The z-component of the vector of angular velocity $\boldsymbol{\omega}$ of the element is therefore

$$\omega_z = \lim_{dt\to 0}\frac{\frac{1}{2}(d\alpha - d\beta)}{dt} = \frac{1}{2}\lim_{dt\to 0}\frac{\tan(d\alpha) - \tan(d\beta)}{dt}$$

$$= \frac{1}{2}\left(\frac{\partial v}{\partial x} - \frac{\partial u}{\partial y}\right). \tag{6–1}$$

Similarly, it can be shown that

$$\omega_y = \frac{1}{2}\left(\frac{\partial u}{\partial z} - \frac{\partial w}{\partial x}\right), \tag{6–2}$$

$$\omega_x = \frac{1}{2}\left(\frac{\partial w}{\partial y} - \frac{\partial v}{\partial z}\right). \tag{6–3}$$

In fluid mechanics, the vector $\boldsymbol{\omega}$ is usually called $\frac{1}{2}\boldsymbol{\xi}$, where the vector $\boldsymbol{\xi}$ is called the *vorticity*. [For the relationship between $\boldsymbol{\xi}$ and $\mathbf{q}$ in vector notation, see Eq. 9–21(a).]

By definition, an *irrotational flow* is one in which $\boldsymbol{\omega}$ is zero at every point. Therefore

$$\xi_z = 2\omega_z = \frac{\partial v}{\partial x} - \frac{\partial u}{\partial y} = 0, \tag{6–4}$$

$$\xi_x = 2\omega_x = \frac{\partial w}{\partial y} - \frac{\partial v}{\partial z} = 0, \tag{6–5}$$

$$\xi_y = 2\omega_y = \frac{\partial u}{\partial z} - \frac{\partial w}{\partial x} = 0. \tag{6–6}$$

In the case of cylindrical coordinates, it can be shown in a similar manner (see Problem 6.2) that for irrotational flows

$$\xi_z = 2\omega_z = \frac{\partial(rv)}{r\,\partial r} - \frac{\partial u}{r\,\partial \theta} = 0, \tag{6–7}$$

$$\xi_r = 2\omega_r = \frac{\partial w}{r\,\partial \theta} - \frac{\partial v}{\partial z} = 0, \tag{6–8}$$

$$\xi_\theta = 2\omega_\theta = \frac{\partial u}{\partial z} - \frac{\partial w}{\partial r} = 0. \tag{6–9}$$

It should be emphasized that rotation indicates the change of orientation of a fluid element and must not be confused with the shape of its path. There are rotational motions where each fluid element, like a wheel of a moving train, is moving in a straight line. There are irrotational motions where the fluid elements move in circles or other curved paths. For example, an irrotational motion with the speed on the inner circular path greater than that on the outer path is shown in Fig. 6–2.

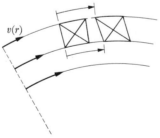

$v(r)$

FIGURE 6–2

The orientation of the diagonals of the fluid element remains the same at the two positions shown. (We have used the notion of a fluid element here to study its motion during a short time dt. We have no interest in following its motion in any longer period of time as it will soon be distorted out of recognition)

PROBLEMS

6-1. Show that, in Fig. 6-1, $d\beta = (\partial u/\partial y)_0 \, dt$.

6-2. Derive Eq. 6-7. [*Hint:* First show that $d\beta/dt = (\partial u/r \, \partial\theta) - (v/r)$ in Fig. 6-3. Use for small $d\theta$, $\sin(d\theta) = d\theta$, $\cos(d\theta) = 1$.]

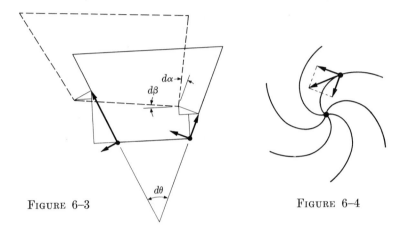

FIGURE 6-3　　　　　　　　　　　　　　　　FIGURE 6-4

6-3. In the flow shown in Fig. 1-1, we have $u = Uy/d$, $v = w = 0$. Is this flow irrotational? Compute the angles corresponding to $d\alpha$ and $d\beta$ in Fig. 6-1.

6-4. In the flow shown in Fig. 6-2, we have $u = w = 0$ and v is a function of r. Show that if the flow is irrotational, v must be inversely proportional to r.

6-5. In the flow shown in Fig. 6-4, u and v are functions of r only, and $w = 0$. Show that if the flow is irrotational and incompressible, both u and v must be inversely proportional to r.

6-2. Velocity potential

It is well known that for any scalar function $\phi(x, y, z, t)$, the order of differentiation is immaterial if the partial derivatives are continuous:

$$\frac{\partial}{\partial y}\left(\frac{\partial\phi}{\partial x}\right) = \frac{\partial}{\partial x}\left(\frac{\partial\phi}{\partial y}\right), \quad \frac{\partial}{\partial z}\left(\frac{\partial\phi}{\partial y}\right) = \frac{\partial}{\partial y}\left(\frac{\partial\phi}{\partial z}\right), \quad \frac{\partial}{\partial x}\left(\frac{\partial\phi}{\partial z}\right) = \frac{\partial}{\partial z}\left(\frac{\partial\phi}{\partial x}\right).$$

An irrotational flow with

$$\frac{\partial u}{\partial y} = \frac{\partial v}{\partial x}, \quad \frac{\partial v}{\partial z} = \frac{\partial w}{\partial y}, \quad \frac{\partial w}{\partial x} = \frac{\partial u}{\partial z}$$

can therefore be described completely by a scalar function $\phi(x, y, z, t)$ with

$$u = -\frac{\partial\phi}{\partial x}, \quad v = -\frac{\partial\phi}{\partial y}, \quad w = -\frac{\partial\phi}{\partial z}. \tag{6-10}$$

This function ϕ is called the *velocity potential* of the flow. It should be clear that a flow can be described by a velocity potential if and only if the flow is irrotational, irrespective of the mechanical properties of the fluid.

In vector notation, Eq. 6–10 can be written as

$$\mathbf{q} = -\nabla\phi \qquad (6\text{–}11)$$

where the vector $\nabla\phi$ is the gradient of the velocity potential ϕ (see Eq. 5–6b). The velocity potential can be visualized as follows. For a function $\phi(x, y, z, t)$, there is a value of ϕ at each point at time t. At any instant, imagine all the points with the same value of ϕ. These points lie on a continuous surface called an *equipotential surface*, as shown by the dashed lines in Fig. 6–5. For an unsteady flow, the equipotential surfaces will move. For a steady flow with $\phi(x, y, z)$, they are stationary. According to Eq. 6–11, the velocity $\mathbf{q}$ is normal to the equipotential surface with a magnitude equal to the rate of change of ϕ in this direction. The negative signs are used in Eqs. 6–10 and 6–11 to conform with the convention that the flow is directed from a higher to a lower potential.

When cylindrical coordinates are used, we have from potential $\phi(r, \theta, z, t)$

$$u = -\frac{\partial\phi}{\partial r},$$

$$v = -\frac{1}{r}\frac{\partial\phi}{\partial\theta}, \qquad (6\text{–}12)$$

$$w = -\frac{\partial\phi}{\partial z}.$$

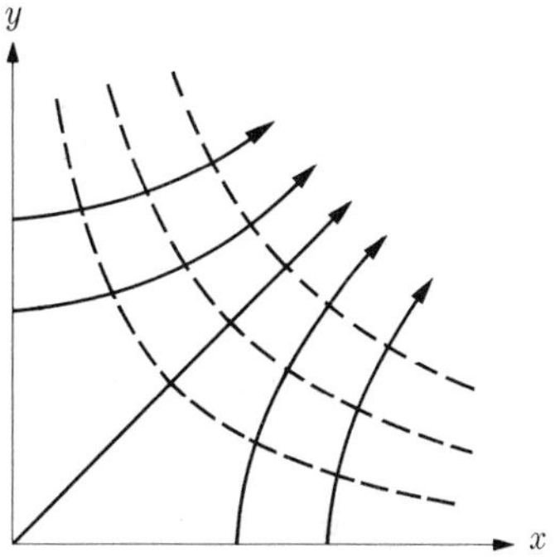

FIGURE 6–5

Here u, v, and w are the velocity components in the r-, θ- and z-directions, respectively.

By introducing the velocity potential, an irrotational flow is completely specified by a scalar function ϕ instead of three functions u, v, and w. Furthermore, the potential function for an incompressible flow is to be determined from a linear equation, as will be shown in the next section.

Example 6–1. A uniform stream of speed U in the x-direction is an irrotational flow. Find its velocity potential.

With $u = U$ and $v = w = 0$ in Eq. 6–10, we have

$$\frac{\partial\phi}{\partial x} = -U, \qquad \frac{\partial\phi}{\partial y} = 0, \qquad \frac{\partial\phi}{\partial z} = 0.$$

Therefore $\phi = -Ux + C$, where C is the constant of integration. Any value of C can be used, since the velocity distribution is related to the spatial rate of change of ϕ, and not to the absolute value of ϕ. Using $C = 0$, we have

$$\phi = -Ux. \qquad (6\text{–}13)$$

Points with the same x form an equipotential surface. Such surfaces are shown in Fig. 6–6. As expected, the streamlines are directed from a higher potential to a lower potential, and are normal to the equipotential surfaces.

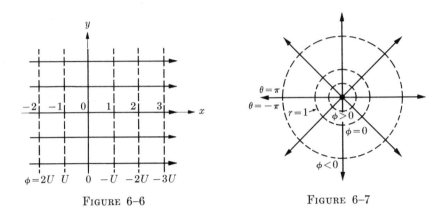

FIGURE 6–6   FIGURE 6–7

Example 6–2. Find the velocity potential of the flow generated by a line source of uniform strength. This flow is an irrotational flow with radial streamlines.

A *line source* is an infinitely long straight line where an incompressible fluid is supposedly being continuously created and sent out evenly in radial directions away from the line. Although it exists only in our imagination, the line source will be found useful in the study of irrotational flow. Let Q be the volume of fluid sent out per unit time per unit length of the line source. It is also called the *strength* of the line source. In a cylindrical coordinate system with the line source along the z-axis, the velocity is in the r-direction as shown in Fig. 6–7. At a distance r, the flow is distributed uniformly over a cylindrical surface with a circumference of $2\pi r$. Thus $u = Q/(2\pi r)$, $v = 0$, and $w = 0$.

It can be verified with Eqs. 6–7 to 6–9 that this flow is irrotational and therefore has a velocity potential that can be found from Eq. 6–12. With $\partial\phi/\partial r = -Q/(2\pi r)$, $\partial\phi/\partial\theta = 0$, and $\partial\phi/\partial z = 0$, we have

$$\phi = -C \ln r, \tag{6–14}$$

where $C = Q/(2\pi)$, and the constant of integration has been set equal to zero for convenience. The equipotential surfaces are therefore cylindrical surfaces with constant r as shown in Fig. 6–7.

A *line sink* is the negative of a line source. If Q is the volume of fluid supposedly absorbed per unit time per unit length, we have

$$\phi = C \ln r, \tag{6–15}$$

where $C = Q/(2\pi)$.

Example 6-3. Find the velocity potential of a *free vortex*, which is an irrotational flow with circular paths around an axis.

Using cylindrical coordinates with the z-axis along the axis of the vortex, we have $u = 0$ and $w = 0$. Since the motion is irrotational, we have from Eqs. 6-7 and 6-8, $\partial(rv)/\partial r = 0$ and $\partial v/\partial z = 0$. From the equation of continuity, Eq. 4-20, we have $\partial v/\partial \theta = 0$. Thus

$$rv = \text{constant} = \frac{\Gamma}{2\pi}, \tag{6-16}$$

where the constant Γ is the circulation around the axis. (For the definition of circulation, see Section 8-3.)

To find the velocity potential, we have from Eq. 6-12, $\partial\phi/\partial r = 0$, $\partial\phi/r\,\partial\theta = -v$, and $\partial\phi/\partial z = 0$. Thus

$$\phi = -vr\theta = -\frac{\Gamma}{2\pi}\,\theta. \tag{6-17}$$

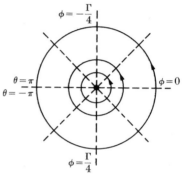

FIGURE 6-8

The equipotential surfaces are radial planes with constant θ, as shown in Fig. 6-8. Since a point at any θ_1 may also be said to have the coordinate $(\theta_1 + 2\pi n)$, where n can be any integer, the potential at this point can have an infinite number of values. To reduce ϕ to a single-valued function, it is necessary to limit the range of θ to 2π. For example, the range of θ in Fig. 6-8 is from $-\pi$ to π.

PROBLEMS

6-6. The velocity potential for the flow shown in Fig. 6-5 is $\phi = -kxy$ with $k = 1$. (a) Using Eq. 6-10, compute the velocity components at the point $(0.8, 1, 0)$. (b) Construct the curves for $\phi = -0.8$ and $\phi = -1.0$ in the interval $0.8 \le x \le 1$. Find these velocity components from the spatial rate of change of ϕ in your graph. (c) From the components in (a) compute the speed q at this point. Check by the greatest rate of change of ϕ in your graph. (d) What are the dimensions of ϕ and k?

6-7. Find the velocity potential of a uniform stream parallel to the xy-plane with speed U directed at an angle α with the x-axis.

6-8. Can a flow with $u = -2x$, $v = 2y$, and $w = 0$ be described with a velocity potential? If so, find the potential.

6-9. A *point source* is a point where an incompressible fluid is imagined to be created and sent out evenly in all directions. Show that its velocity potential is $\phi = k/r$, where k is a constant and r is a spherical coordinate with the source as the origin.

6–3. Irrotational flow of homogeneous frictionless fluids

Although velocity potentials can be used whenever a flow has been shown to be irrotational irrespective of the fluid properties, we shall consider only homogeneous frictionless fluids in this chapter. With the introduction of the velocity potential, the problem is reduced to finding the velocity potential ϕ and pressure p in accordance with the momentum equations and the equation of continuity.

We consider Eq. 5–1, the momentum equation for the x-direction:

$$\frac{\partial u}{\partial t} + u \frac{\partial u}{\partial x} + v \frac{\partial u}{\partial y} + w \frac{\partial u}{\partial z} = -\frac{1}{\rho} \frac{\partial p}{\partial x} - g \frac{\partial h}{\partial x}.$$

With the irrotational conditions $\partial u/\partial y = \partial v/\partial x$ and $\partial u/\partial z = \partial w/\partial x$, and $u = -\partial \phi/\partial x$, we have

$$-\frac{\partial^2 \phi}{\partial t\, \partial x} + u \frac{\partial u}{\partial x} + v \frac{\partial v}{\partial x} + w \frac{\partial w}{\partial x} = -\frac{1}{\rho} \frac{\partial p}{\partial x} - g \frac{\partial h}{\partial x}.$$

For a homogeneous incompressible fluid with constant ρ, this equation can be written as

$$\frac{\partial}{\partial x}\left[-\frac{\partial \phi}{\partial t} + \frac{1}{2}(u^2 + v^2 + w^2) + \frac{p}{\rho} + gh \right] = \frac{\partial}{\partial x}\left(-\frac{\partial \phi}{\partial t} + \frac{q^2}{2} + \frac{p}{\rho} + gh \right) = 0.$$

In a similar manner, we obtain from the momentum equations for the y- and z-directions,

$$\frac{\partial}{\partial y}\left(-\frac{\partial \phi}{\partial t} + \frac{q^2}{2} + \frac{p}{\rho} + gh \right) = 0, \qquad \frac{\partial}{\partial z}\left(-\frac{\partial \phi}{\partial t} + \frac{q^2}{2} + \frac{p}{\rho} + gh \right) = 0.$$

The quantity in parentheses does not vary with x, y, or z. Thus

$$-\frac{\partial \phi}{\partial t} + \frac{q^2}{2} + \frac{p}{\rho} + gh = f(t). \tag{6–18}$$

For a steady flow, $\partial \phi/\partial t = 0$. If there is no overall change of pressure, $f(t)$ is a constant. Then

$$\frac{q^2}{2} + \frac{p}{\rho} + gh = \text{constant}. \tag{6–19}$$

This is *Bernoulli's equation* for steady irrotational flow of a homogeneous frictionless fluid. The constant is the same throughout the flow. (The reader will recall that the constant may vary from one streamline to another in a steady, incompressible, frictionless but rotational flow.)

Next consider the equation of continuity for incompressible flow

$$\frac{\partial u}{\partial x} + \frac{\partial v}{\partial y} + \frac{\partial w}{\partial z} = 0.$$

Substituting in Eq. 6–10, we have

$$\frac{\partial^2 \phi}{\partial x^2} + \frac{\partial^2 \phi}{\partial y^2} + \frac{\partial^2 \phi}{\partial z^2} = 0. \tag{6–20}$$

This is the *Laplace equation* from which the velocity potential is to be determined. The solution must also satisfy the boundary conditions of the case under consideration. For example, at a stationary solid boundary, we must have $\partial \phi / \partial n = 0$, where n is the direction normal to the solid boundary, so that there is no normal velocity component.

When cylindrical coordinates are used, the Laplace equation is obtained by substituting Eq. 6–12 into Eq. 4–21 for incompressible flow:

$$\frac{\partial \phi}{r \, \partial r} + \frac{\partial^2 \phi}{\partial r^2} + \frac{\partial^2 \phi}{r^2 \, \partial \theta^2} + \frac{\partial^2 \phi}{\partial z^2} = 0. \tag{6–21}$$

Equations 6–20 and 6–21 express the same physical fact and are written in vector notation as $\nabla \cdot \mathbf{q} = -\nabla \cdot \nabla \phi = -\nabla^2 \phi = 0$ (see Eqs. 4–29, 4–30, and 6–11). Thus

$$\nabla^2 \phi = 0. \tag{6–22}$$

When the velocity potential has been obtained from the Laplace equation, satisfying the boundary conditions, the velocity components can be computed from Eqs. 6–10 and 6–12. The pressure distribution can then be obtained from Eqs. 6–18 and 6–19 for unsteady and steady flows, respectively. The problem is therefore completely solved once the velocity potential is found. The Laplace equation has been extensively studied since it is encountered in many fields of study, such as electricity, magnetism, and heat conduction. However, many methods of solution involve mathematics higher than the level intended for this book. Several of the simpler methods are presented in the following sections. For more advanced methods, consult books on hydrodynamics.

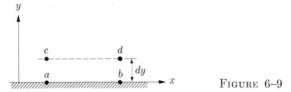

FIGURE 6–9

Note that in an irrotational flow we cannot specify zero tangential velocity relative to a solid boundary. Consider the irrotational flow parallel to the xy-plane as shown in Fig. 6–9. Since we specify $v = 0$ at the wall along the x-axis, we have $\partial v / \partial x = 0$ there. Since the flow is irrotational, we also have $\partial u / \partial y = \partial v / \partial x = 0$ at points a and b. Suppose that we also specify the tangential velocity $u = 0$ at the wall. Then $\partial u / \partial x = 0$ there. Due to continuity, we also have $\partial v / \partial y = -\partial u / \partial x = 0$ at a and b. With $\partial u / \partial y = 0$ and $\partial v / \partial y = 0$

at these points, the particles at a short distance from the wall, such as c and d, would also have $u = 0$ and $v = 0$. These particles would behave as if they were in contact with a solid wall. In fact, it can be proved that the velocity would be zero everywhere. In other words, no irrotational flow is possible if we specify no tangential velocity at a solid surface. We therefore allow in an irrotational flow a tangential velocity relative to a solid boundary, although all real fluids have been observed to have no relative velocity there. The irrotational flow is intended to be an approximation of the flow outside the boundary layer at high Reynolds number. This tangential relative velocity represents therefore the velocity at the edge of the thin boundary layer. This topic will be further discussed in Section 6–7 and subsequent chapters.

Example 6–4. Investigate the flow with the velocity potential

$$\phi = -k(x^2 + y^2 - 2z^2).$$

First, verify that this function is a solution of the Laplace equation so that the flow satisfies the continuity requirement of an incompressible flow:

$$\frac{\partial^2 \phi}{\partial x^2} + \frac{\partial^2 \phi}{\partial y^2} + \frac{\partial^2 \phi}{\partial z^2} = -2k - 2k + 4k = 0.$$

The velocity components are $u = -\partial\phi/\partial x = 2kx$, $v = 2ky$, and $w = -4kz$; and

$$q^2 = u^2 + v^2 + w^2 = 4k^2(x^2 + y^2 + 4z^2).$$

Since the normal velocity to the xy-plane is zero (i.e., $w = 0$ at $z = 0$), this flow may be regarded as a flow against the xy-plane, as shown in Fig. 6–10(a). Note that there is a tangential velocity component along this plane. As $u = 0$ at $x = 0$, this flow can also be regarded as the flow shown in Fig. 6–10(b). In fact, this ϕ gives the flow over any boundary surface formed by the streamlines.

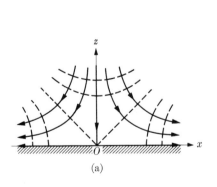
(a)

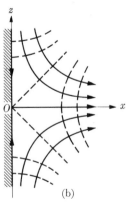
(b)

FIGURE 6–10

To find the pressure distribution, let the z-axis be pointing upward, and let $h = z$. From Bernoulli's equation

$$2k^2(x^2 + y^2 + 4z^2) + \frac{p}{\rho} + gz = \text{constant.}$$

Let p_0 be the pressure at the origin, so that the constant is p_0. Then

$$p = p_0 - 2\rho k^2(x^2 + y^2 + 4z^2) - \rho gz.$$

The origin, where the velocity is zero, is called a *stagnation point.*

PROBLEMS

6–10. Does $\phi = -kxy$ describe a possible flow of an incompressible fluid? If so, find the pressure difference between the points (x_1, y_1, z_1) and (x_2, y_2, z_2), the y-axis being vertically upward.

6–11. Verify that the potentials in Eqs. 6–13, 6–14, and 6–17 are solutions of the Laplace equation.

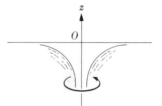

FIGURE 6–11

6–12. A liquid moves irrotationally about the vertical z-axis as shown in Fig. 6–11. Show that the equation of the free surface is $z = -\Gamma^2/8\pi^2 gr^2$, where Γ is the constant circulation.

6–4. Superposition of solutions of Laplace equation

An important property of linear equations is that solutions can be superposed. If $\phi_1(x, y, z, t)$ and $\phi_2(x, y, z, t)$ are individually solutions of the Laplace equation, which is linear, $\phi(=\phi_1 + \phi_2)$ is also a solution:

$$\frac{\partial^2 \phi}{\partial x^2} + \frac{\partial^2 \phi}{\partial y^2} + \frac{\partial^2 \phi}{\partial z^2} = \left(\frac{\partial^2 \phi_1}{\partial x^2} + \frac{\partial^2 \phi_1}{\partial y^2} + \frac{\partial^2 \phi_1}{\partial z^2}\right) + \left(\frac{\partial^2 \phi_2}{\partial x^2} + \frac{\partial^2 \phi_2}{\partial y^2} + \frac{\partial^2 \phi_2}{\partial z^2}\right)$$

$$= 0 + 0 = 0.$$

Since ϕ_1 and ϕ_2 individually satisfy the Laplace equation, the value of each of the parentheses is zero. Thus ϕ also satisfies the equation. In fact, the velocity given by ϕ at any point is the vector sum of those given by ϕ_1 and ϕ_2. Let u,

u_1, and u_2 be the x-components of the velocities given by ϕ, ϕ_1, and ϕ_2, respectively. Then

$$u = -\frac{\partial \phi}{\partial x} = -\frac{\partial}{\partial x}(\phi_1 + \phi_2) = -\frac{\partial \phi_1}{\partial x} - \frac{\partial \phi_2}{\partial x} = u_1 + u_2.$$

Similar results can be obtained for the y- and z-components.   Thus, scalar addition of potentials is equivalent to vectorial addition of the corresponding velocities.

We shall study some results obtained by superposition of known solutions of the Laplace equation. This method of solution is of considerable practical value.

Example 6–5. Given $\phi_1 = -Ux$ and $\phi_2 = (Q/2\pi) \ln r$, find the velocity at the point $(1, 1, 0)$ due to the velocity potential ϕ which is $\phi_1 + \phi_2$.

From

$$\phi = \phi_1 + \phi_2 = -Ux + \frac{Q}{2\pi} \ln r = -Ux + \frac{Q}{2\pi} \ln \sqrt{x^2 + y^2},$$

we have in general

$$u = -\frac{\partial \phi}{\partial x} = U - \frac{Q}{2\pi}\frac{x}{x^2 + y^2} = U - \frac{Q}{2\pi r}\frac{x}{r},$$

$$v = -\frac{\partial \phi}{\partial y} = -\frac{Q}{2\pi}\frac{y}{x^2 + y^2} = -\frac{Q}{2\pi r}\frac{y}{r}.$$

At $x = 1$, $y = 1$, we have $u = U - (Q/4\pi)$, $v = -Q/4\pi$.

The potential ϕ_1 gives a uniform flow with $q_1 = U$ in the x-direction (see Eq. 6–13), and the potential ϕ_2 gives the flow toward a line sink with $q_2 = Q/(2\pi r)$ (see Eq. 6–15). Note that the velocity $\mathbf{q}$ due to ϕ is the vector sum of the velocities $\mathbf{q}_1$ and $\mathbf{q}_2$ due to ϕ_1 and ϕ_2, respectively (see Fig. 6–12).

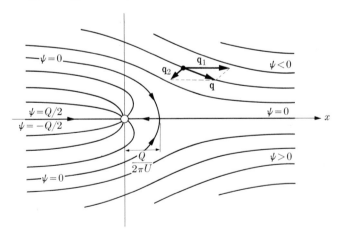

FIGURE 6–12

Example 6–6. (a) Find the velocity potential due to a line source together with a line sink of equal strength. (b) If the distance between them approaches zero, the result is called a *two-dimensional doublet*. Find the velocity potential of this doublet.

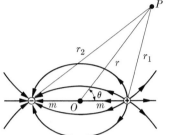

FIGURE 6–13

(a) When the line source and line sink are placed as shown in Fig. 6–13, we have, from Eqs. 6–14 and 6–15, for any point P,

$$\phi_1 = -\frac{Q}{2\pi} \ln r_1 \quad \text{(source)},$$

$$\phi_2 = \frac{Q}{2\pi} \ln r_2 \quad \text{(sink)},$$

where r_1 and r_2 are measured from the source and the sink, respectively, and can be expressed in terms of the cylindrical coordinates r and θ of P:

$$r_1^2 = (r \sin \theta)^2 + (r \cos \theta - m)^2 = r^2 - 2rm \cos \theta + m^2,$$

$$\ln r_1 = \tfrac{1}{2} \ln r_1^2 = \tfrac{1}{2} \ln \left\{ r^2 \left[1 - 2 \cos \theta \, \frac{m}{r} + \left(\frac{m}{r} \right)^2 \right] \right\}$$

$$= \tfrac{1}{2} \ln r^2 + \tfrac{1}{2} \ln \left[1 - 2 \cos \theta \, \frac{m}{r} + \left(\frac{m}{r} \right)^2 \right].$$

Similarly,

$$\ln r_2 = \tfrac{1}{2} \ln r^2 + \tfrac{1}{2} \ln \left[1 + 2 \cos \theta \, \frac{m}{r} + \left(\frac{m}{r} \right)^2 \right].$$

Therefore, for the line source and line sink together,

$$\phi = \phi_1 + \phi_2 = -\frac{Q}{4\pi} \left\{ \ln \left[1 - 2 \cos \theta \, \frac{m}{r} + \left(\frac{m}{r} \right)^2 \right] \right.$$

$$\left. - \ln \left[1 + 2 \cos \theta \, \frac{m}{r} + \left(\frac{m}{r} \right)^2 \right] \right\}.$$

(b) If $m = 0$, we have $\phi = 0$; i.e., there is no flow, with the discharge from the source being absorbed directly by the sink. However, if m approaches

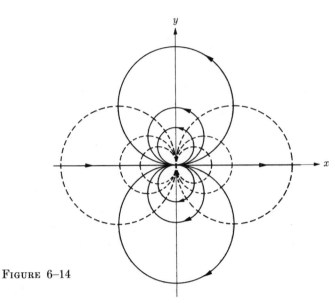

FIGURE 6–14

zero but is not exactly equal to zero, we have the flow shown in Fig. 6–14. As $m \to 0$, m/r becomes infinitesimal for any finite value of r. To find ϕ in this case, we have to find an expression for $\ln (1 + x)$ as $x \to 0$. Taylor's expansion of a function about $x = 0$ is

$$f(x) = f(0) + f'(0)x + \tfrac{1}{2}f''(0)x^2 + \cdots,$$

where $f'(0)$, $f''(0)$, ... are the values of df/dx, d^2f/dx^2, ... at $x = 0$, respectively. With $f(x) = \ln (1 + x)$, we have $f'(x) = (1 + x)^{-1}$, $f''(x) = -(1 + x)^{-2}$, ..., and $f(0) = 0, f'(0) = 1, f''(0) = -1, \ldots$. Thus

$$\ln (1 + x) = 0 + x - \tfrac{1}{2}x^2 + \cdots.$$

As $x \to 0$, the higher-order terms vanish by comparison. Thus $\ln (1 + x) = x$ as $x \to 0$. As $m/r \to 0$,

$$\lim_{(m/r) \to 0} \left\{ \ln \left[1 - 2 \cos \theta \, \frac{m}{r} + \left(\frac{m}{r} \right)^2 \right] - \ln \left[1 + 2 \cos \theta \, \frac{m}{r} + \left(\frac{m}{r} \right)^2 \right] \right\}$$

$$= -2 \cos \theta \, \frac{m}{r} - 2 \cos \theta \, \frac{m}{r} = -4 \cos \theta \, \frac{m}{r}.$$

Here the higher-order terms of m/r have disappeared by comparison. Therefore, for a two-dimensional doublet, as shown in Fig. 6–14,

$$\phi = \frac{Qm}{\pi} \frac{\cos \theta}{r} = k \frac{\cos \theta}{r}, \qquad (6\text{--}23)$$

where the constant k is called the *strength* of the doublet.

Example 6–7. Investigate the flow with a two-dimensional doublet opposing a uniform stream.

For a doublet pointing in the x-direction, we have, from Eq. 6–23, $\phi_1 = k\cos\theta/r$. For a uniform stream in the negative x-direction, we have $\phi_2 = Ux$. Thus for the specified flow

$$\phi = \phi_1 + \phi_2 = k\,\frac{\cos\theta}{r} + Ux = k\,\frac{\cos\theta}{r} + Ur\cos\theta.$$

With cylindrical coordinates, the radial velocity component is

$$u = -\frac{\partial\phi}{\partial r} = \left(\frac{k}{r^2} - U\right)\cos\theta.$$

Note that $u = 0$ at $r = \sqrt{k/U}$. This flow can therefore be considered as a uniform stream passing a circular cylinder of radius $R = \sqrt{k/U}$, as shown in Fig. 6–15. For given radius R, use $k = R^2U$. Then

$$\phi = U\cos\theta\left(r + \frac{R^2}{r}\right), \tag{6–24}$$

$$u = U\cos\theta\left(\frac{R^2}{r^2} - 1\right),$$

$$v = -\frac{\partial\phi}{r\,\partial\theta} = U\sin\theta\left(1 + \frac{R^2}{r^2}\right).$$

There are stagnation points (with $u = 0$ and $v = 0$) at the point $r = R$, $\theta = 0$, and the point $r = R$, $\theta = \pi$.

Note that the circumferential velocity component v is not zero at the surface of the cylinder as shown in Fig. 6–15. As explained at the end of Section 6–3, the velocity relative to the solid surface represents the velocity just outside the boundary layer. The question is whether the boundary layer will remain thin compared with the radius R so that Fig. 6–15 can represent the flow outside the boundary layer. This problem will be discussed in Section 6–7.

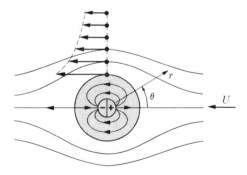

FIGURE 6–15

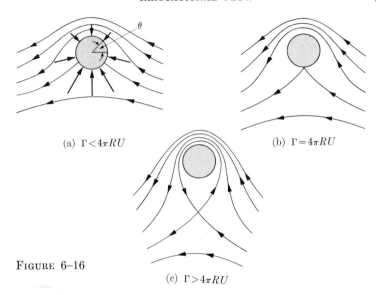

(a) $\Gamma < 4\pi R U$ (b) $\Gamma = 4\pi R U$

FIGURE 6–16

(c) $\Gamma > 4\pi R U$

Example 6–8. The flow given in Example 6–7 is not the only possible irrotational flow around a circular cylinder. Investigate the flow with circulation around the cylinder.

Superpose a free vortex with potential $\phi_1 = -\Gamma\theta/(2\pi)$ to the potential in Eq. 6–24:

$$\phi = -\frac{\Gamma}{2\pi}\,\theta + U\cos\theta\left(r + \frac{R^2}{r}\right).$$

For this flow, we have

$$u = -\frac{\partial\phi}{\partial r} = U\cos\theta\left(1 - \frac{R^2}{r^2}\right),$$

$$v = -\frac{\partial\phi}{r\,\partial\theta} = \frac{\Gamma}{2\pi r} + U\sin\theta\left(1 + \frac{R^2}{r^2}\right).$$

To find the stagnation points, set $u = 0$ and $v = 0$. One possible solution is

$$r = R \qquad \text{and} \qquad \sin\theta = -\frac{\Gamma}{4\pi R U}.$$

This solution gives two stagnation points on the cylinder if $\Gamma/(4\pi R U) < 1$, as shown in Fig. 6–16(a). Another solution for $u = 0$ and $v = 0$ is $\cos\theta = 0$ and

$$\frac{\Gamma}{2\pi r} \pm U\left(1 + \frac{R^2}{r^2}\right) = 0, \qquad \text{or} \qquad r = R\left[\frac{\Gamma}{4\pi R U} \pm \sqrt{\left(\frac{\Gamma}{4\pi R U}\right)^2 - 1}\right].$$

For $\Gamma/(4\pi R U) > 1$, there is one stagnation point outside the cylinder ($r > R$) as shown in Fig. 6–16(c). For $\Gamma/(4\pi R U) = 1$, both solutions give one stagnation point on the cylinder, as shown in Fig. 6–16(b).

PROBLEMS

6–13. Given that ϕ_1 and ϕ_2 are solutions of the Laplace equation, show that $\phi \, (= c_1\phi_1 + c_2\phi_2)$ is also a solution, where c_1 and c_2 are arbitrary constants.

6–14. Compute the components of $\mathbf{q}_1$, $\mathbf{q}_2$, and $\mathbf{q}$ in Fig. 6–12 from the potentials ϕ_1, ϕ_2, and ϕ in Example 6–5 with $Q = 2\pi U$. Since $\phi = \phi_1 + \phi_2$, verify $\mathbf{q} = \mathbf{q}_1 + \mathbf{q}_2$. Also find the location of the stagnation point.

6–15. If in Fig. 6–13 there are a point source $(\phi = k/r_1)$ and a point sink of equal strength $(\phi_2' = -k/r_2)$, show that as $m \to 0$, the potential of the *three-dimensional doublet* is $\phi = 2mk \cos \theta / r^2$, where r and θ are spherical coordinates. [*Hint:* Show that $r_2 - r_1 = 2m \cos \theta$, as $m \to 0$.]

6–16. Show that the potential of an irrotational flow passing a sphere of radius R is

$$\phi = U \cos \theta \, [r + (R^3/2r^2)],$$

where r and θ are spherical coordinates (see Fig. 6–15).

6–5. Dynamic pressure and lift

According to Bernoulli's equation for steady irrotational flow of incompressible frictionless fluids,

$$\frac{p}{\rho} + gh + \frac{q^2}{2} = \text{constant},$$

the pressure difference between two points is due to the differences of elevation h and speed q. It is often convenient to separate the effects of h and q by writing

$$p = p_d + p_s,$$

where p_d is called the *dynamic pressure*, and p_s the *hydrostatic pressure*. The latter is defined as

$$\frac{p_s}{\rho} + gh = \text{constant}. \qquad (6\text{–}25)$$

Bernoulli's equation can then be written as

$$\frac{p_d}{\rho} + \frac{q^2}{2} = \text{constant}. \qquad (6\text{–}26)$$

The difference of p_s is then due to the difference of elevation, while p_d is influenced only by the speed.

For a given p, the values of p_d and p_s have not been uniquely defined, since the constant in Eq. 6–25 is arbitrary. This constant is to be so chosen that the advantage of the division of p into p_d and p_s is enhanced. For example, in the case of a uniform stream of speed U passing an object, as shown in Fig.

6–15, we have Bernoulli's equation

$$\frac{p_d}{\rho} + \frac{p_s}{\rho} + gh + \frac{q^2}{2} = \frac{p_0}{\rho} + gh_0 + \frac{U^2}{2},$$

where p_0 and h_0 are the pressure and elevation, respectively, of any point in the undisturbed stream. It is advantageous to write Eq. 6–25 as

$$\frac{p_s}{\rho} + gh = \frac{p_0}{\rho} + gh_0,$$

so that

$$\frac{p_d}{\rho} + \frac{q^2}{2} = \frac{U^2}{2}. \tag{6–27}$$

Then dynamic pressure p_d is zero in the undisturbed flow, where $q = U$, and the value of p_d at any point represents the change of pressure due to the difference between U and q at this point. In particular, at a stagnation point where $q = 0$,

$$p_d = \frac{\rho U^2}{2}.$$

This dynamic pressure is called the *stagnation pressure*. Sometimes this value is called the *dynamic pressure* of the flow system.

The division of p into p_s and p_d also facilitates the discussion of the force acting on an object by the surrounding fluid. In the case of frictionless fluids, this force can be found by summing up the forces due to p on the surface of the object. The sum of the forces due to p_s is equal to the buoyant force that would be exerted by the incompressible fluid if at rest (compare Eq. 6–25 with Eq. 3–2). The sum of the forces due to p_d is the *dynamic force* on the object. Its component along the direction of motion is called the *drag*, and the other component perpendicular to this direction is called the *lift*.

Example 6–9. Find the dynamic force acting on a circular cylinder in a uniform stream.

When there is no circulation around the cylinder, as shown in Fig. 6–15, the distribution of q is symmetrical with respect to two perpendicular diameters. According to Eq. 6–27, the distribution of p_d is also symmetrical with respect to these diameters. There is therefore no net dynamic force acting on the cylinder. (There is, however, a buoyant force due to the variation of p_s over the surface.)

For a flow with a circulation as shown in Fig. 6–16, there is no drag due to symmetry of the flow about the vertical diameter. However, since the speed is higher above the cylinder than below, p_d is higher underneath. There is therefore a lift which can be computed as follows. To find p_d on the surface of the cylinder, first find the speed q there. From Example 6–8, we have at $r = R$,

$u = 0$, and

$$q = v = 2U \sin \theta + \frac{\Gamma}{2\pi R}.$$

From Eq. 6–27, we have at $r = R$,

$$p_d = \rho \left(\frac{U^2}{2} - \frac{q^2}{2} \right) = \frac{\rho U^2}{2} \left[1 - \left(2 \sin \theta + \frac{\Gamma}{2\pi R U} \right)^2 \right].$$

The lift L per unit length of the cylinder is therefore (see Fig. 6–16a)

$$L = \int -\sin \theta \, dF = \int_0^{2\pi} -\sin \theta \, p_d R \, d\theta$$

$$= -\frac{\rho R U^2}{2} \int_0^{2\pi} \sin \theta \left[1 - \left(2 \sin \theta + \frac{\Gamma}{2\pi R U} \right)^2 \right] d\theta = \rho U \Gamma. \qquad (6\text{–}28)$$

PROBLEMS

6–17. For the flow around a circular cylinder without circulation, compute p_d at the stagnation points, and at the top and bottom of the cylinder. Also find the four points where p_d is zero. Does $p_d = 0$ mean vacuum, atmospheric pressure, or neither?

6–18. Verify that the drag of the cylinder in Fig. 6–15 is zero.

6–6. Methods of approximate solution of Laplace equation

While there are exact methods of solution of the Laplace equation for special types of boundary conditions, technical problems are often of such complexity that solution is difficult if not impossible. One must then have recourse to approximate methods. Two methods are presented in this section, and two more can be found in Section 7–4 and Example 11–7.

(a) Electrical conductor analog. The Laplace equation is encountered in problems of many physical systems other than fluid flow, e.g., in the flow of electricity and the conduction of heat. It is therefore possible to construct an analog (a model of a different physical system) in which measurements can be more easily made. Because of the relative ease of electrical measurement, the electrical conductor analog is often used.

The flow of electric charges in a conductor is in the direction normal to surfaces of constant voltage, and the intensity of flow at any point is proportional to the gradient of the voltage:

$$\mathbf{J} = -k\nabla V,$$

where $\mathbf{J}$ is the current intensity in charges per unit time per unit area of the constant voltage surface, k is the electrical conductivity of the material, and V is the voltage. Compare this equation with Eq. 6–11. In this potential flow, the components of $\mathbf{J}$ along the coordinate axes are

$$J_x = -k\,\frac{\partial V}{\partial x}, \qquad J_y = -k\,\frac{\partial V}{\partial y}, \qquad J_z = -k\,\frac{\partial V}{\partial z}.$$

Due to the conservation of electric charges, we have from a derivation similar to that of the equation of continuity,

$$\frac{\partial J_x}{\partial x} + \frac{\partial J_y}{\partial y} + \frac{\partial J_z}{\partial z} = 0.$$

Thus, for a homogeneous conductor with constant k,

$$\frac{\partial^2 V}{\partial x^2} + \frac{\partial^2 V}{\partial y^2} + \frac{\partial^2 V}{\partial z^2} = 0,$$

which is the Laplace equation. Thus the distribution of voltage in a homogeneous conductor is analogous to the distribution of velocity potential if the boundary conditions are similar.

The analog should be geometrically similar to the fluid system. For a solid wall, the analog should have an insulated surface. For a boundary with a constant velocity potential, a good conductor along this boundary should be used in the analog. In Fig. 6–17 is shown an analog for the determination of the velocity potential around an airfoil. When the Wheatstone bridge is balanced, we have the ratio of the voltage drop ΔV_1, from electrode A to the pointer P, to the total voltage drop ΔV from A to B:

$$\frac{\Delta V_1}{\Delta V} = \frac{r_1}{r_1 + r_2}.$$

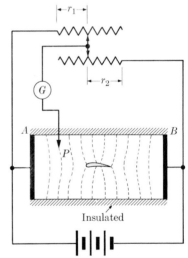

FIGURE 6–17

Thus an equipotential surface can be traced with one setting of r_1 and r_2 by balancing the bridge at various positions of the pointer P. In a similar manner, equipotential surfaces with equal increments of potential can be traced. The velocity vectors are normal to these equipotential surfaces. The relative magnitude of the velocity is indicated by the spatial rate of decrease of the potential, and is therefore inversely proportional to the spacing of these equipotential surfaces.

(b) Method of finite differences. In the *method of finite differences*, a differential equation is replaced by approximate algebraic equations. Take for

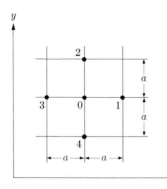

FIGURE 6–18

example the Laplace equation for two-dimensional flow in the xy-plane. This equation specifies that at any point 0, the second derivatives of the potential must be such that

$$\left(\frac{\partial^2 \phi}{\partial x^2}\right)_0 + \left(\frac{\partial^2 \phi}{\partial y^2}\right)_0 = 0, \tag{6–29}$$

where the subscript 0 indicates values at point 0. These derivatives can be expressed approximately in terms of the values of ϕ at points near 0, such as ϕ_1, ϕ_2, etc. (see Fig. 6–18). In this way, Eq. 6–29 is replaced by an approximate algebraic equation involving these values of ϕ. To do this, express $\phi(x, y)$ in terms of values at 0 by Taylor's series. For a function of two variables, we have

$$\phi(x, y) = \phi_0 + \left[\left(\frac{\partial \phi}{\partial x}\right)_0 (x - x_0) + \left(\frac{\partial \phi}{\partial y}\right)_0 (y - y_0)\right] + \frac{1}{2!}\left[\left(\frac{\partial^2 \phi}{\partial x^2}\right)_0 (x - x_0)^2\right.$$
$$\left. + 2\left(\frac{\partial^2 \phi}{\partial x\, \partial y}\right)_0 (x - x_0)(y - y_0) + \left(\frac{\partial^2 \phi}{\partial y^2}\right)_0 (y - y_0)^2\right] + \cdots.$$

Thus for point 1 at $x = x_0 + a$, $y = y_0$,

$$\phi_1 = \phi_0 + \left(\frac{\partial \phi}{\partial x}\right)_0 a + \frac{1}{2}\left(\frac{\partial^2 \phi}{\partial x^2}\right)_0 a^2 + \frac{1}{3!}\left(\frac{\partial^3 \phi}{\partial x^3}\right)_0 a^3 + \frac{1}{4!}\left(\frac{\partial^4 \phi}{\partial x^4}\right)_0 a^4 + \cdots, \tag{6–30}$$

and for point 3 at $x = x_0 - a$, $y = y_0$,

$$\phi_3 = \phi_0 - \left(\frac{\partial \phi}{\partial x}\right)_0 a + \frac{1}{2}\left(\frac{\partial^2 \phi}{\partial x^2}\right)_0 a^2 - \frac{1}{3!}\left(\frac{\partial^3 \phi}{\partial x^3}\right)_0 a^3 + \frac{1}{4!}\left(\frac{\partial^4 \phi}{\partial x^4}\right)_0 a^4 - \cdots.$$

Adding these two equations, we have

$$\left(\frac{\partial^2 \phi}{\partial x^2}\right)_0 = \frac{1}{a^2}(\phi_1 + \phi_3 - 2\phi_0) + \frac{1}{12}\left(\frac{\partial^4 \phi}{\partial x^4}\right)_0 a^2 + \cdots.$$

Similarly,

$$\left(\frac{\partial^2 \phi}{\partial y^2}\right)_0 = \frac{1}{a^2}(\phi_2 + \phi_4 - 2\phi_0) + \frac{1}{12}\left(\frac{\partial^4 \phi}{\partial y^4}\right)_0 a^2 + \cdots. \tag{6–31}$$

With all terms of the order of a^2 and higher neglected, the simplest approximation of Eq. 6–29 is

$$\phi_1 + \phi_2 + \phi_3 + \phi_4 - 4\phi_0 = 0. \tag{6–32}$$

For each interior grid point, considered as point 0, there is therefore one approximate algebraic equation.

At a point on a solid boundary, the normal velocity relative to the boundary must be zero. This boundary condition can be expressed as follows. For point 0 in Fig. 6–19, we must have $(\partial\phi/\partial x)_0 = 0$. Making use of Eqs. 6–30, 6–29, and 6–31, we have

$$\left(\frac{\partial\phi}{\partial x}\right)_0 = \frac{\phi_1 - \phi_0}{a} - \frac{1}{2}\left(\frac{\partial^2\phi}{\partial x^2}\right)_0 a - \frac{1}{3!}\left(\frac{\partial^3\phi}{\partial x^3}\right)_0 a^2 - \cdots$$

$$= \frac{\phi_1 - \phi_0}{a} + \frac{1}{2}\left(\frac{\partial^2\phi}{\partial y^2}\right)_0 a - \frac{1}{3!}\left(\frac{\partial^3\phi}{\partial x^3}\right)_0 a^2 - \cdots$$

$$= \frac{\phi_1 - \phi_0}{a} + \frac{\phi_2 + \phi_4 - 2\phi_0}{2a} - \frac{1}{3!}\left(\frac{\partial^2\phi}{\partial x^3}\right)_0 a^2 - \cdots.$$

Neglecting all terms of order of a^2 and higher, so as to be consistent with Eq. 6–32, we have for $(\partial\phi/\partial x)_0 = 0$,

$$2\phi_1 + \phi_2 + \phi_4 - 4\phi_0 = 0. \tag{6–33}$$

Equation 6–33 would be the same as Eq. 6–32 if we use the value ϕ_1 at point 3 in Fig. 6–19.

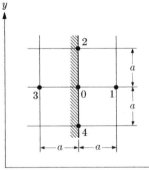

FIGURE 6–19

In applying this method to a two-dimensional potential flow, a grid is constructed over the area of the flow. For each grid point where the potential is to be found, an algebraic equation can be written. The values of the potential at these points can therefore be determined by solving these algebraic equations simultaneously. The solution of these equations may be performed by an electronic computer, or by a method of successive approximations devised by Southwell (see *Relaxation Methods in Theoretical Physics*, by Southwell, Oxford University Press, 1946).

PROBLEMS

6–19. In dealing with grid points near a solid boundary, some of the adjacent points may not be equidistant away, as shown in Fig. 6–20. Show that at point 0, the Laplace equation is approximated by

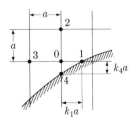

FIGURE 6–20

$$\frac{\phi_1}{k_1} + \phi_2 + \phi_3 + \frac{\phi_4}{k_4} - \left(2 + \frac{1}{k_1} + \frac{1}{k_4}\right)\phi_0 = 0.$$

6–20. For a potential flow symmetrical with respect to the z-axis, the potential $\phi(r, z)$ must obey Eq. 6–21:

$$\frac{1}{r}\frac{\partial\phi}{\partial r} + \frac{\partial^2\phi}{\partial r^2} + \frac{\partial^2\phi}{\partial z^2} = 0.$$

Show that at point 0 in Fig. 6–21, this equation can be approximated by

$$\phi_1 + \phi_2 + \phi_3 + \phi_4 - 4\phi_0$$
$$+ \frac{a}{2r_0}(\phi_1 - \phi_3) = 0.$$

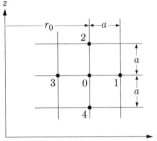

FIGURE 6–21

6–21. In Fig. 6–22 is shown the boundary of a contraction of a two-dimensional potential flow. Given the values of the potential at the upstream and downstream ends, write the algebraic equations for those at the other six grid points.

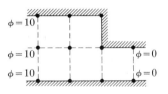

FIGURE 6–22

6–7. Comparison with observations

For mathematical simplicity, the flow of a fictitious frictionless fluid has been studied in the hope that the result can serve as an approximation of the flow of real fluids outside a boundary layer at a high Reynolds number. It is important to compare some typical results with observations to ascertain the practical value of this approach.

Consider the flow around a circular cylinder without circulation. In the irrotational frictionless flow, the pressure drops from the stagnation point at the front to the mid-point at $\theta = \pi/2$, and then rises to the same stagnation pressure at the rear (see Example 6–9). It has been found experimentally that except when the Reynolds number is low, the observed and the theoretical values of pressure agree fairly well in the front of the cylinder but differ considerably at the rear. This discrepancy is explained by the separation of the flow from the cylinder in

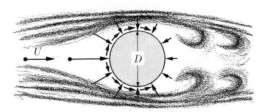

FIGURE 6–23

the case of a real fluid (see Figs. 5–11 and 6–23). The velocity of a real fluid, no matter how small the internal friction, must be zero at the surface of the cylinder. While the velocity outside the boundary layer is described closely by the irrotational flow solution, the velocity in the thin boundary layer is much lower than that in the irrotational flow near the solid surface. In the back of the cylinder where the fluid elements are decelerated, the velocity of the fluid in the boundary layer may be reduced to zero and reversed (see Fig. 5–11). This phenomenon of separation changes the flow pattern completely. In this respect, the irrotational flow of a frictionless fluid is not a good approximation of the flow of real fluids, even if the internal friction is small.

However, when the solid body is *streamlined*, i.e., so shaped that the deceleration is very small so that there is no separation or separation takes place near the end of the body, the observed pressure distribution agrees with the irrotational frictionless flow solution. Since the lift of a body depends mainly on the pressure distribution, it can be predicted closely by assuming the fluid to be frictionless. Thus, for an airfoil at an angle of inclination of a few degrees to the stream, the lift can be estimated with reasonable accuracy from an irrotational frictionless flow with a circulation such that a streamline leaves the trailing edge smoothly, as shown in Fig. 6–24. [In the same figure are also shown cases where the circulation is either too large or too small. There is one circulation that yields the flow shown in Fig. 6–24(b) which is assumed to be representative of the actual flow.]

The drag of a cylinder in a steady irrotational frictionless flow has been shown to be zero. This can be shown also to be the case with bodies of other shapes. In a real fluid, the drag of a body is not zero for three reasons. First, there are tangential shearing stresses on the surface, as shown in Fig. 6–23.

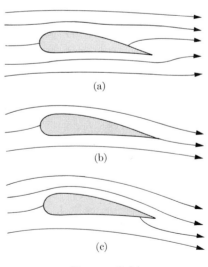

FIGURE 6–24

These tangential forces contribute a drag, generally known as *skin friction*. Secondly, if the body is not streamlined, there will be separation (unless the Reynolds number is very low), and the pressure is lower at the rear than in the front. This difference of pressure gives rise to a drag generally known as *form drag*. For a blunt body with separation, the form drag is much greater than skin friction. For a streamlined body, the drag is due to skin friction only and is therefore much smaller than that of a blunt body of about the same thickness (D in Fig. 6–23) with separation. To calculate skin friction, internal friction must be considered (see Chapters 12 and 13). The computation of form drag is a very difficult problem, and experimental results must be relied upon in practice. Thirdly, finite streamlined bodies may have *induced drag* due to the presence of vorticity in the wake, as will be discussed in Section 8–4.

7 Streamlines and Stream Functions

7-1. Equations of streamlines

It is often desirable to find the equations describing the streamlines of a flow. When the velocity components $u(x, y, z, t)$, $v(x, y, z, t)$, and $w(x, y, z, t)$ are known, differential equations for the streamlines can be written. Since the velocity vector q at any point is tangential to the streamline passing this point, we have for any length dl along a streamline, its components $dx = dl \cdot u/q$, $dy = dl \cdot v/q$, and $dz = dl \cdot w/q$. (See Fig. 7–1 for a case with $w = 0$.) Thus

$$dx : dy : dz = u : v : w, \tag{7–1}$$

from which two independent differential equations for the streamline can be obtained. From such differential equations, two equations for the streamline may be obtained, such as $y(x, t)$ and $z(x, t)$, which together describe a curve in space at time t. For a steady flow, these equations are independent of time, and they become also the equations of a pathline.

When cylindrical coordinates are used, we have for the streamline

$$dr : r\, d\theta : dz = u : v : w. \tag{7–2}$$

Example 7–1. In the steady flow shown in Fig. 6–10, we have $u = 2kx$, $v = 2ky$, and $w = -4kz$. Find the equation of the streamline passing the point $(1, 0, 1)$.

The streamlines are described by

$$\frac{dy}{dx} = \frac{v}{u} = \frac{2ky}{2kx} = \frac{y}{x}$$

and

$$\frac{dz}{dx} = \frac{w}{u} = \frac{-4kz}{2kx} = -\frac{2z}{x}.$$

Separating the variables and integrating, we obtain

$$y = c_1 x \quad \text{and} \quad z = \frac{c_2}{x^2}.$$

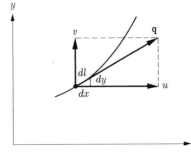

FIGURE 7–1

118

These two equations, with different values of c_1 and c_2, describe all the streamlines in this flow.

For the streamline passing the point $(1, 0, 1)$, c_1 and c_2 must be such that $y = 0$ and $z = 1$ at $x = 1$; that is, $c_1 = 0$ and $c_2 = 1$. This streamline is therefore described by

$$y = 0 \quad \text{and} \quad z = \frac{1}{x^2}.$$

PROBLEMS

7–1. For the flow in Fig. 6–10, we have $u = 2kr$, $v = 0$, and $w = -4kz$ when cylindrical coordinates are used. Find the equations of the streamlines. In particular, find the streamline passing $r = 1$, $z = 1$, and $\theta = 0$.

7–2. For a viscous laminar flow in a pipe of radius R along the z-axis, we have $u = 0, v = 0$, and $w = U[1 - (r/R)^2]$. Find the equation of the streamlines.

7–3. For the two-dimensional doublet shown in Fig. 6–14, we have from Eq. 6–23, $u = k \cos \theta / r^2$, and $v = k \sin \theta / r^2$. Find the equation of the streamline passing the point $r = 1$ and $\theta = \pi/2$.

7–2. Stream function of two-dimensional incompressible flow

In the case of irrotational flow, we have seen a great mathematical simplification of the problem when the velocity potential is introduced. In place of three velocity components, the velocity potential alone is adequate to describe the velocity distribution. In this chapter, stream functions are introduced, which alone also describe completely the velocity distribution, whether the flow is irrotational or not. These functions exist by virtue of the equation of continuity.

In this section we consider *two-dimensional* incompressible flows. The velocity is everywhere parallel to a plane, and the velocity distribution is identical in each plane of flow. For a two-dimensional flow parallel to the xy-plane, we have $u(x, y, t)$, $v(x, y, t)$, and $w = 0$. Some actual flows are approximately two dimensional; e.g., the flow passing a long cylinder. For an incompressible two-dimensional flow, the equation of continuity is

$$\frac{\partial u}{\partial x} + \frac{\partial v}{\partial y} = 0.$$

A function $\psi(x, y, t)$ can therefore be defined by

$$u = -\frac{\partial \psi}{\partial y} \quad \text{and} \quad v = \frac{\partial \psi}{\partial x}, \tag{7–3}$$

which will automatically satisfy the equation of continuity:

$$\frac{\partial u}{\partial x} + \frac{\partial v}{\partial y} = -\frac{\partial^2 \psi}{\partial x \, \partial y} + \frac{\partial^2 \psi}{\partial y \, \partial x} = 0.$$

The function ψ is called the *Lagrange stream function* of a two-dimensional incompressible flow which, through Eq. 7–3, describes the velocity distribution.

It can be shown immediately that the value of $\psi(x, y, t)$ on a streamline is constant. Consider the instantaneous difference $d\psi$ between two points at dx and dy apart. Since

$$d\psi = \frac{\partial \psi}{\partial x} \, dx + \frac{\partial \psi}{\partial y} \, dy = v \, dx - u \, dy,$$

the difference $d\psi$ depends on the local values of u and v. If these two points lie on a streamline, then according to Eq. 7–1,

$$v \, dx = u \, dy,$$

and therefore $d\psi = 0$. Thus the value of ψ is constant along a streamline.

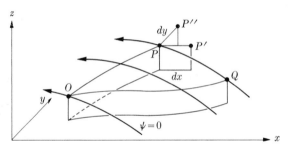

FIGURE 7–2

There is a physical interpretation of the stream function. Let O be an arbitrary fixed point, and OP be any surface perpendicular to the plane of flow. It will be seen that the value of ψ_P at P may be regarded as the discharge through unit depth of the surface OP (see Fig. 7–2). This discharge is considered positive if the net flow across the surface is from right to left when viewed from O toward P. For example, ψ_P is positive in Fig. 7–2. Consider point P' at a distance dx from P. The value of $\psi_{P'}$ will then be the discharge through surface OPP'. Calling $d\psi$ the change of ψ from P to P', we have

$$d\psi = \psi_{P'} - \psi_P = \text{discharge through } OPP' - \text{discharge through } OP$$
$$= \text{discharge through } PP' = v \, dx,$$

where v is the y-component of the velocity in this dx. Thus, with y held constant, we have $d\psi = v \, dx$, or

$$v = \frac{\partial \psi}{\partial x},$$

which is the second of Eqs. 7–3. In a similar manner, we can show that the first of Eqs. 7–3 is also satisfied by considering point P'' at a distance dy from P. Thus the value of ψ at a point may be interpreted as the discharge through unit depth of a surface passing this point and a fixed point O. The dimensional formula of ψ is that of discharge per unit depth or L^2T^{-1}.

It can also be seen from this interpretation of ψ that the value of ψ is constant along a streamline. Consider point Q on the same streamline as P. For continuity, the discharge of an incompressible flow through OP must be the same as that through OQ. Thus $\psi_P = \psi_Q$. To each streamline can therefore be designated a value of ψ. For the streamline passing the reference point O, we have $\psi = 0$.

When cylindrical coordinates are used, the equation of continuity for two-dimensional incompressible flows is

$$\frac{\partial(ur)}{\partial r} + \frac{\partial v}{\partial \theta} = 0.$$

This equation is automatically satisfied by using a stream function $\psi(r, \theta, t)$ defined by

$$ur = -\frac{\partial \psi}{\partial \theta} \quad \text{and} \quad v = \frac{\partial \psi}{\partial r}. \tag{7-4}$$

This stream function can also be interpreted as the discharge per unit depth, and has the same value along a streamline.

Example 7–2. Find the stream function of the steady two-dimensional flow shown in Fig. 7–3, where $u = Uy/d$ and $v = 0$.

From Eqs. 7–3, we have

$$-\frac{\partial \psi}{\partial y} = u = \frac{U}{d} y \quad \text{and} \quad \frac{\partial \psi}{\partial x} = v = 0.$$

These equations require that

$$\psi = -\frac{U}{2d} y^2 + f_1(x) \quad \text{and} \quad \psi = f_2(y),$$

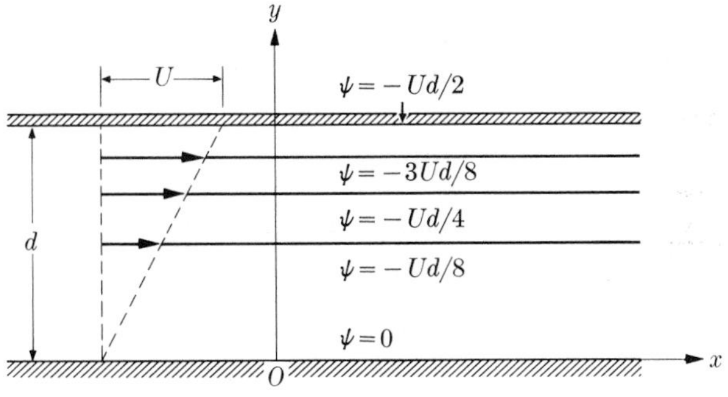

Figure 7–3

respectively. The stream function ψ that satisfies both of these requirements is

$$\psi = -\frac{U}{2d} y^2 + C.$$

Since the velocity is described by the derivatives of ψ, the value of the constant C is immaterial. The value of C depends on the choice of the streamline for $\psi = 0$. It is convenient in this case to use the streamline along the bottom for $\psi = 0$. Then C must be zero so that $\psi = 0$ at $y = 0$:

$$\psi = -\frac{U}{2d} y^2.$$

In the entire flow field, the value of ψ is negative. This means that when viewed from point O (where $\psi = 0$) toward any point P in the field, the discharge across OP is from left to right.

The equation of a streamline is obtained by assigning a proper value for ψ. For example, the equation of the streamline with $\psi = -UD/4$ is

$$-\frac{Ud}{4} = -\frac{U}{2d} y^2, \quad \text{or} \quad y = \frac{d}{\sqrt{2}}.$$

Streamlines with equal increments of ψ are shown in Fig. 7–3. The discharge between each pair of adjacent streamlines is the same. The spacing of these streamlines is therefore inversely proportional to the local speed.

Example 7–3. Find the stream function of a line source.

For a line source, we have $u = Q/(2\pi r)$ and $v = 0$. Therefore

$$-\frac{1}{r}\frac{\partial\psi}{\partial\theta} = u = \frac{Q}{2\pi r} \quad \text{and} \quad \frac{\partial\psi}{\partial r} = v = 0.$$

From these, we obtain

$$\psi = -\frac{Q}{2\pi}\theta + C.$$

The streamlines are radial lines with constant θ. In this case, the stream function is multivalued unless the range of θ is limited to 2π, for example, $-\pi \leq \theta \leq \pi$ in Fig. 6–7.

PROBLEMS

7–4. Show that ψ in Eqs. 7–4 is constant along a streamline. Also show that it can be interpreted as the discharge per unit depth through OP in Fig. 7–2.

7–5. If the streamline along the upper wall is designated by $\psi = 0$, find the stream function for the flow in Fig. 7–3. What is then the value of ψ for the lowest streamline?

7–6. Find the stream functions for the following flows, and in each case, plot three or more streamlines with equal increments of ψ: (a) a uniform flow in the

xy-plane of speed U directed at an angle α with the x-axis; (b) rotation about an axis as if the fluid were a rigid body (see Fig. 5–2); (c) a free vortex (see Eq. 6–16).

7–7. Given $\phi = x^2 - y^2$ for a two-dimensional irrotational incompressible flow, (a) find the stream function; (b) find the equation of the streamline passing $x = 1$, $y = 1$. (c) Can you tell from the stream function that it describes a flow near the corner of two walls along the x- and y-axes?

7–8. For a steady two-dimensional compressible flow, the equation of continuity is

$$\frac{\partial(\rho u)}{\partial x} + \frac{\partial(\rho v)}{\partial y} = 0.$$

A stream function $\chi(x, y)$ can be defined by $\rho u = -\partial\chi/\partial y$ and $\rho v = \partial\chi/\partial x$. Show that χ is constant along a streamline, and that $\Delta\chi$ between two streamlines is the mass flow between them per unit time per unit depth.

7–3. Solution of two-dimensional irrotational incompressible flow

For any incompressible two-dimensional flow, the functions u, v, and p must satisfy the equation of continuity, two momentum equations, and the boundary conditions. By introducing the stream function ψ, the continuity equation is automatically satisfied. The functions ψ and p are therefore determined by the two momentum equations and the boundary conditions. Note that the number of unknown functions has been thus reduced by one.

Although the stream function ψ is applicable to two-dimensional incompressible flows in general (see Problem 7–13), attention will be confined to frictionless fluids in irrotational motion in this chapter. It has been shown that such a motion can be determined from the Laplace equation $\nabla^2\phi = 0$, where ϕ is the velocity potential. The use of the stream function for the solution of an irrotational flow is therefore not absolutely necessary. In fact, both being related to u and v, the stream function can be computed from the velocity potential and vice versa if desired. However, by using ψ instead of ϕ, we obtain directly the equations of the streamlines.

The momentum equations have been shown to yield for a steady irrotational flow of a homogeneous incompressible fluid

$$\frac{p}{\rho} + gh + \frac{q^2}{2} = \text{constant}$$

throughout the field. In an irrotational flow, the stream function must be such that $\partial u/\partial y = \partial v/\partial x$, i.e.

$$\frac{\partial}{\partial y}\left(-\frac{\partial\psi}{\partial y}\right) = \frac{\partial}{\partial x}\left(\frac{\partial\psi}{\partial x}\right),$$

$$\frac{\partial^2\psi}{\partial x^2} + \frac{\partial^2\psi}{\partial y^2} = 0, \qquad (7\text{–}5)$$

which is the Laplace equation. This equation is to be solved with the proper boundary conditions. For example, ψ must be constant along a stationary solid wall, since it is a streamline. With the stream function and therefore the velocity determined, the pressure distribution can then be determined for a steady flow from Bernoulli's equation.

The mathematical solution of the problem can be seen to be the same whether the stream function or the velocity potential is used. The only difference lies in the boundary conditions. While $\partial\phi/\partial n = 0$ at a fixed solid wall (n normal to the wall), $\partial\psi/\partial l = 0$ in the direction l along the wall. In the following example, we shall use the method of superposition which has been shown in Section 6–4 to be applicable since the Laplace equation is linear.

Example 7–4. Find the resultant flow of a line sink of strength Q and a uniform stream of speed U in the x-direction, as shown in Fig. 6–12.

For the uniform stream with

$$-\frac{\partial\psi_1}{\partial y} = U \qquad \text{and} \qquad \frac{\partial\psi_1}{\partial x} = 0,$$

we have

$$\psi_1 = -Uy + C_1.$$

For the line sink with

$$-\frac{\partial\psi_2}{r\,\partial\theta} = -\frac{Q}{2\pi r} \qquad \text{and} \qquad \frac{\partial\psi_2}{\partial r} = 0,$$

we have

$$\psi_2 = \frac{Q}{2\pi}\theta + C_2.$$

Both of these flows are irrotational. The functions ψ_1, ψ_2, and therefore their sum ψ are solutions of the Laplace equation. The flow for ψ as shown in Fig. 6–12 is therefore also irrotational:

$$\psi = \psi_1 + \psi_2 = -Uy + \frac{Q}{2\pi}\theta + C_1 + C_2 = -Ur\sin\theta + \frac{Q}{2\pi}\theta + C.$$

The value of C depends on the choice of the streamline for $\psi = 0$. There is a streamline with constant ψ along the line $\theta = 0$. Setting $\psi = 0$ on this streamline, we have $C = 0$ and

$$\psi = -Ur\sin\theta + \frac{Q}{2\pi}\theta.$$

Streamlines with constant values of ψ are shown in Fig. 6–12. The function ψ_2 and therefore ψ are multivalued unless θ is limited to a range of 2π, for example, $-\pi \le \theta \le \pi$ in Fig. 6–12.

It is interesting to find the equation of the streamline separating the flow into the sink from the rest. This streamline is a part of that with $\psi = 0$. The equation of this streamline is therefore

$$r = \frac{Q}{2\pi U}\frac{\theta}{\sin\theta}.$$

As $\theta \to 0$, we have $\sin\theta \to \theta$. Therefore the stagnation point is located at $r = Q/(2\pi U)$, $\theta = 0$.

PROBLEMS

7-9. Verify that the stream functions in Example 7-3 and Problems 7-6(a), 7-6(c), and 7-7 satisfy the Laplace equation. Also verify that the stream functions in Example 7-2 and Problem 7-6(b) are not solutions of the Laplace equation. Why?

7-10. Given $\psi_1 = (\Gamma/2\pi)\ln r$ for a free vortex, and $\psi_2 = Q\theta/2\pi$ for a line sink, show that the streamlines of a two-dimensional irrotational flow toward a line sink are the spirals

$$re^{Q\theta/\Gamma} = \text{constant}$$

as shown in Fig. 6-4.

7-11. (a) Given $\psi_1 = Uy$ for a uniform stream in the negative x-direction and $\psi_2 = -k\sin\theta/r$ for an opposing two-dimensional doublet (see Fig. 6-15), find the stream function of the flow around a circular cylinder without circulation. Verify that the lines of $\theta = 0$, $\theta = \pi$, and $r = \sqrt{k/U}$ form one streamline. (b) Find the equation of the streamline passing the point at $\theta = \pi/2$, $r = \frac{3}{2}R$, where $R = \sqrt{k/U}$. Plot a few points of this streamline.

7-12. Show that scalar addition of ψ_1 and ψ_2 is equivalent to vectorial addition of their velocities. Also show that if ψ_1 and ψ_2 represent irrotational flows, their sum also represents one.

7-13. Show that for a two-dimensional frictionless but rotational incompressible flow, ψ must satisfy the nonlinear equation

$$\frac{\partial}{\partial t}(\nabla^2\psi) + \frac{\partial\psi}{\partial y}\cdot\frac{\partial}{\partial x}(\nabla^2\psi) - \frac{\partial\psi}{\partial x}\cdot\frac{\partial}{\partial y}(\nabla^2\psi) = 0.$$

[*Hint:* Eliminate the terms involving p and g in Eqs. 5-1 and 5-2 by differentiating with respect to y and x, respectively.]

7-4. Graphical solution of two-dimensional irrotational flow

Two approximation methods of solution of the Laplace equation have been presented in Section 6-6. In this section, two graphical methods of solution are described which are applicable to two-dimensional irrotational incompressible flows.

(a) Superposition of stream functions. If ψ_1 and ψ_2 are stream functions of irrotational flows and are therefore solutions of the Laplace equation (Eq. 7-5), their sum ψ is also a solution of the linear equation. A simple graphical method

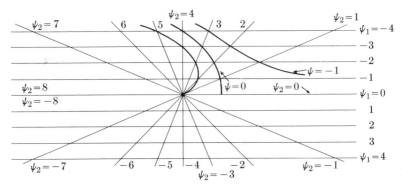

FIGURE 7–4

for the construction of the streamlines of constant ψ can be devised. If stream-lines with equal increments of ψ_1 and ψ_2 are constructed, the desired stream-lines can be obtained by joining the intersections of the two sets of streamlines as shown in Fig. 7–4. In this drawing, streamlines of a uniform stream with $\psi_1 = -Uy$ and of a line sink with $\psi_2 = Q\theta/2\pi$ are drawn with $U = 1$ and $Q = 16$ with increments $\Delta\psi_1 = \Delta\psi_2 = 1$. Several streamlines with constant ψ $(= \psi_1 + \psi_2)$ have been obtained by joining the opposite corners of the four-sided figures with equal values of $(\psi_1 + \psi_2)$. (The reader may complete Fig. 7–4 as an exercise.)

(b) The flow net. According to the definitions of the velocity potential ϕ and the stream function ψ, the speed q at any point in a two-dimensional irrotational flow may be written as

$$-\frac{\partial\phi}{\partial l} = q = \frac{\partial\psi}{\partial n},$$

where l is the direction of the velocity $\mathbf{q}$, and n is a direction along the equipotential line, as shown in Fig. 7–5. Therefore, if equipotential lines with increments $\Delta\phi$ and streamlines with increments $\Delta\psi$ are drawn, their spacings Δl and Δn are related by

$$\frac{\Delta\phi}{\Delta l} \doteq q \doteq \frac{\Delta\psi}{\Delta n}. \qquad (7\text{–}6)$$

Thus, if $\Delta\phi$ and $\Delta\psi$ are chosen to be equal, we have $\Delta l \doteq \Delta n$. Since the equipotential lines are perpendicular to the streamlines (except at stagnation points where $q = 0$), these two sets of curves will form a net of squares if Δl and Δn are infinitesimal. This net is called a *flow net*. With finite spacings, however, these squares will not be perfect.

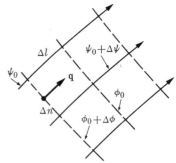

FIGURE 7–5

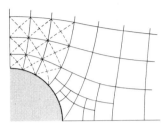

FIGURE 7–6

In this graphical method, the flow net is sketched by trials according to the boundary conditions until a net of near-squares are obtained. In Fig. 7–6, the flow net of a flow passing a circular cylinder is shown. In sketching this flow net, the solid surface has to be a streamline, and the streamlines must be evenly spaced in the uniform approach flow. Near the stagnation points, a finer net may be used for a more accurate result. To check the correctness of a sketch, lines joining the corners of the squares may be drawn as shown in Fig. 7–6. These lines should also form a net of near-squares. When a satisfactory flow net is obtained, the relative magnitude of the speed can be determined from the size of the squares, in accordance with Eq. 7–6. It must be pointed out, however, that this method can yield only a qualitative description of the flow pattern.

7–5. Stokes' stream function

It has been shown in the previous sections that a two-dimensional motion can be described with a stream function. For a three-dimensional motion, it is generally necessary to use two stream functions for its description. However, when the flow is symmetrical with respect to an axis, it is again possible to describe the motion with one stream function.

With cylindrical coordinates, we have, for a flow symmetrical with respect to the z-axis, $v = 0$ and u and w as functions of r and z only. Since the flow is identical in all the meridian planes, it is necessary to study only the flow in one meridian plane at a constant θ, as shown in Fig. 7–7. For an incompressible flow with $v = 0$, we have the equation of continuity from Eq. 4–21:

$$\frac{\partial(ur)}{r\,\partial r} + \frac{\partial w}{\partial z} = 0, \qquad \text{or} \qquad \frac{\partial(ur)}{\partial r} + \frac{\partial(wr)}{\partial z} = 0.$$

Thus it is possible to describe the flow with a function $\Psi(r, z, t)$ defined by

$$ur = -\frac{\partial \Psi}{\partial z} \qquad \text{and} \qquad wr = \frac{\partial \Psi}{\partial r}. \tag{7–7}$$

The flow described by such a function will automatically satisfy the equation of continuity:

$$\frac{\partial(ur)}{\partial r} + \frac{\partial(wr)}{\partial z} = -\frac{\partial^2 \Psi}{\partial r\,\partial z} + \frac{\partial^2 \Psi}{\partial z\,\partial r} = 0.$$

This function Ψ is called the *Stokes' stream function* for axially symmetric incompressible flow. Note that unlike the velocity potential ϕ and the Lagrange stream function ψ which have the dimensions L^2/T, the Stokes' stream function Ψ is dimensionally L^3/T.

Stokes' stream function can be shown to be constant along a streamline. For any two points distances dr and dz apart, the instantaneous difference of $\Psi(r, z, t)$ is

$$d\Psi = \frac{\partial \Psi}{\partial r}\, dr + \frac{\partial \Psi}{\partial z}\, dz = wr\, dr - ur\, dz.$$

If these two points are situated on a streamline, we have from Eq. 7–2,

$$u:w = dr:dz \qquad \text{or} \qquad w\, dr = u\, dz.$$

Thus $d\Psi = 0$, that is, Ψ is constant along a streamline.

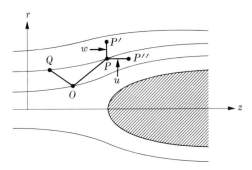

FIGURE 7–7

The Stokes' stream function may also be defined in another manner. Let O be a fixed point and P be a point in the same meridian plane, as shown in Fig. 7–7. The quantity $2\pi\Psi$ at P is defined as the discharge through the surface of revolution generated by a line OP about the z-axis. This discharge is considered positive when the flow is from left to right when viewed from O toward P, as is the case shown in Fig. 7–7. At point P' a distance dr from P, the increase of discharge is then

$$2\pi\, d\Psi = (2\pi r\, dr)\, w \qquad \text{(at constant } z),$$

where $(2\pi r\, dr)$ is the surface of revolution generated by the line PP'. Thus $\partial \Psi/\partial r = wr$, which is the same as Eq. 7–7. Similarly, we can obtain $\partial \Psi/\partial z = -ur$. If points P and Q are located on the same streamline of an incompressible flow, the discharges $2\pi\Psi_P$ and $2\pi\Psi_Q$ through the surfaces of revolution generated by OP and OQ must be the same. Thus $\Psi_P = \Psi_Q$, that is, Ψ as defined in this manner is constant along a streamline.

Equations 7–7 are valid for axially symmetric incompressible flows in general. If the flow is irrotational with $\partial u/\partial z = \partial w/\partial r$ (see Eq. 6–9), Ψ must satisfy

$$\frac{\partial^2 \Psi}{\partial r^2} - \frac{1}{r}\frac{\partial \Psi}{\partial r} + \frac{\partial^2 \Psi}{\partial z^2} = 0. \tag{7–8}$$

While it is not a Laplace equation, this equation is linear. Its solutions may therefore be obtained by superposition.

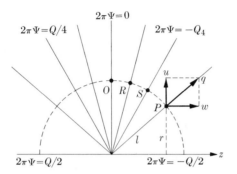

FIGURE 7–8

Example 7–5. Find the Stokes' stream function of a point source.

With a point source of discharge Q at the origin, the speed q at any point P is (see Fig. 7–8) $q = Q/(4\pi l^2)$, where $4\pi l^2$ is the surface area of a sphere of radius l. From Eq. 7–7,

$$-\frac{1}{r}\frac{\partial \Psi}{\partial z} = u = q\,\frac{r}{l} = \frac{Qr}{4\pi l^3} = \frac{Q}{4\pi}\frac{r}{(r^2 + z^2)^{3/2}}$$

and

$$\frac{1}{r}\frac{\partial \Psi}{\partial r} = w = q\,\frac{z}{l} = \frac{Qz}{4\pi l^3} = \frac{Q}{4\pi}\frac{z}{(r^2 + z^2)^{3/2}}.$$

Integrating, we have

$$\Psi = C - \frac{Q}{4\pi}\frac{z}{\sqrt{r^2 + z^2}}.$$

Let $\Psi = 0$ along $z = 0$. Then $C = 0$, and

$$\Psi = -\frac{Q}{4\pi}\frac{z}{\sqrt{r^2 + z^2}}. \tag{7–9}$$

Several streamlines at equal $\Delta\Psi$ apart are shown in Fig. 7–8. The same discharge flows through the surfaces of revolution generated by lines OR, RS, SP, etc. Note that although the particle speed is the same at all points on these lines, the spacings of these streamlines are not equal.

PROBLEMS

7-14. Find the Stokes' stream function Ψ of a uniform stream in the z-direction. Plot several streamlines at equal $\Delta\Psi$ apart. Are they evenly spaced?

7-15. Show that if Ψ_1 and Ψ_2 are the Stokes' stream functions of two irrotational flows, $\Psi = \Psi_1 + \Psi_2$ is the Stokes' stream function of another irrotational flow.

7-16. By placing a point source in a uniform stream, the pattern of flow passing a semiinfinite body of revolution is obtained (see Fig. 7–9). (a) Find the Stokes' stream function such that $\Psi = 0$ along OP. (b) Find the equation $f(r, z) = 0$ of the body for a point source of discharge Q in a stream of speed U.

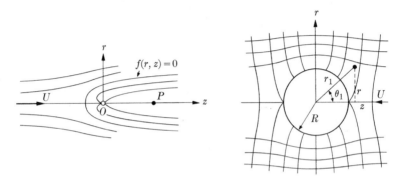

FIGURE 7–9   FIGURE 7–10

7-17. The velocity potential of an irrotational flow passing a sphere of radius R is

$$\phi = U \cos \theta_1 [r_1 + (R^3/2r_1^2)],$$

where r_1 and θ_1 are spherical coordinates (see Problem 6–16 and Fig. 7–10). In terms of the cylindrical coordinates,

$$\phi = Uz + \frac{R^3 Uz}{2(r^2 + z^2)^{3/2}}.$$

Find the velocity components u and w, and the Stokes' stream function in the cylindrical coordinate system. Plot one streamline and compare with those in Fig. 7–6 for the flow passing a cylinder.

8 Vorticity

8-1. Introduction

In Chapter 6, we studied fluid flow problems under the assumption that flows were irrotational. Obviously, not all flows are irrotational. When we are faced with a practical problem in fluid flow, we must somehow decide whether the flow is irrotational before we can apply the analysis presented in Chapter 6. We may also ask at this point whether it is very important to differentiate irrotational flows from rotational flows. In this chapter, we shall first formally present certain kinematic relations for a rotational flow field, demonstrate that the lift and drag experienced by a body in a steady flow is intimately related to the concept of rotational flow, and derive the conditions or criteria with which to judge whether a given fluid flow is irrotational.

In a given flow field, it is possible that some parts of the flow field are irrotational, while other parts are not. The rotation of a fluid element shall be denoted by the rotation vector $\boldsymbol{\omega}$, which is defined as the average angular velocity of two originally perpendicular lines conceptually attached to the fluid element of interest (see Section 6–1). It is customary, however, to deal with the vorticity vector $\boldsymbol{\xi}$, which is simply defined as $\boldsymbol{\xi} = 2\boldsymbol{\omega}$. In this and the following two sections, we shall study some kinematic properties of $\boldsymbol{\xi}$.

In rectangular coordinates, the components of the vorticity vector ξ_x, ξ_y, ξ_z have been derived in Section 6–1 and are given by Eqs. 6–4, 6–5, and 6–6. They are

$$\xi_x = \frac{\partial w}{\partial y} - \frac{\partial v}{\partial z}, \qquad \xi_y = \frac{\partial u}{\partial z} - \frac{\partial w}{\partial x}, \qquad \xi_z = \frac{\partial v}{\partial x} - \frac{\partial u}{\partial y}, \qquad (8\text{–}1)$$

where u, v, and w are the velocity components in the x-, y-, and z-directions, respectively. In cylindrical polar coordinates, its components ξ_r, ξ_θ, and ξ_z are given in Eqs. 6–7, 6–8, and 6–9 as

$$\xi_r = \frac{1}{r}\frac{\partial w}{\partial \theta} - \frac{\partial v}{\partial z}, \qquad \xi_\theta = \frac{\partial u}{\partial z} - \frac{\partial w}{\partial r}, \qquad \xi_z = \frac{1}{r}\frac{\partial}{\partial r}(rv) - \frac{1}{r}\frac{\partial u}{\partial \theta}, \qquad (8\text{–}2)$$

where u, v, and w are velocity components in the r-, θ-, z-directions, respectively. We see that given a velocity field $\mathbf{q}$, the vorticity field $\boldsymbol{\xi}$ is obtained by a certain combination of the spatial partial derivatives of $\mathbf{q}$, provided that such

derivatives exist. If other coordinate systems are used, such as spherical coordinates, one must derive the corresponding expressions for ξ in that coordinate system starting again from the definition. In Section 8–3, we shall give
an alternative definition for ξ which is more suitable for this purpose (see
Example 8–2). For the moment, let it be understood that in different coordinate systems, the formulas for the components of vorticity are quite different.
(The relationship between ξ and $\mathbf{q}$ in vector notations will be given in Eq.
9–21a.)

Let us first calculate the quantity

$$\frac{\partial}{\partial x}\,\xi_x + \frac{\partial}{\partial y}\,\xi_y + \frac{\partial}{\partial z}\,\xi_z. \tag{8-3}$$

Using Eq. 8–1, we have

$$\frac{\partial}{\partial x}\,\xi_x = \frac{\partial^2 w}{\partial x\,\partial y} - \frac{\partial^2 v}{\partial z\,\partial x}, \quad \frac{\partial}{\partial y}\,\xi_y = \frac{\partial^2 u}{\partial y\,\partial z} - \frac{\partial^2 w}{\partial x\,\partial y}, \quad \frac{\partial}{\partial z}\,\xi_z = \frac{\partial^2 v}{\partial z\,\partial x} - \frac{\partial^2 u}{\partial y\,\partial z}.$$

Summing the results, we find

$$\frac{\partial}{\partial x}\,\xi_z + \frac{\partial}{\partial y}\,\xi_y + \frac{\partial}{\partial z}\,\xi_z = 0. \tag{8-4}$$

This is really an astonishing result, for we see from Eq. 8–4 that the three
components of the vorticity vector ξ at any point are not independent. For
example, $\xi_x = x$, $\xi_y = y$, $\xi_z = z$ is not a permissible vorticity vector field!
Furthermore, Eq. 8–4 reminds us of the incompressible continuity equation,
Eq. 4–19; if ξ_x, ξ_y, and ξ_z were replaced by u, v, and w, the velocity components in the x-, y-, z-directions, respectively, Eqs. 8–4 and 4–19 would become identical. This observation prompts us to try out this idea in the
cylindrical polar coordinate system by calculating the quantity

$$\frac{1}{r}\frac{\partial}{\partial r}\,(r\xi_r) + \frac{1}{r}\frac{\partial}{\partial \theta}\,\xi_\theta + \frac{\partial}{\partial z}\,\xi_z,$$

which is suggested by the form of Eq. 4–21, the incompressible continuity
equation in cylindrical polar coordinates. A simple calculation using Eq. 8–2
shows that

$$\frac{1}{r}\frac{\partial}{\partial r}\,(r\xi_r) + \frac{1}{r}\frac{\partial}{\partial \theta}\,\xi_\theta + \frac{\partial}{\partial z}\,\xi_z = 0. \tag{8-5}$$

We shall see from Eq. 9–21b that in any coordinate system, the vorticity
vector ξ indeed satisfies the corresponding incompressible continuity equation
if $\mathbf{q}$ is replaced by ξ. In vector form, the above result can be written as

$$\nabla \cdot \xi = 0, \tag{8-6}$$

and we say that the divergence of the vorticity is always zero. In integral form, Eq. 8–6 is equivalent to

$$\int_S \mathbf{n} \cdot \boldsymbol{\xi} \, dS = 0, \qquad (8\text{–}7)$$

where S is a closed surface enclosing an arbitrary volume of fluid of interest, and $\mathbf{n}$ is the unit outward normal of the surface element dS, so that $\mathbf{n} \cdot \boldsymbol{\xi}$ is the outward normal component of the vector $\boldsymbol{\xi}$ on the surface element dS. (See Eq. 4–26 for the definition of divergence.) Physically, Eq. 8–7 states that for an arbitrary volume of fluid, the algebraic sum of the vorticity flux coming out of the surface of the volume is zero. In other words, whatever goes in must come out. This kinematic property of vorticity is generally referred to as the *conservation of vorticity*.

PROBLEMS

8–1. Compute the vorticity field of the following flows and show that for each case $\nabla \cdot \boldsymbol{\xi} = 0$. (a) In rectangular coordinates $u = ay$, $v = 0$, $w = 0$, where a is a constant. (b) In rectangular coordinates, $u = ax$, $v = -ay$, $w = 1$. (c) In cylindrical polar coordinates, $u = 0$, $v = ar$, $w = 0$.

8–2. Using Eq. 8–2, show that

$$\frac{1}{r} \frac{\partial}{\partial r} (r\xi_r) + \frac{1}{r} \frac{\partial}{\partial \theta} \xi_\theta + \frac{\partial}{\partial z} \xi_z = 0.$$

8–3. Given in cylindrical polar coordinates

$$u = 0, \quad v = arz, \quad \text{and} \quad w = 0$$

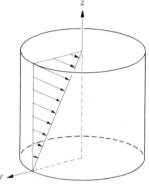

for the flow in a cylindrical container as shown in Fig. 8–1. (a) Find the vorticity vector field $\boldsymbol{\xi}$. (b) Draw on an rz-plane at several points the vorticity vectors. (c) Verify Eq. 8–7 by evaluating the surface integral over any convenient surface.

FIGURE 8-1

8–2. Vortex lines and vortex tubes

Given a velocity field $\mathbf{q}$, the corresponding vorticity field $\boldsymbol{\xi}$ is determined and can be calculated. A line in the flow field which is everywhere tangent to the local vorticity vector is called a *vortex line*. Given a small closed curve in the vorticity field, all the vortex lines which touch this closed curve will form a tube. Such tubes are called *vortex tubes* or *vortex filaments*. The definitions of vortex lines and vortex tubes are entirely analogous to those of streamlines and streamtubes.

For a given streamtube in an incompressible flow, the volume flow rate Q is a constant at any cross section:

$$Q = \int_S (\mathbf{n} \cdot \mathbf{q})\, dS,$$

where the surface integral is taken over any cross section of the tube, and $\mathbf{n} \cdot \mathbf{q}$ is the component of the velocity vector normal to the surface element. This constancy of Q is a consequence of the conservation of mass, and therefore is derivable from the incompressible continuity equation. Now, since the vorticity vector $\boldsymbol{\xi}$ satisfies the same equation as $\mathbf{q}$, one can expect that a similar property exists for $\boldsymbol{\xi}$, such as

$$\Gamma = \int_S (\mathbf{n} \cdot \boldsymbol{\xi})\, dS,$$

where S now denotes any cross-sectional surface of a given vortex tube. The quantity Γ, called circulation, is then expected to be a constant for this vortex tube.

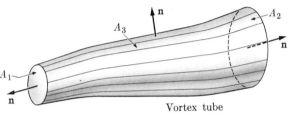

Vortex tube

FIGURE 8–2

To demonstrate that the above heuristic arguments are correct, we proceed as follows. Consider the volume bounded by a small vortex tube and two flat ends A_1 and A_2, as shown in Fig. 8–2. Performing the surface integral Eq. 8–7 over the surface of this volume, we have

$$\int_{A_1} (\mathbf{n} \cdot \boldsymbol{\xi})\, dS + \int_{A_2} (\mathbf{n} \cdot \boldsymbol{\xi})\, dS + \int_{A_3} (\mathbf{n} \cdot \boldsymbol{\xi})\, dS = 0,$$

where A_3 is the lateral surface area of the vortex tube. The third integral in the above equation is identically zero, since $\mathbf{n} \cdot \boldsymbol{\xi}$ on the surface of a vortex tube is zero by definition. Hence we have

$$\int_{A_1} (\mathbf{n} \cdot \boldsymbol{\xi})\, dS + \int_{A_2} (\mathbf{n} \cdot \boldsymbol{\xi})\, dS = 0. \tag{8–8a}$$

Equation 8–8a is of course valid for any *given* vortex tube. We can rewrite it as

$$\Gamma = \int_{A_2} (\mathbf{n} \cdot \boldsymbol{\xi})\, dS = \int_{A_1} (-\mathbf{n} \cdot \boldsymbol{\xi})\, dS, \tag{8–8b}$$

where Γ is, for the moment, the circulation of surface A_1 or A_2. However, Eq. 8–8b is valid for a given vortex tube for any choice of A_1 and A_2. Consequently, Γ must be a constant, for once it is evaluated at any cross section such as A_2, the same value must result from any other cross section. Hence the proof is complete.

Note that for a given vortex tube, the circulation Γ is essentially the product of the *mean* vorticity and the cross-sectional area. Hence with constant Γ, the smaller the cross-sectional area, the stronger will be the vorticity. A tornado is a giant vortex tube. Thus the air particles in the center of a tornado rotate with a much higher angular velocity near the ground where the funnel is narrowest. The high-speed circular flow outside the tornado, however, is essentially irrotational.

Equation 8–8b implies the interesting result that a vortex tube cannot end in a fluid. Since the circulation of a vortex tube is a constant, it is clear that if we suppose it ends at a certain cross section, we would arrive at the contradiction that the circulation there is zero. This property is again analogous to the fact that streamtubes cannot end in a fluid. A vortex tube in a fluid must therefore either form a ring, such as a smoke ring, or extend to infinity. It can, however, attach itself to a solid surface, as a tornado attaches itself to the earth. Note that the above properties of vorticity have been arrived at from purely kinematic considerations, without reference to the physical properties of the fluid, such as compressibility and viscosity. For example, the vorticity field of a compressible flow still satisfies Eq. 8–6, even though the continuity equation for $\mathbf{q}$ is no longer that given by $\nabla \cdot \mathbf{q} = 0$.

PROBLEMS

8–4. For the flow described in Problem 8–3, sketch several vortex lines in an rz-plane. Show that the equations for these lines are $r^2 = \text{constant}/z$. [*Hint:* The equation for a vortex line is derived analogously as the equation for a streamline. See Section 7–1.]

8–5. Given a vortex tube which forks into two separate tubes as shown in Fig. 8–3. If $\int_{A_1} (\mathbf{n}_1 \cdot \boldsymbol{\xi})\, dS = 3$ and $\int_{A_2} (\mathbf{n}_2 \cdot \boldsymbol{\xi})\, dS = 1$, find the value of $\int_{A_3} (\mathbf{n}_3 \cdot \boldsymbol{\xi})\, dS$, where $\mathbf{n}_1$, $\mathbf{n}_2$, $\mathbf{n}_3$ are unit normal vectors to surfaces A_1, A_2, A_3, respectively, as shown in the figure.

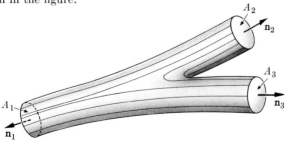

FIGURE 8–3

8–3. Circulation

In the previous section, the surface integral

$$\Gamma = \int_S (\mathbf{n} \cdot \boldsymbol{\xi}) \, dS \tag{8–9}$$

taken over a certain surface S of interest is called *circulation*. Consider an arbitrary surface S bounded by a closed curve C as shown in Fig. 8–4. Let us draw lines on this surface and divide it into many small pieces which may be of various sizes, but each one is approximately rectangular and flat. We shall construct on each small surface a local rectangular coordinate system (x, y, z), the z-axis being normal to the surface, as shown in Fig. 8–4. We shall now calculate the contribution of $\int_s (\mathbf{n} \cdot \boldsymbol{\xi}) dS$ over any one of these small surfaces. The value of $\mathbf{n} \cdot \boldsymbol{\xi}$ at any point on this surface is approximately ξ_z, since $\mathbf{n}$ and the z-axis are approximately parallel everywhere. We can write

$$\int_s (\mathbf{n} \cdot \boldsymbol{\xi}) \, dS = \iint_s \xi_z \, dx \, dy = \iint_s \left(\frac{\partial v}{\partial x} - \frac{\partial u}{\partial y} \right) dx \, dy, \tag{8–10}$$

where u and v are, as usual, velocity components in the local x- and y-directions. Now integrating over the surface $abcd$, we have,

$$\iint_s \frac{\partial v}{\partial x} \, dx \, dy = \int (v_{ab} - v_{cd}) \, dy,$$

where v_{ab}, v_{cd} denote the values of v along the sides ab and cd, respectively. Similarly, we have

$$\iint_s \frac{\partial u}{\partial y} \, dx \, dy = \int (u_{bc} - u_{da}) \, dx.$$

Thus Eq. 8–10 becomes

$$\int_s (\mathbf{n} \cdot \boldsymbol{\xi}) \, dS = \int_a^b v_{ab} \, dy - \int_b^c u_{bc} \, dx - \int_c^d v_{cd} \, dy + \int_d^a u_{da} \, dx.$$

The right-hand side is known as a line integral; it is an integral taken along

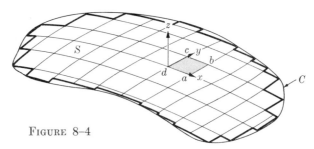

FIGURE 8–4

a line. In this case, the integrand is the velocity component along the line element. Thus it can be written as

$$\int_s (\mathbf{n} \cdot \boldsymbol{\xi})\, dS = \oint_c \mathbf{q} \cdot d\mathbf{l}, \tag{8–11}$$

where $d\mathbf{l}$ is a differential displacement vector along the sides of the rectangle, and is considered positive in the direction of the bent fingers of the right hand when the thumb is pointed along the vector $\mathbf{n}$. The approximate relation, Eq. 8–11, becomes exact in the limit of an infinitesimal area. In order to calculate the value of $\int_S (\mathbf{n} \cdot \boldsymbol{\xi})\, dS$ for the whole surface S, we must sum up the values of $\oint \mathbf{q} \cdot d\mathbf{l}$ of all the small rectangles. However, it is easy to see that for adjacent rectangles, the contributions from the common sides precisely cancel each other. When the results of all the individual line integrals are summed, we shall obtain a line integral taken over the bold jagged line in Fig. 8–4. In the limit of infinitely many divisions, the bold line will tend to the bounding curve C. Hence we have

$$\int_S (\mathbf{n} \cdot \boldsymbol{\xi})\, dS = \oint_C \mathbf{q} \cdot d\mathbf{l}, \tag{8–12}$$

where the line integral is taken around the closed curve C. Equation 8–12 is generally referred to as *Stokes' theorem*. Since the left-hand side of Eq. 8–12 is the circulation Γ, we can now give an alternative definition for *circulation* as

$$\Gamma = \oint_C \mathbf{q} \cdot d\mathbf{l}. \tag{8–13}$$

Note that for a given velocity vector field, the magnitude of the circulation of a surface S depends *only* on the boundary curve C of S, and can be computed from Eq. 8–13 without reference to $\boldsymbol{\xi}$ or the precise shape of the surface.

Example 8–1. In Example 6–3, we found that the velocity distribution for a free vortex is

$$u = 0, \qquad v = \frac{\Gamma}{2\pi r}, \qquad w = 0,$$

where u, v, and w are velocity components in the r-, θ-, and z-directions, respectively. Find the circulation using any closed curve C surrounding the z-axis.

Instead of calculating directly $\oint_C \mathbf{q} \cdot d\mathbf{l}$, let us first consider the surface integral $\int_S (\mathbf{n} \cdot \boldsymbol{\xi})\, dS$ taken over the shaded surface

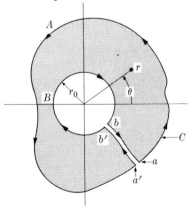

FIGURE 8–5

S, as shown in Fig. 8–5. This shaded surface is bounded by the curve C and a circle of radius r_0 and the two lines ab and $a'b'$ as shown. Since the flow is irrotational everywhere on the shaded surface S, the magnitude of ξ is zero everywhere. Hence

$$\int_S (\mathbf{n} \cdot \boldsymbol{\xi})\, dS = 0.$$

Using Eq. 8–12, we then have

$$\oint_C \mathbf{q} \cdot d\mathbf{l} = \int_{aAa'} \mathbf{q} \cdot d\mathbf{l} + \int_{a'b'} \mathbf{q} \cdot d\mathbf{l} + \int_{b'Bb} \mathbf{q} \cdot d\mathbf{l} + \int_{ba} \mathbf{q} \cdot d\mathbf{l} = 0.$$

Now, if the lines ab and $a'b'$ are brought adjacent to each other, the integrals along these lines will cancel each other. The curves aAa' and $b'Bb$ become closed curves C and a complete circle, respectively. Hence we have

$$\int_{aAa'} \mathbf{q} \cdot d\mathbf{l} = -\int_{b'Bb} \mathbf{q} \cdot d\mathbf{l}.$$

That is,

$$\oint_C \mathbf{q} \cdot d\mathbf{l} = -\int_{b'Bb} \mathbf{q} \cdot d\mathbf{l}.$$

Now, along the circle $b'Bb$, we have

$$\mathbf{q} \cdot d\mathbf{l} = \frac{\Gamma}{2\pi r_0}\,(-r_0\, d\theta).$$

Thus

$$\oint_C \mathbf{q} \cdot d\mathbf{l} = \int_0^{2\pi} \frac{\Gamma}{2\pi}\, d\theta = \Gamma.$$

Hence, the circulation computed around *any* closed curve C surrounding the origin is Γ. This was, of course, the reason for choosing the symbol Γ in Eq. 6–16.

Using Stokes' theorem, it is seen that the surface integral $\int_S (\mathbf{n} \cdot \boldsymbol{\xi})\, dS$ for the whole surface bounded by any closed curve surrounding the origin must also be equal to Γ. The value of $\mathbf{n} \cdot \boldsymbol{\xi}$ on the surface is given by

$$\mathbf{n} \cdot \boldsymbol{\xi} = \xi_z = \frac{1}{r}\frac{\partial}{\partial r}(rv) - \frac{1}{r}\frac{\partial u}{\partial \theta},$$

which is zero everywhere except at $r = 0$, where it is indeterminate. Such points are called *singular points* and must be treated with great care and should be avoided if possible. In the calculation presented in this example, this singular point was excluded for precisely this reason.

Example 8–2. By using Stokes' theorem given in Eq. 8–12, find the expressions for the vorticity components in cylindrical coordinates.

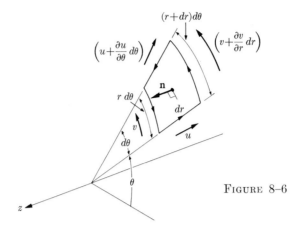

FIGURE 8–6

In Section 6–1, the expressions for vorticity components were derived by considering the rotation of a fluid element. It is often simpler to arrive at these expressions by means of Stokes' theorem. Consider an infinitesimal plane area normal to the z-axis as shown in Fig. 8–6. The left-hand side of Eq. 8–12 is evaluated directly as

$$\int_S (\mathbf{n} \cdot \boldsymbol{\xi})\, dS = \xi_z r\, d\theta\, dr, \tag{8–13a}$$

where $\mathbf{n}$ is chosen to be in the z-direction, so that $\mathbf{n} \cdot \boldsymbol{\xi} = \xi_z$. The line integral on the right-hand side of Eq. 8–12 is then taken in the counterclockwise direction according to the right-hand rule. We obtain

$$\oint_C \mathbf{q} \cdot d\mathbf{l} = u\, dr + \left(v + \frac{\partial v}{\partial r}\, dr\right)(r + dr)\, d\theta - \left(u + \frac{\partial u}{\partial \theta}\, d\theta\right) dr - vr\, d\theta, \tag{8–13b}$$

where u, v, etc., are the main values of velocity components along the sides as shown in Fig. 8–6. Equating Eqs. 8–13a and Eq. 8–13b and solving for ξ_z, we obtain

$$\xi_z = \frac{1}{r}\left[\frac{\partial}{\partial r}(rv) - \frac{\partial u}{\partial \theta}\right], \tag{8–14}$$

in agreement with Eq. 6–7. In similar manner, the expressions for ξ_r and ξ_θ can be derived.

PROBLEMS

8–6. Verify the formulas for ξ_r and ξ_θ in Eqs. 6–8 and 6–9 using Stokes' theorem.

8–7. Prove by using Stokes' theorem, 8–12, that the surface integral $\int_S (\mathbf{n} \cdot \boldsymbol{\xi})\, dS$ is zero if the surface S under consideration is any closed surface. [*Hint:* Divide the closed surface into two separate surfaces with the same boundary.]

8–8. Consider a cylindrical tank of radius R filled with a viscous fluid, spinning steadily about its axis with constant angular velocity ω. For a viscous fluid, the fluid adjacent to a solid surface has no relative velocity with respect to the surface. We assume that the flow is in a steady state so that the fluid rotates as a solid body with the tank, as shown in Fig. 5–2. (a) Find $\int_S (\mathbf{n} \cdot \boldsymbol{\xi})\, dS$, where S is a horizontal plane surface in the fluid bounded by the wall of the tank. (b) The tank then stops spinning. Find again the value of $\int_S (\mathbf{n} \cdot \boldsymbol{\xi})\, dS$ before the fluid motion has ceased.

8–4. Role of vorticity in lift and drag

In many cases, we are interested in the forces experienced by solid bodies moving in a fluid, e.g., the lift and drag of the wing of an airplane. In Chapter 6, we have studied flows over simple bodies such as a circular cylinder and a sphere under the assumption that the flow was irrotational everywhere. It was found that there was no drag; and for the case of the cylinder, unless circulation was present, there was no lift. Actually, for a *finite* body of arbitrary shape in a *steady, frictionless,* and incompressible flow *completely free from vorticity*, it can be shown quite generally that that body will experience no net force. This surprising result is generally known as *D'Alembert's paradox.* The reader may recall that in Example 6–9, a lift force was obtained for a circular cylinder with circulation. The D'Alembert paradox does not apply in this case, because a two-dimensional body is laterally infinite and is not a finite body. In any case, forces are closely related to circulations and vorticities, as we shall see presently.

For this purpose, let us consider a finite thin flat wing moving with constant velocity U in a fluid otherwise at rest, and let us suppose that the pressure on the lower surface of the wing is on the average higher than that on the upper surface, so that a net lift is experienced by the wing. When this is the case, the fluid beneath the wing near the wing tips will tend to spill over the tip, as shown in Fig. 8–7. Now at a certain time, let us consider the circulation around the dotted loop C somewhere behind the wing tip. We see that if this spilling flow is present, the circulation Γ computed around C using Eq. 8–13 will be nonzero. Consequently, there must be vorticity in the fluid, since from Stokes' theorem the surface integral $\int_S (\mathbf{n} \cdot \boldsymbol{\xi})\, dS$ using any surface in the fluid having the dotted loop C as a boundary will yield $\Gamma = \oint_C \mathbf{q} \cdot d\mathbf{l}$, which is nonzero.

The vortex tube generated from loop C must obey the kinematic properties derived earlier; i.e., it cannot end in the fluid, but must attach itself to the solid body (the wing), or extend to infinity. Consequently, we see that when lift is present, the flow field as a whole cannot be completely free from vorticity. As a consequence of the presence of vorticity in the fluid, a drag force must be experienced by the wing, as can be seen from the following arguments.

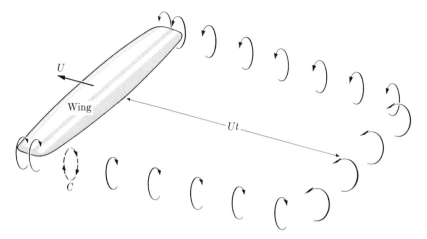

FIGURE 8–7

Suppose that the body has been moving with a constant velocity for some time t in a fluid at rest. If the flow is completely irrotational, the flow field can be studied by methods presented in Chapter 6. An outstanding feature of such irrotational solution is that the flow pattern relative to the body (observed by an observer moving with the body) is independent of time. Thus the kinetic energy of the fluid in the flow field as a whole must be independent of time. If the body experiences a drag D in this motion, then per unit time an amount of energy equal to DU has been expended. Since for an irrotational flow the total kinetic energy of the flow in the whole flow field did not change, the drag experienced by the body must be zero. On the other hand, when vorticity discussed previously is present, such vorticities will be found behind the body in a wake roughly of length Ut. Since vortex tubes have constant circulations and cannot end in the fluid, they must take the shape of horseshoes, attaching both ends to the wing, as shown in Fig. 8–7. Within this wake, the mean kinetic energy of the fluid per unit length of the wake shall be denoted by K. Consequently, the vorticity wake as a whole contains kinetic energy which increases with time at a rate equal to KU. This increase in kinetic energy must have come from the rate of work done by the drag force experienced by the wing. Hence a nonzero drag force, called the *induced drag*, must always be present when there is lift, and its magnitude must equal K. By a much more elaborate argument, we can also show that in the absence of vorticity the net lift experienced by a finite body in steady flow is also zero. It should now be clear that vorticity is closely related to both lift and drag on bodies in steady flows. This is one of the reasons why vorticity is always considered an important topic in the study of fluid mechanics, especially for aeronautical engineers. It must be mentioned that this conceptual model of lift and drag based on vorticity was due to L. Prandtl and formed the basis for his celebrated lifting line theory for finite wings.

PROBLEM

8-9. A long but *finite* circular cylinder is in a uniform flow normal to its axis. The flow is assumed frictionless and incompressible. If the flow about the center portion of the cylinder is well approximated by

$$\phi = -\frac{\Gamma}{2\pi}\,\theta + U\cos\theta\left(r + \frac{R^2}{r}\right),$$

where r, θ, and z are cylindrical polar coordinates with the z-axis coincident with the axis of the cylinder, ϕ is the velocity potential, and R is the radius of the cylinder, show that the flow field as a whole cannot be completely free from vorticity. [*Hint:* Find the circulation around the middle section of the cylinder and then apply Stokes' theorem.]

8–5. Rate of change of vorticity

We may recall that so far we have given no formal criterion to judge whether a flow can be assumed irrotational. As a matter of fact, one may have already noticed that the Laplace equation for the velocity potential was apparently derived without the benefit of any momentum considerations (Section 6–3). In this section, we shall devote our attention to the derivation of the conditions under which a flow can be assumed irrotational.

To this end, let us calculate in this section formally the rate change of vorticity following a given element of fluid. Since vorticity is a vector, we shall need to compute this rate for each of its three components. First, let us write down the momentum equations in the x-, y-, and z-directions as follows:

$$\frac{\partial u}{\partial t} + u\frac{\partial u}{\partial x} + v\frac{\partial u}{\partial y} + w\frac{\partial u}{\partial z} = -\frac{1}{\rho}\frac{\partial p}{\partial x} + f_x, \qquad (8\text{--}15a)$$

$$\frac{\partial v}{\partial t} + u\frac{\partial v}{\partial x} + v\frac{\partial v}{\partial y} + w\frac{\partial v}{\partial z} = -\frac{1}{\rho}\frac{\partial p}{\partial y} + f_y, \qquad (8\text{--}15b)$$

$$\frac{\partial w}{\partial t} + u\frac{\partial w}{\partial x} + v\frac{\partial w}{\partial y} + w\frac{\partial w}{\partial z} = -\frac{1}{\rho}\frac{\partial p}{\partial z} + f_z. \qquad (8\text{--}15c)$$

These equations are the same as Eqs. 5–1, 5–2, and 5–3 except that, for generality, we write f_x, f_y, and f_z to represent either body forces or frictional forces per unit mass in the x-, y-, and z-directions, respectively. Let us first calculate $D\xi_z/Dt$, the rate of change of ξ_z of a particle. Since

$$\frac{D\xi_z}{Dt} = \frac{\partial \xi_z}{\partial t} + u\frac{\partial \xi_z}{\partial x} + v\frac{\partial \xi_z}{\partial y} + w\frac{\partial \xi_z}{\partial z}$$

and

$$\xi_z = \frac{\partial v}{\partial x} - \frac{\partial u}{\partial y},$$

we have, by differentiating Eqs. 8–15a and 8–15b with respect to y and x, respectively, and subtracting from each other, the following expression:

$$\frac{D\xi_z}{Dt} = \xi_z \left(\frac{1}{\rho} \frac{D\rho}{Dt} \right) + \left(\xi_x \frac{\partial w}{\partial x} + \xi_y \frac{\partial w}{\partial y} + \xi_z \frac{\partial w}{\partial z} \right)$$

$$- \left[\frac{\partial}{\partial x} \left(\frac{1}{\rho} \frac{\partial p}{\partial y} \right) - \frac{\partial}{\partial y} \left(\frac{1}{\rho} \frac{\partial p}{\partial x} \right) \right] + \left(\frac{\partial f_y}{\partial x} - \frac{\partial f_x}{\partial y} \right). \qquad (8\text{–}16a)$$

Similar expressions can readily be written down for ξ_y and ξ_x:

$$\frac{D\xi_y}{Dt} = \xi_y \left(\frac{1}{\rho} \frac{D\rho}{Dt} \right) + \left(\xi_x \frac{\partial v}{\partial x} + \xi_y \frac{\partial v}{\partial y} + \xi_z \frac{\partial v}{\partial z} \right)$$

$$- \left[\frac{\partial}{\partial z} \left(\frac{1}{\rho} \frac{\partial p}{\partial x} \right) - \frac{\partial}{\partial x} \left(\frac{1}{\rho} \frac{\partial p}{\partial z} \right) \right] + \left(\frac{\partial f_x}{\partial z} - \frac{\partial f_z}{\partial x} \right), \qquad (8\text{–}16b)$$

$$\frac{D\xi_x}{Dt} = \xi_x \left(\frac{1}{\rho} \frac{D\rho}{Dt} \right) + \left(\xi_x \frac{\partial u}{\partial x} + \xi_y \frac{\partial u}{\partial y} + \xi_z \frac{\partial u}{\partial z} \right)$$

$$- \left[\frac{\partial}{\partial y} \left(\frac{1}{\rho} \frac{\partial p}{\partial z} \right) - \frac{\partial}{\partial z} \left(\frac{1}{\rho} \frac{\partial p}{\partial y} \right) \right] + \left(\frac{\partial f_z}{\partial y} - \frac{\partial f_y}{\partial z} \right). \qquad (8\text{–}16c)$$

We are now ready to discuss the rate of change of the vorticity vector $\boldsymbol{\xi}$. Consider an element of fluid which at some instant $t = 0$ had zero vorticity ($\xi_x = 0$, $\xi_y = 0$, $\xi_z = 0$). The first two parentheses on the right-hand side of Eqs. 8–16 will be zero at that instant. We see that this element of fluid will continue to have no vorticity if the following conditions are true (see Problem 8–13):

$$\rho = \rho(p), \qquad (8\text{–}17)$$

and

$$f_x = \frac{\partial F}{\partial x}, \qquad f_y = \frac{\partial F}{\partial y}, \qquad f_z = \frac{\partial F}{\partial z}, \qquad (8\text{–}18)$$

where Eq. 8–17 expresses the requirement that density is a function of pressure only, and Eqs. 8–18 express the requirement that the force $\mathbf{f}$ must be the gradient of some scalar function $F(x, y, z)$. When Eqs. 8–17 and 8–18 are true, the vorticity of the fluid particle in question at time $t = 0$ will have zero rate of change, i.e.:

$$\frac{D\xi_z}{Dt} = 0, \qquad \frac{D\xi_y}{Dt} = 0, \qquad \frac{D\xi_x}{Dt} = 0,$$

and hence it will continue to have no vorticity.

Equation 8–17 is called the *barotropic condition*, and a flow satisfying such a condition is called a *barotropic flow*. A constant density flow is a barotropic

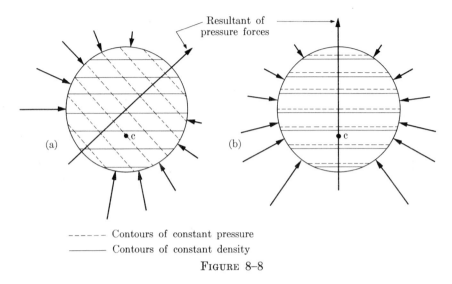

Contours of constant pressure
Contours of constant density

FIGURE 8–8

flow. Another important barotropic flow is constant entropy flow, which for a perfect gas satisfies

$$p \propto \rho^\gamma,$$

where γ is the ratio of specific heats (see Eq. 1–5 and Chapters 14 and 15). In general, the judgment of whether a flow is barotropic or not must come from thermodynamic or other considerations. It cannot be deduced from continuity or momentum considerations.

Condition 8–18 is called the conservative force field condition. In other words, when Eq. 8–18 is satisfied, the force $\mathbf{f}$ ($= \nabla F$) is said to be *conservative*. It will suffice to state here that frictional forces are never conservative (see Eqs. 11–30 to 11–32), and that gravitational force is conservative (see Eq. 5–6a). Thus in flows where frictional forces are not expected to be negligible, the assumption of irrotational flow cannot be used.

When a flow is nonbarotropic, vorticity can be either generated or destroyed. The physical mechanism can be seen in the following. Consider a given small element of fluid of nonconstant density as shown in Fig. 8–8(a). The net resultant pressure force acting on the surface of this fluid element may not always pass through its center of mass, since density and pressure are not related. A moment therefore results about the center of mass. Consequently, the rotation vector or vorticity of this fluid element will change with time. However, if the flow is barotropic, the surfaces of constant density will also be surfaces of constant pressure as shown in Fig. 8–8(b). The resultant vector of the surface pressure forces will now always pass through the center of mass of the small parcel of fluid. Hence, no moment will result and the rotation vector of the fluid element will not change with time. Note that even if the flow is barotropic, the fluid element in Fig. 8–8(b) can still rotate if there are frictional forces acting on the surface of the fluid element. Except on rare occasions this results in a net torque.

We thus see that in order to decide whether a given flow field is irrotational, we must first find out whether both Eqs. 8–17 and 8–18 are satisfied. If they are satisfied, then we must examine the flow field *particle by particle* in a conceptual manner. Any particle which at some earlier time had no vorticity will continue to have none. Any particle which at some earlier time had nonzero vorticity will generally be rotational. As an example, consider an incompressible, frictionless, uniform flow over an obstacle such as a wing. For this case we have ρ = constant and $F = 0$, and thus Eqs. 8–17 and 8–18 are satisfied. Far upstream of the obstacle, the vorticity of every fluid particle is zero, since the flow is uniform there. Consequently, we can *conclude* that the flow is irrotational. It is seen that irrotationality is no longer an assumption itself, but is now a consequence of the incompressible, frictionless, upstream uniform flow assumptions.

The next question which naturally arises is, if such a flow is irrotational, then how can vorticity be present in the wake of a wing, as discussed in the previous section? The answer to this question is rather subtle. Immediately adjacent to the wing surface, the frictionless assumption is not valid, and fluid particles gain vorticity there from frictional processes. Once they gain some vorticity and flow off the wing, the kinematic properties of vorticity—conservation of vorticity, etc.—demand the vorticity wake picture described in the previous section. Thus the flow about a lifting finite wing is almost irrotational everywhere, except in the wake, where vorticity must be present. It must be emphasized that fluid friction, or viscosity, only acts as an agent for the existence of the drag force on the wing. The magnitude of the drag due to the vortex wake is independent of the viscosity of the fluid. The thickness of the vortex wake, however, depends on viscosity. For air, viscosity is very small; such vortex wake is extremely thin, and is generally referred to as *vortex sheet*. If there exists a fluid which is absolutely frictionless and its density is absolutely constant, then the D'Alembert's paradox will apply, and no net force will be experienced by any finite body in a steady uniform flow.

PROBLEMS

8–10. Show that for a frictionless, steady, two-dimensional, rotational flow with constant density and conservative body forces, vorticity is constant along a streamline; i.e., ξ_z is a function of ψ only. [*Hint:* Use x and ψ as independent variables instead of x and y.]

8–11. Consider a steady, frictionless, two-dimensional flow in a duct with a smooth contraction. Far upstream the velocity U_1 is uniform across the channel. The density at this section, however, is a function of y:

$$\rho_1 = \rho_0 e^{-y/L},$$

where ρ_0 is a constant and L is the width of the duct far upstream. The fluid is assumed incompressible. Far downstream, the flow again becomes parallel. Show that the velocity distribution U_2 at distances far downstream satisfies the differential equation

$$2LU_2 \frac{dU_2}{dy} = \left(\frac{U_2^2}{U_1^2} - 1\right) U_1^2.$$

[*Hint:* First find the rate of change of Bernoulli's constant with respect to y in the downstream section.]

8-12. Using momentum equations in cylindrical polar coordinates, show that

$$\frac{D}{Dt}\left(\frac{\xi_\theta}{\rho r}\right) = 0$$

if the flow is axisymmetric ($\partial/\partial\theta = 0$) and is barotropic, and if f is a conservative force field.

8-13. Show that

$$\frac{\partial}{\partial z}\left(\frac{1}{\rho}\frac{\partial p}{\partial x}\right) - \frac{\partial}{\partial x}\left(\frac{1}{\rho}\frac{\partial p}{\partial z}\right) = 0$$

if the barotropic condition, Eq. 8–17, is satisfied. [*Hint:* $\partial p/\partial x$ can be rewritten as $(dp/d\rho)(\partial\rho/\partial x)$.]

9 The Momentum Theorem

9–1. Motion of a system of particles

The momentum equations for infinitesimal fluid elements have been used in previous chapters to study the velocity distribution in fluid motion. These equations are often found to be difficult to solve. Corresponding equations can be derived for finite volumes of fluid which contain a system of particles. Although these equations describe the group behavior and not the details of the flow, they are often easier to solve and can yield valuable information.

Consider a system of particles of masses $m_1, m_2, \ldots$. Let $\mathbf{F}_1$ be the forces from sources outside the system acting on mass m_1, and $\mathbf{f}_1$ be the forces acting on m_1 by the other particles in the system. According to Newton's law of motion, we have for m_1

$$\mathbf{F}_1 + \mathbf{f}_1 = \frac{d}{dt}(m_1\mathbf{q}_1),$$

where $\mathbf{q}_1$ is the velocity of m_1. Similarly, we have for the other particles $m_2, \ldots$ in the system

$$\mathbf{F}_2 + \mathbf{f}_2 = \frac{d}{dt}(m_2\mathbf{q}_2)$$
$$\vdots$$

Summing up all these equations, we obtain

$$\mathbf{F}_1 + \mathbf{F}_2 + \cdots + \mathbf{f}_1 + \mathbf{f}_2 + \cdots = \frac{d}{dt}(m_1\mathbf{q}_1) + \frac{d}{dt}(m_2\mathbf{q}_2) + \cdots$$

$$= \frac{d}{dt}(m_1\mathbf{q}_1 + m_2\mathbf{q}_2 + \cdots).$$

The forces $\mathbf{f}_1, \mathbf{f}_2, \ldots$ are the actions and reactions among the particles of the system. By the law of actions and reactions, the sum $(\mathbf{f}_1 + \mathbf{f}_2 + \cdots)$ must be zero. This is true whether there is internal friction or not. Thus

$$\sum \mathbf{F} = \frac{d\mathbf{M}}{dt}, \tag{9–1}$$

where $\sum \mathbf{F}$ is the vector sum of the forces acting on the system from sources external to the system, and $\mathbf{M} = \sum m\mathbf{q}$ is the vector sum of the momenta of the particles and is called the *total momentum* of the system. The derivative $d\mathbf{M}/dt$ is the rate of change of the total momentum of these particles. The reader will recall that in the study of rigid-body mechanics, this equation is reduced to the convenient form $\sum \mathbf{F} = (\sum m)\mathbf{a}_c$, where $\sum m$ is the total mass of the system, and $\mathbf{a}_c$ is the acceleration of the center of mass of the system. However, this form is not convenient in the study of fluid motion. Equation 9–1 must be rewritten in a more useful form.

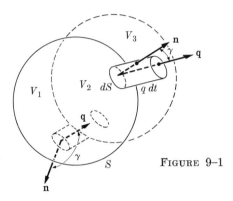

FIGURE 9–1

Consider a volume of fluid bounded by a closed surface S at t_0, occupying the position of volume V, which is the sum of V_1 and V_2 in Fig. 9–1. Equation 9–1 states that the external forces acting on this volume are equal to the rate of change of the total momentum of the particles inside. Let us follow these particles for a short time dt and find the change of their total momentum $\mathbf{M}$. At time $t_0 + dt$, they will occupy the position of volumes V_2 and V_3. The change of their total momentum during dt is therefore

$$d\mathbf{M} = [\mathbf{M} \text{ in } V_2 \text{ at } (t_0 + dt) + \mathbf{M} \text{ in } V_3 \text{ at } (t_0 + dt)]$$
$$- [\mathbf{M} \text{ in } V_1 \text{ at } t_0 + \mathbf{M} \text{ in } V_2 \text{ at } t_0]. \tag{9–2}$$

Now the total momentum vector $\mathbf{M}$ in a volume V can be expressed as

$$\mathbf{M} = \sum m\mathbf{q} = \lim_{\Delta V \to 0} \sum (\rho \, \Delta V)\mathbf{q} = \int_V \rho \mathbf{q} \, dV.$$

Here the integral is the sum of the infinite number of momentum vectors $\rho \, dV\mathbf{q}$. For $\mathbf{M}$ in V_3, we have, for every infinitesimal surface dS of S, a volume $dV = (q \, dt)(dS \cos \gamma)$, where q is the speed at dS, $q \, dt$ is the length of the cylinder dV as shown in Fig. 9–1, γ is the angle between the vector $\mathbf{q}$ and the outward normal unit vector $\mathbf{n}$ at dS, and $dS \cos \gamma$ is the cross-sectional area of the cylinder. In terms of the scalar product of two vectors (see Eq. 4–22),

$q \cos \gamma = \mathbf{n} \cdot \mathbf{q}$, $dV = (\mathbf{n} \cdot \mathbf{q}) \, dS \, dt$, and

$$\mathbf{M} \text{ in } V_3 = \int_{V_3} \rho \mathbf{q} \, dV = \int_{S_3} \rho \mathbf{q} (\mathbf{n} \cdot \mathbf{q}) \, dS \, dt,$$

where S_3 is the surface separating V_2 and V_3. Similarly, for $\mathbf{M}$ in V_1, we have $dV = (q \, dt)(-dS \cos \gamma) = -(\mathbf{n} \cdot \mathbf{q}) \, dS \, dt$. Here the negative sign is necessary because $\cos \gamma$ is negative, while the volume dV is positive:

$$\mathbf{M} \text{ in } V_1 = \int_{V_1} \rho \mathbf{q} \, dV = -\int_{S_1} \rho \mathbf{q} (\mathbf{n} \cdot \mathbf{q}) \, dS \, dt,$$

where S_1 is equal to $S - S_3$. As $dt \to 0$, we have for Eq. 9–2,

$$(\mathbf{M} \text{ in } V_3) - (\mathbf{M} \text{ in } V_1) = \int_{S_3} \rho \mathbf{q} (\mathbf{n} \cdot \mathbf{q}) \, dS \, dt + \int_{S_1} \rho \mathbf{q} (\mathbf{n} \cdot \mathbf{q}) \, dS \, dt$$

$$= \int_{S} \rho \mathbf{q} (\mathbf{n} \cdot \mathbf{q}) \, dS \, dt.$$

The other two terms in Eq. 9–2 are the local change of $\mathbf{M}$ in V_2 during dt:

$$[\mathbf{M} \text{ in } V_2 \text{ at } (t_0 + dt)] - [\mathbf{M} \text{ in } V_2 \text{ at } t_0] = \frac{d}{dt} \left[\int_{V_2} \rho \mathbf{q} \, dV \right] dt.$$

Here we use d/dt because the integral over a fixed V is a function of t only. Since dt is infinitesimal, V_2 approaches V. Thus Eq. 9–2 can be written as

$$\frac{d\mathbf{M}}{dt} = \frac{d}{dt} \left[\int_{V} \rho \mathbf{q} \, dV \right] + \int_{S} \rho \mathbf{q} (\mathbf{n} \cdot \mathbf{q}) \, dS.$$

Note that $d\mathbf{M}/dt$ is a particle derivative, since we have followed the same particles in finding the change $d\mathbf{M}$. The rate of change of the total momentum of a system of particles occupying momentarily the fixed volume V bounded by surface S can be seen to be the sum of two quantities, namely the local rate of change of momentum in V and the rate of flow of momentum out of the surface S.

Equation 9–1 can therefore be written as

$$\sum \mathbf{F} = \frac{d}{dt} \left[\int_{V} \rho \mathbf{q} \, dV \right] + \int_{S} \rho \mathbf{q} (\mathbf{n} \cdot \mathbf{q}) \, dS, \tag{9–3}$$

where $\sum \mathbf{F}$ is the sum of the external forces acting on the fluid in the control volume V. This is Newton's law of motion written for a finite volume of fluid, and is usually called *Euler's momentum theorem*. It is applicable to all fluids.

For steady flow, the distribution of ρ and $\mathbf{q}$ in the space of the control volume V remains the same. The volume integral in Eq. 9–3 is therefore constant.

Thus, for steady flows,

$$\sum \mathbf{F} = \int_S \rho \mathbf{q}(\mathbf{n} \cdot \mathbf{q}) \, dS. \tag{9-4}$$

In application, it is convenient to use the components of the vectors. With **i**, **j**, and **k** as unit vectors in the directions of the rectangular coordinate axes and $\mathbf{n} \cdot \mathbf{q} = q \cos \gamma$, Eq. 9-4 can be written as

$$\mathbf{i}\sum F_x + \mathbf{j}\sum F_y + \mathbf{k}\sum F_z$$

$$= \int_S \rho(\mathbf{i}u + \mathbf{j}v + \mathbf{k}w)q \cos \gamma \, dS$$

$$= \mathbf{i}\int_S \rho u q \cos \gamma \, dS + \mathbf{j}\int_S \rho v q \cos \gamma \, dS + \mathbf{k}\int_S \rho w q \cos \gamma \, dS.$$

Hence

$$\sum F_x = \int_S \rho u q \cos \gamma \, dS, \tag{9-5}$$

$$\sum F_y = \int_S \rho v q \cos \gamma \, dS, \tag{9-6}$$

$$\sum F_z = \int_S \rho w q \cos \gamma \, dS. \tag{9-7}$$

The momentum theorem is valid for fluid flows in general. In using these equations, the control volume in S must first be chosen. These equations are usually used for computing the external forces acting on this volume when the velocities at the surface S can be measured or estimated, and vice versa. Being the momentum equation for a system of particles, this theorem cannot yield the details of a flow pattern. However, with supporting information, it can give valuable, although often only approximate, solutions.

Example 9-1. Find the discharge from a large tank of a liquid with little internal friction through a *Borda's mouthpiece*, as shown in Fig. 9-2.

From Bernoulli's equation, we obtain the speed of the jet $q_j = \sqrt{2gH}$ (see Example 5-3). The problem is to find the area A_j of the jet. As indicated in Example 5-3, A_j can be obtained theoretically by solving the equations of motion. However, if the Borda's mouthpiece is sufficiently long such that the velocity of the fluid is negligible near the walls of the container, A_j can be estimated as follows. Since the flow in this case can be approximated by an irrotational flow where Bernoulli's equation is applicable,

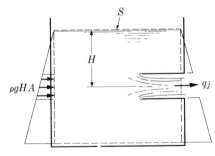

FIGURE 9-2

we have the pressure at the walls given by

$$\frac{p}{\rho} + gh + \frac{q^2}{2} = \frac{p}{\rho} + gh = \text{constant.}$$

Thus the pressure on the walls is hydrostatically distributed, as shown in Fig. 9–2.

We take the control volume in S shown in Fig. 9–2, and make use of the fact that the liquid gains momentum in the x-direction. The net external force acting on S in this direction is $\rho g H A$ due to the unbalanced pressure on the left wall, since the pressure over the opening A is atmospheric. The only contribution to the integral of flow of momentum in Eq. 9–5 is from the jet of (unknown) area A_j. For this area, the angle γ between its outward normal and the velocity vector is zero, and u is equal to q_j. Thus

$$\int_S \rho u q \cos \gamma \, dS = \rho q_j q_j A_j = \rho (2gH) A_j.$$

Substituting this and the force $\rho g H A$ into Eq. 9–5, we obtain

$$\rho g H A = 2 \rho g H A_j.$$

Therefore the *coefficient of contraction* A_j/A is equal to $\frac{1}{2}$, and the discharge Q is

$$Q = A_j q_j = \tfrac{1}{2} A \sqrt{2gH}.$$

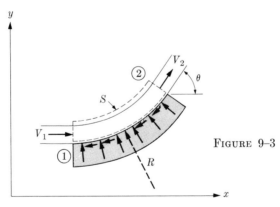

FIGURE 9–3

Example 9–2. A jet of liquid with speed V_1 and discharge Q is deflected in a horizontal plane without splashing through an angle θ by a stationary bucket, as shown in Fig. 9–3. The speed after deflection is V_2. Find the resultant R of the distributed forces acting on the jet by the bucket.

Choose the control volume in S as shown. To evaluate the flow of momentum in Eqs. 9–5 and 9–6, we have at section 1: $\gamma = \pi$, $u = q = V_1$, and $v = 0$. At section 2, we have: $\gamma = 0$, $q = V_2$, $u = V_2 \cos \theta$, and $v = V_2 \sin \theta$.

Thus

$$\int_S \rho u q \cos \gamma \, dS = -\rho V_1 V_1 A_1 + \rho V_2 \cos \theta \, V_2 A_2 = \rho Q(V_2 \cos \theta - V_1),$$

$$\int_S \rho v q \cos \gamma \, dS = \rho V_2 \sin \theta \, V_2 A_2 = \rho Q V_2 \sin \theta,$$

where $Q = A_1 V_1 = A_2 V_2$. Let R_x and R_y be the components of the resultant of the distributed forces acting on this volume of liquid by the bucket. From Eqs. 9–5 and 9–6, we have

$$R_x = \rho Q(V_2 \cos \theta - V_1), \qquad R_y = \rho Q V_2 \sin \theta.$$

The force $\mathbf{F}$ acting on the bucket by the liquid is the reaction of this resultant $\mathbf{R}$, that is, $\mathbf{F} = -\mathbf{R}$, or

$$F_x = -R_x = \rho Q(V_1 - V_2 \cos \theta), \qquad F_y = -R_y = -\rho Q V_2 \sin \theta.$$

This result is valid whether there is internal friction in the liquid or not. The internal friction affects only the magnitude of V_2 in these equations. For a frictionless fluid, Bernoulli's equation gives $V_2 = V_1$.

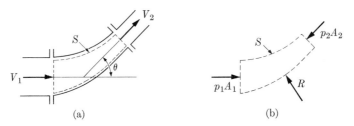

FIGURE 9–4

Example 9–3. At a horizontal bend of a pipeline, the cross-sectional area is reduced from A_1 to A_2, as shown in Fig. 9–4. Find the forces acting on the two joints when the discharge is Q under pressure p_1 at section 1. The bend is so gradual that there is no separation of flow from the wall and the velocity distribution over a section may be assumed to be practically uniform.

Choose the control volume in S as shown. The forces acting on this volume are $p_1 A_1$, $p_2 A_2$, and R, the resultant of the distributed forces from the walls of the bend. From the equation of continuity for one-dimensional analysis,

$$A_1 V_1 = A_2 V_2 = Q.$$

From Bernoulli's equation, which is assumed to be applicable here,

$$\frac{p_2}{\rho} = \frac{p_1}{\rho} + \frac{V_1^2}{2} - \frac{V_2^2}{2} = \frac{p_1}{\rho} + \frac{Q^2}{2}\left(\frac{1}{A_1^2} - \frac{1}{A_2^2}\right).$$

From Eqs. 9–5 and 9–6, we have

$$p_1 A_1 - p_2 A_2 \cos \theta + R_x = -\rho V_1 V_1 A_1 + \rho V_2 \cos \theta V_2 A_2$$

$$= -\rho Q^2 \left(\frac{1}{A_1} - \frac{\cos \theta}{A_2} \right)$$

and

$$- p_2 A_2 \cos \theta + R_y = \rho V_2 \sin \theta V_2 A_2 = \rho Q^2 \frac{\sin \theta}{A_2}.$$

Substituting p_2 from Bernoulli's equation above, we have R_x and R_y in terms of the given quantities.

The force $\mathbf{F}$ acting on the bend by the fluid is the reaction of $\mathbf{R}$. The joints must be designed to withstand the force $\mathbf{F} = -\mathbf{R}$.

Example 9–4. A propeller with a projected area A travels at a constant speed. The speed of the fluid relative to the propeller is increased from V_1 to V_4 within the *slipstream boundary*, as shown in Fig. 9–5(a). Find the thrust and the relative speed V across the propeller.

To find the thrust, we consider the volume of fluid in S_b, as shown in Fig. 9–5(b). The forces acting on this volume are $p_2 A$, $p_3 A$, and R, the resultant of the distributed forces acting on the fluid by the propeller. The flow of momentum across S_b in Eq. 9–5 is

$$-\rho V_2 V_2 A + \rho V_3 V_3 A = -\rho V^2 A + \rho V^2 A = 0.$$

Thus, from Eq. 9–5, we have

$$p_2 A - p_3 A + R = 0, \quad \text{or} \quad R = (p_3 - p_2) A.$$

To compute R, it is necessary to find $p_3 - p_2$. Since the flow across the propeller is unsteady, Bernoulli's equation is not applicable between points 2 and

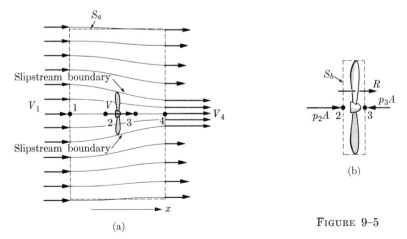

(a)

FIGURE 9–5

3. However, the equation can be applied along the streamline between points 1 and 2, and between points 3 and 4:

$$\frac{p_1}{\rho} + gh_1 + \frac{V_1^2}{2} = \frac{p_2}{\rho} + gh_2 + \frac{V_2^2}{2},$$

$$\frac{p_3}{\rho} + gh_3 + \frac{V_3^2}{2} = \frac{p_4}{\rho} + gh_4 + \frac{V_4^2}{2},$$

where $h_1 = h_2 = h_3 = h_4$ and $V_2 = V_3$. Since the streamlines are practically parallel at sections 1 and 4, it can be shown with Eq. 5–10 that $p_1 = p_4$. Thus

$$p_3 - p_2 = \frac{\rho}{2}(V_4^2 - V_1^2) \qquad \text{and} \qquad R = \frac{\rho A}{2}(V_4^2 - V_1^2).$$

To find V across the propeller, we consider the large volume enclosed by S_a in Fig. 9–5(a). Since the streamlines at some distance from the propeller are practically undisturbed, there is very little flow across the cylindrical surface of S_a. With $p_1 = p_4$, the net force acting on this volume is R from the propeller. Neglecting the change of momentum of the fluid outside the slipstream boundary, we have from Eq. 9–5,

$$R = -\rho V_1 V_1 A_1 + \rho V_4 V_4 A_4,$$

where A_1, and A_4 are the cross-sectional areas within the slipstream boundary. By continuity

$$A_1 V_1 = A_4 V_4 = AV.$$

Therefore

$$\frac{\rho A}{2}(V_4^2 - V_1^2) = \rho AV(V_4 - V_1), \qquad V = \tfrac{1}{2}(V_4 + V_1).$$

Several approximations have been employed in this analysis. The influx of momentum across the cylindrical surface of S_a and the effect of the deficiency of speed around the jet have been neglected for simplicity in presentation. However, it can be shown by a more refined analysis that this result is correct.

PROBLEMS

9–1. Show that for the control volume V, shown in Fig. 9–1, bounded by the surface S, the equation of continuity can be written as

$$\int_S \rho \mathbf{q} \cdot \mathbf{n}\, dS = \frac{d}{dt}\left[\int_V \rho\, dV\right].$$

Also show that for the incompressible flow shown in Fig. 5–14, this equation is reduced to Eq. 5–17.

9–2. In Fig. 5–7 is shown the flow of a thin liquid through a sharp-edged orifice of area A. Show that the discharge is expected to be greater than $\tfrac{1}{2}A\sqrt{2gH}$.

9–3. (a) A jet of water with a discharge of 10 ft³/sec and a speed of 20 ft/sec impinges normally on a large stationary flat plate. Assuming that the water leaves the plate in directions parallel to the plate, find the force required to hold the plate stationary. (b) If the plate moves at 5 ft/sec in the same direction as the jet, what is the force acting on the plate by the jet? [*Hint:* Let the coordinate system move with the plate to obtain a steady flow.]

9–4. (a) A jet of liquid with a discharge Q_1 and a speed V_1 is deflected by a small block, as shown in Fig. 9–6. The discharge of the jet going upward is Q_2. Assuming that there is negligible frictional effect, find the resultant force acting on the block. (b) A jet of liquid of discharge Q_1 and speed V_1 impinges on a large plate at an angle θ, as shown in Fig. 9–7. Assuming that there is negligible frictional effect, find the resultant force acting on the plate. Since the surface friction is negligible, this force may be assumed to act in the direction normal to the plate.

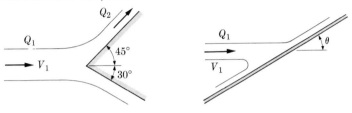

FIGURE 9–6   FIGURE 9–7

9–5. A jet of liquid with speed U strikes a bucket similar to that shown in Fig. 9–3. The bucket moves with constant u in the x-direction. Assuming negligible frictional effect, show that $R_y = \rho Q U \sin \theta \, [1 - (u/U)^2]$, where $Q = U A_1$. [*Hint:* Move the coordinate system with the bucket. Then $V_1 = V_2 = U - u$ in Fig. 9–3.]

9–6. A nozzle is connected to a hose with four bolts, as shown in Fig. 9–8. The water pressure at the joint is observed to be 144 psi. Assuming negligible frictional effect in the nozzle, find the tensile force in each bolt.

9–7. To determine experimentally the drag of a circular cylinder in an incompressible flow, the velocity and pressure are measured at the control surface S in the two-dimensional flow shown in Fig. 9–9. The pressure is observed to be constant over S. With the observed velocities shown, compute the drag coefficient of the cylinder as defined in Eq. 2–13. The frictional force on S can be assumed to be negligible.

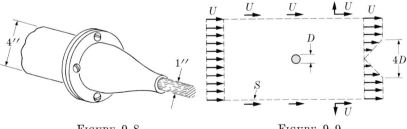

FIGURE 9–8   FIGURE 9–9

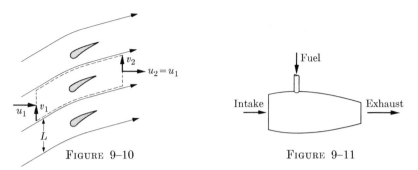

FIGURE 9–10   FIGURE 9–11

9–8. In Fig. 9–10 is shown the two-dimensional flow of an incompressible fluid being deflected by a row of turbine blades spaced a distance L apart. The velocity components u_1, v_1, u_2, and v_2 relative to the blades have been observed. Assuming that the flow is irrotational and that therefore Bernoulli's equation is applicable throughout the fluid, find the force acting on each blade.

9–9. A jet engine burns 5 lb of fuel per second. The fuel enters the engine vertically as shown in Fig. 9–11. At the intake, the speed of air relative to the engine is 300 ft/sec. The density is 0.0023 slug/ft³, and the intake area is 4 ft². At the exhaust, the relative speed is 1800 ft/sec, and the exhaust area is 2 ft². (a) Find the density of the exhaust gas. (b) Find the thrust developed by the engine.

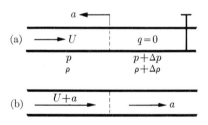

FIGURE 9–12

9–10. A fluid flowing at speed U in a rigid pipe is stopped by sudden closure of a valve as shown in Fig. 9–12(a). The pressure and the density of the fluid near the valve are suddenly increased and a pressure wave propagates up the pipe with speed a. (a) Reduce the flow to the steady flow shown in Fig. 9–12(b), and show that the increase of pressure Δp and the wave speed a are related by

$$\Delta p = \rho U(U + a), \qquad a(U + a) = \Delta p / \Delta \rho.$$

(b) For water, the bulk modulus $K(= \rho \cdot dp/d\rho)$ is 43×10^6 lb/ft². Compute the wave speed a in a rigid pipe and Δp due to the sudden stoppage of U of 1 ft/sec. (The pressure waves created by rapid change of flow in a water line are referred to as *water-hammers*.) (c) Repeat part (a) using instead the unsteady flow in Fig. 9–12(a).

9–2. Momentum theorem in one-dimensional analysis

In many cases of channel and pipe flow, one-dimensional analysis can furnish answers of interest (see Section 5–7). Although the velocity $\mathbf{q}$ is not absolutely uniform over a cross section, the flow is described in one-dimensional analysis with the mean speed V at the cross section which is defined as

$$V = \frac{Q}{A} = \frac{1}{A} \int_A v_l \, dA,$$

where v_l is the velocity component normal to the area dA of the cross section. In applying the momentum theorem in one-dimensional analysis, it is necessary to express the flow of momentum across a cross section $\int \rho v_l q \cos \gamma \, dA$ in the direction of the flow in terms of the mean speed V. If this cross section is the portion of the control surface S where the fluid is flowing out of the control volume, such as section 2 in Fig. 9–13, we have $q \cos \gamma = v_l$ and

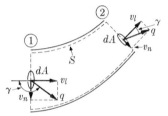

$$\int_A \rho v_l q \cos \gamma \, dA = \int_A \rho v_l^2 \, dA$$

$$= \beta \rho V^2 A = \beta \rho Q V, \quad (9\text{–}8)$$

where

$$\beta = \frac{1}{A V^2} \int_A v_l^2 \, dA. \quad (9\text{–}9)$$

Figure 9–13

is called the *momentum correction factor*. If the cross section is the portion of S where the fluid is flowing into the control volume, such as section 1 in Fig. 9–13, we have $q \cos \gamma = -v_l$, and

$$\int_A \rho v_l q \cos \gamma \, dA = -\beta \rho Q V.$$

Since the velocity vector $\mathbf{q}$ at a cross section are not all directed along the axis of the channel, there is also a flow of momentum across the section in a direction normal to the axis. For an n-direction normal to the axis of the channel (see Fig. 9–13), the flow of momentum across cross section A is

$$\int_A \rho v_n q \cos \gamma \, dA = \pm \int_A \rho v_n v_l \, dA. \quad (9\text{–}10)$$

This quantity may not be zero. For example, at section 1 in Fig. 9–14 where a fluid flows into a branch from a main, v_n over the whole cross section is of the same sign. In this case, the flow of momentum across section 1 in the n-direction may be a significant quantity. However, in most cases where one-dimensional analysis is suitable, v_n is zero or of different sign over different parts of the

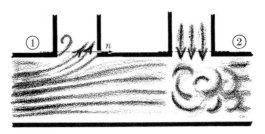

FIGURE 9–14

cross section, so that the component of the flow of momentum in the n-direction is negligible. It can be seen that this is true at section 2 in Fig. 9–14 where a fluid is discharged from a branch into the main. In applying the momentum theorem in one-dimensional analysis, it is necessary to choose cross sections where the quantity in Eq. 9–10 is negligible or can be estimated.

When the quantity in Eq. 9–10 is negligible, the flow of momentum across a cross section can be considered to be $\pm\beta\rho Q\mathbf{V}$, a vector directed along the axis of the channel. The x-component of this vector is $\pm\beta\rho Q V_x$, where V_x is the x-component of $\mathbf{V}$. Then for the system shown in Fig. 9–15, we have from Eqs. 9–5 to 9–7,

$$\sum F_x = \rho Q(\beta_2 V_{x2} - \beta_1 V_{x1}), \qquad (9\text{–}11)$$

$$\sum F_y = \rho Q(\beta_2 V_{y2} - \beta_1 V_{y1}), \qquad (9\text{–}12)$$

$$\sum F_z = \rho Q(\beta_2 V_{z2} - \beta_1 V_{z1}). \qquad (9\text{–}13)$$

Since the velocity distribution is unknown in one-dimensional analysis, the value of β is usually estimated. Its value is always less than the kinetic energy correction factor α for nonuniform velocity distribution. For laminar flow in a circular pipe, β can be shown to be $\frac{4}{3}$. For turbulent flow where the velocity

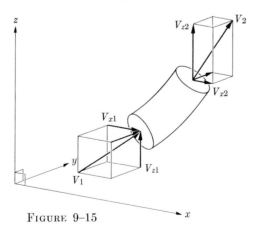

FIGURE 9–15

is more uniformly distributed, β may be less than 1.05. In such cases, it is often assumed to be unity for convenience. However, when the velocity distribution is highly nonuniform, β may be exceedingly large. In applying the momentum theorem in one-dimensional analysis, efforts should be made to select for the control surface S cross sections where the velocity distribution is expected to be reasonably uniform.

The momentum theorem can be used to great advantage in cases where violent disturbances of the flow occur within a short length, so that the unknown force due to the shearing stresses acting on the control surface S is negligible within this short length. In these cases, Bernoulli's equation cannot be used due to the increase of turbulence in this region (see Section 5–6). It is necessary to apply a correction E_L to Bernoulli's equation:

$$\frac{p_1}{\rho} + gh_1 + \frac{\alpha_1 V_1^2}{2} = \frac{p_2}{\rho} + gh_2 + \frac{\alpha_2 V_2^2}{2} + E_L, \qquad (9\text{--}14)$$

where E_L represents mainly the energy per unit mass converted into energy of turbulence which will be dissipated through internal friction into heat. The energy represented by E_L is no longer available to the main flow, and E_L is therefore called the average *loss of energy* per unit mass of the main flow between sections 1 and 2.

Example 9–5. At a sudden expansion of a pipe, the cross-sectional area changes from A_1 to A_2. It is observed that at a short distance downstream of the expansion, the velocity distribution is reasonably uniform over the cross section. Find the loss of main-flow energy per unit mass at the expansion when the discharge is Q.

We consider the control surface shown in Fig. 9–16. The forces in the x-direction acting on this surface are $p_1 A_2$, $p_2 A_2$, and the shearing forces at the wall. Since the length of this volume has been observed to be short, these shearing forces are negligible compared with $(p_1 - p_2)A_2$. Assuming that $\beta_1 = \beta_2 = 1$, we have from Eq. 9–11,

$$(p_1 - p_2)A_2 = \rho Q(V_2 - V_1).$$

Note that at section 1, the flow of momentum takes place over A_1, while mean pressure p_1 acts over A_2. By continuity,

$$A_1 V_1 = A_2 V_2 = Q.$$

Thus

$$p_1 - p_2 = \frac{\rho Q}{A_2} (V_2 - V_1)$$

$$= \rho V_2(V_2 - V_1).$$

FIGURE 9–16

Since $V_2 < V_1$, p_2 is higher than p_1. Thus, in slowing down, the fluid gains in pressure. There is an amount of energy E_L of the main flow being converted into energy of turbulence at the expansion. This quantity E_L can be found by substituting the last equation into Eq. 9–14. With $\alpha_1 = \alpha_2 = 1$, we obtain

$$E_L = \frac{p_1 - p_2}{\rho} + \frac{V_1^2 - V_2^2}{2} = V_2(V_2 - V_1) + \frac{V_1^2 - V_2^2}{2} = \frac{(V_1 - V_2)^2}{2}.$$

Example 9–6. Under suitable conditions, an open-channel flow may expand to flow at a greater depth as shown in Fig. 9–17. (See also Fig. 10–11b.) This phenomenon is called a *hydraulic jump*. It has been observed that within a short distance downstream of the jump, the velocity distribution over the cross section is reasonably uniform. Given V_1 and the depth Y_1 in a horizontal channel with a rectangular cross section, find the depth Y_2 after the jump.

We choose the control surface as shown. The forces in the x-direction acting on this volume are the forces due to pressure on sections 1 and 2, and the frictional forces at the bottom and the sidewalls. In the short distance, the resultant shearing force is negligible. Since the streamlines at sections 1 and 2 are practically parallel, the pressure distribution at these sections may be considered to be hydrostatic (see Eq. 5–10). Thus with $\beta_1 = \beta_2 = 1$, we have from Eq. 9–11,

$$\tfrac{1}{2}(\rho g Y_1)(Y_1 B) - \tfrac{1}{2}(\rho g Y_2)(Y_2 B) = \rho Q(V_2 - V_1),$$

where B is the width of the channel, and $\rho g Y/2$ is the mean pressure at a cross section. By continuity,

$$Q = BY_1 V_1 \qquad \text{and} \qquad V_2 = \frac{V_1 Y_1}{Y_2}.$$

Substituting these into the momentum equation, and simplifying, we have

$$\tfrac{1}{2}g(Y_1^2 - Y_2^2)Y_2 = V_1^2 Y_1(Y_1 - Y_2).$$

Obviously, one solution of this cubic equation of Y_2 is $Y_2 = Y_1$, i.e., a uni-

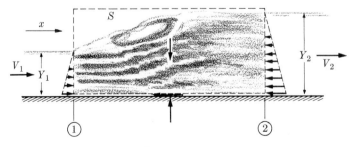

FIGURE 9–17

form flow without a jump. For solutions with $Y_2 \neq Y_1$, this equation can be reduced to a quadratic equation of Y_2

$$\tfrac{1}{2}g(Y_1 + Y_2)Y_2 = V_1^2 Y_1,$$

which yields

$$Y_2 = \frac{Y_1}{2}(\sqrt{1 + 8F_1^2} - 1),$$

where $F_1^2 = V_1^2/(gY_1)$. The corresponding values of Y_1 and Y_2 are called the *conjugate depths* of the jump.

The hydraulic jump is similar to the pipe expansion in Fig. 9–16 in many aspects. In going through the jump, or expansion, some energy of the main flow is converted into energy of turbulence which is subsequently dissipated into heat. To compute this loss of main-flow energy per unit mass, Eq. 9–14 may be written for this case as

$$gY_1 + \alpha_1 \frac{V_1^2}{2} = gY_2 + \alpha_2 \frac{V_2^2}{2} + E_L.$$

With $\alpha_1 \doteq \alpha_2 \doteq 1$, and $V_2 = V_1 Y_1/Y_2 = 2V_1/(\sqrt{1 + 8F_1^2} - 1)$,

$$\frac{E_L}{gY_1} = 1 - \tfrac{1}{2}(\sqrt{1 + 8F_1^2} - 1) + \tfrac{1}{2}F_1^2\left[1 - \frac{4}{(\sqrt{1 + 8F_1^2} - 1)^2}\right].$$

When $F_1 = 1$, we have $Y_2 = Y_1$ and $E_L = 0$. If $F_1 < 1$, we would have $Y_2 < Y_1$ and $E_L < 0$; i.e., there would be an increase of energy of the main flow, a situation considered physically impossible. Thus a jump can occur only when $F_1 > 1$ with $Y_2 > Y_1$ and $E_L > 0$. It can be easily verified that $F_2 (= V_2/\sqrt{gY_2})$ is always less than unity.

PROBLEMS

9–11. Find the momentum correction factor for laminar flow and turbulent flow in circular pipes as described in Problem 5–14.

9–12. In a sudden contraction of pipe, the flow forms a *vena contracta* as shown at section c in Fig. 9–18. The loss of main-flow energy occurs mainly in the expansion from the *vena contracta* to A_2. For the case with $A_1 = 2A_2$, it is observed that at the *vena contracta*, $A_c = 0.68A_2$. Find the loss of energy per unit mass in terms of V_2.

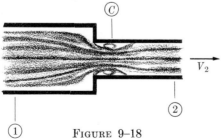

FIGURE 9–18

9-13. If Y_2 and V_2 are given for the horizontal channel in Fig. 9-17, what must be the depth Y_1 for a jump to occur?

FIGURE 9-19

9-14. Water is flowing in a horizontal pipe 12 in. in diameter. The mean speed at section 1 is 2 ft/sec. Within a short distance, two small pipes add 0.1 cfs each to the main flow as shown in Fig. 9-19. Find the pressure difference $p_1 - p_2$.

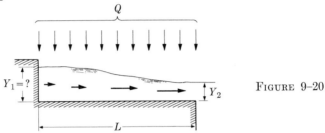

FIGURE 9-20

9-15. A discharge Q of a thin liquid drops vertically into a short horizontal rectangular channel of width B, as shown in Fig. 9-20. The depth at the outlet is Y_2. Find the depth Y_1 at the upstream end.

9-3. Vector product and curl of vectors

In the following discussion, the vector product of two vectors is involved. The *vector product* of vectors **A** and **B** is defined as a vector with a magnitude $AB \sin \theta$ (where θ is the smaller angle between **A** and **B**) and having the direction in which a right-handed screw would advance when turned from **A** toward **B** through the angle θ, as shown in Fig. 9-21. Let this direction be represented by a unit vector **e**. The vector product can then be written as

$$\mathbf{A} \times \mathbf{B} = AB \sin \theta \mathbf{e}. \qquad (9\text{-}15)$$

According to this definition, the following expressions are self-evident:

$$\mathbf{A} \times \mathbf{A} = 0, \qquad (9\text{-}16)$$

$$\mathbf{B} \times \mathbf{A} = -(\mathbf{A} \times \mathbf{B}), \qquad (9\text{-}17)$$

$$(k\mathbf{A}) \times \mathbf{B} = k(\mathbf{A} \times \mathbf{B}). \qquad (9\text{-}18)$$

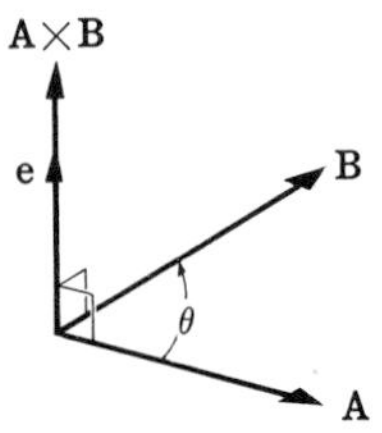

It can also be shown that the vector product is distributive:

$$\mathbf{A} \times (\mathbf{B} + \mathbf{C}) = (\mathbf{A} \times \mathbf{B}) + (\mathbf{A} \times \mathbf{C}). \qquad (9\text{-}19)$$

FIGURE 9-21

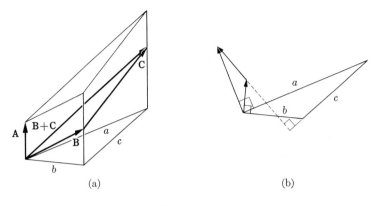

(a)　　　　　　　　　　　　　　　　　　　　(b)

FIGURE 9–22

Take any three vectors, **A**, **B**, and **C**, as shown in Fig. 9–22(a). The vector products **A** × (**B** + **C**), **A** × **B**, and **A** × **C** are all lying in the plane perpendicular to **A**. Let the lengths a, b, and c be the components of the vectors (**B** + **C**), **B**, and **C**, respectively, on this plane. The vector product **A** × (**B** + **C**) is perpendicular to the parallelogram with sides **A** and (**B** + **C**), and according to Eq. 9–15, has a magnitude equal to the area of the parallelogram. Thus, **A** × (**B** + **C**) can be shown in Fig. 9–22(b) with a magnitude Aa. Similarly, **A** × **B** is perpendicular to the line of length b with a magnitude Ab, and **A** × **C** is perpendicular to the line of length c with a magnitude Ac. Therefore the three vector products form a triangle similar to the one with sides a, b, and c. Thus Eq. 9–19 holds.

The vector product can be computed from the components of the vectors as follows:

$$\mathbf{A} \times \mathbf{B} = (\mathbf{i}A_x + \mathbf{j}A_y + \mathbf{k}A_z) \times (\mathbf{i}B_x + \mathbf{j}B_y + \mathbf{k}B_z)$$
$$= \mathbf{i}(A_yB_z - A_zB_y) + \mathbf{j}(A_zB_x - A_xB_z) + \mathbf{k}(A_xB_y - A_yB_x). \quad (9\text{–}19a)$$

Here we have made use of Eq. 9–19 and the following relations:

$$\mathbf{i} \times \mathbf{i} = 0, \quad \mathbf{i} \times \mathbf{j} = \mathbf{k}, \quad \mathbf{j} \times \mathbf{i} = -\mathbf{k},$$

etc., for a right-hand coordinate system, as shown in Fig. 4–7. Equation 9–19a can also be written in the following convenient form:

$$\mathbf{A} \times \mathbf{B} = \begin{vmatrix} \mathbf{i} & \mathbf{j} & \mathbf{k} \\ A_x & A_y & A_z \\ B_x & B_y & B_z \end{vmatrix}. \quad (9\text{–}19b)$$

The vector product of two vectors has been so defined as to be useful in the analysis of vector quantities. In particular, we are interested in the moment of a vector about a point. For example, the *moment of a force* **F** about a point a is defined as

$$\mathbf{T} = \mathbf{R} \times \mathbf{F}, \tag{9-20}$$

where **R** is the vector directed from *moment center* a to the point of application of **F**, as shown in Fig. 9–23. The magnitude of **T** is therefore equal to $RF \sin \theta$. The distance $R \sin \theta$ is the distance from the moment center to the line of action of **F**, and is called the *moment arm* of **F** about a. The magnitude T can be seen to be equal to the product of F and the moment arm. Similarly, the *moment of the momentum* $m\mathbf{q}$ of a particle at **R** from a point a is defined as $\mathbf{R} \times (m\mathbf{q})$.

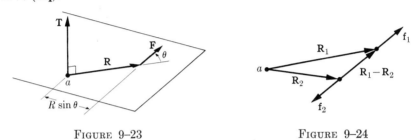

FIGURE 9–23   FIGURE 9–24

From Eqs. 9–16 to 9–19, we can immediately draw many conclusions about moments of vectors. For example, we can show that the sum of the moments of two equal and opposite colinear vectors about a point is zero. If $\mathbf{f}_2 = -\mathbf{f}_1$, then the sum of their moments about a in Fig. 9–24 is

$$(\mathbf{R}_1 \times \mathbf{f}_1) + (\mathbf{R}_2 \times \mathbf{f}_2) = (\mathbf{R}_1 \times \mathbf{f}_1) - (\mathbf{R}_2 \times \mathbf{f}_1)$$
$$= (\mathbf{R}_1 - \mathbf{R}_2) \times \mathbf{f}_1 = k\mathbf{f}_1 \times \mathbf{f}_1 = 0, \tag{9-21}$$

where $\mathbf{R}_1 - \mathbf{R}_2$ is in the direction of $\mathbf{f}_1$ and therefore can be written as $k\mathbf{f}_1$. Several other useful conclusions are left to the student in Problem 9–16.

Corresponding to the definition of gradient in Eq. 4–31 and that of divergence in Eq. 4–26, we have the definition of the *curl* of a vector field **q**:

$$\text{curl } \mathbf{q} = \lim_{V \to 0} \frac{1}{V} \int_S \mathbf{n} \times \mathbf{q} \, dS,$$

where V is the volume enclosed by the surface S, as shown in Fig. 4–16. Using the differential volume in Fig. 4–13 with rectangular coordinates, and evaluating the surface integral, we can show that

$$\text{curl } \mathbf{q} = \mathbf{i} \left(\frac{\partial w}{\partial y} - \frac{\partial v}{\partial z} \right) + \mathbf{j} \left(\frac{\partial u}{\partial z} - \frac{\partial w}{\partial x} \right) + \mathbf{k} \left(\frac{\partial v}{\partial x} - \frac{\partial u}{\partial y} \right),$$

where u, v, and w are the x-, y-, and z-components of **q**, respectively. Thus when **q** is the velocity, curl **q** is the vorticity $\boldsymbol{\xi}$ of the flow (see Eqs. 6–4 to 6–6).

Since in rectangular coordinates

$$\nabla = \mathbf{i}\, \frac{\partial}{\partial x} + \mathbf{j}\, \frac{\partial}{\partial y} + \mathbf{k}\, \frac{\partial}{\partial z},$$

we have from Eq. 9–19a

$$\nabla \times \mathbf{q} = \mathbf{i}\left(\frac{\partial w}{\partial y} - \frac{\partial v}{\partial z}\right) + \mathbf{j}\left(\frac{\partial u}{\partial z} - \frac{\partial w}{\partial x}\right) + \mathbf{k}\left(\frac{\partial v}{\partial x} - \frac{\partial u}{\partial y}\right).$$

Thus in general

$$\boldsymbol{\xi} = \operatorname{curl} \mathbf{q} = \nabla \times \mathbf{q}. \tag{9–21a}$$

It can be verified with Eq. 4–25 that

$$\nabla \cdot (\nabla \times \mathbf{q}) = 0, \tag{9–21b}$$

which is Eq. 8–6.

PROBLEMS

9–16. Using Eqs. 9–16 to 9–19, show the following. (a) The moment of force $\mathbf{F}$ about a point on its line of action is zero. (b) The sum of moments of concurrent forces $\mathbf{F}_1$, $\mathbf{F}_2$, ... about a point is equal to the moment of their resultant $(\mathbf{F}_1 + \mathbf{F}_2 + \cdots)$ about the same point. (c) The sum of the moments of two equal and opposite but noncolinear forces about any moment center is the same. This sum is equal to $\mathbf{s} \times \mathbf{F}$ (see Fig. 9–25).

FIGURE 9–25

9–17. Given any vector field $\mathbf{q}$, show that $\nabla \cdot (\nabla \times \mathbf{q}) = 0$. (If $\mathbf{q}$ is velocity, this shows that div $\boldsymbol{\xi} = 0$.)

9–18. Given any scalar field ϕ, show that $\nabla \times (\nabla\phi) = 0$. (Thus for any velocity potential, $\boldsymbol{\xi} = \nabla \times \mathbf{q} = 0$; i.e., the flow is irrotational.)

9–4. Theorem of moment of momentum

In this section, we study the moments of forces and momenta to arrive at a theorem analogous to Euler's momentum theorem in Eq. 9–3. This theorem of moment of momentum is found particularly useful in the study of *turbomachinery*, such as turbines and pumps.

Consider a particle of mass m located instantaneously at $\mathbf{R}$ from a fixed moment center a, as shown in Fig. 9–26. This particle has a momentum $m\mathbf{q}$ and is acted upon by force $\mathbf{F}$. Take the moments of the vectors in Newton's equation of motion for this particle:

$$\mathbf{R} \times \mathbf{F} = \mathbf{R} \times \frac{d(m\mathbf{q})}{dt}. \tag{9–22}$$

During Δt, the particle will move through $\Delta\mathbf{R}$ with a change of momentum

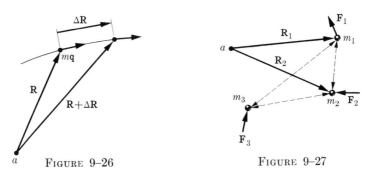

FIGURE 9–26   FIGURE 9–27

$\Delta m\mathbf{q}$. According to the definition of the derivative (see Eq. 4–4), we have the rate of change of $(R \times m\mathbf{q})$ of this particle:

$$\frac{d}{dt}(\mathbf{R} \times m\mathbf{q}) = \lim_{\Delta t \to 0} \frac{(\mathbf{R} + \Delta\mathbf{R}) \times (m\mathbf{q} + \Delta m\mathbf{q}) - (\mathbf{R} \times m\mathbf{q})}{\Delta t}$$

$$= \lim_{\Delta t \to 0} \frac{(\mathbf{R} \times \Delta m\mathbf{q}) + (\Delta\mathbf{R} \times m\mathbf{q})}{\Delta t} = \mathbf{R} \times \frac{d(m\mathbf{q})}{dt} + \frac{d\mathbf{R}}{dt} \times m\mathbf{q}.$$

Since $\Delta\mathbf{R}$ is the displacement of the particle during Δt, the derivative $d\mathbf{R}/dt$ is the velocity vector $\mathbf{q}$. Therefore

$$\frac{d\mathbf{R}}{dt} \times m\mathbf{q} = \mathbf{q} \times m\mathbf{q} = 0$$

and

$$\mathbf{R} \times \frac{d(m\mathbf{q})}{dt} = \frac{d}{dt}(\mathbf{R} \times m\mathbf{q}).$$

Equation 9–22 can therefore be written as

$$\mathbf{R} \times \mathbf{F} = \frac{d}{dt}(\mathbf{R} \times m\mathbf{q}). \tag{9–23}$$

Thus the rate of change of the moment of momentum $\mathbf{R} \times m\mathbf{q}$ of a particle is equal to the moment of the forces acting on it.

Now consider a system of particles of masses $m_1, m_2, \ldots$ (see Fig. 9–27). Let $\mathbf{F}_1$ be the forces acting on m_1 from sources outside the system, and $\mathbf{f}_1$ be the forces acting on m_1 by the other particles of the system. Let $\mathbf{R}_1$ be the location vector to m_1 from a fixed moment center. From Eq. 9–23, we have

$$\mathbf{R}_1 \times \mathbf{F}_1 + \mathbf{R}_1 \times \mathbf{f}_1 = \frac{d}{dt}(\mathbf{R}_1 \times m_1\mathbf{q}_1).$$

Similarly, for the other particles,

$$\mathbf{R}_2 \times \mathbf{F}_2 + \mathbf{R}_2 \times \mathbf{f}_2 = \frac{d}{dt}(\mathbf{R}_2 \times m_2\mathbf{q}_2).$$

$$\vdots$$

Adding these equations, we obtain

$$\mathbf{R}_1 \times \mathbf{F}_1 + \mathbf{R}_2 \times \mathbf{F}_2 + \cdots + \mathbf{R}_1 \times \mathbf{f}_1 + \mathbf{R}_2 \times \mathbf{f}_2 + \cdots$$

$$= \frac{d}{dt} (\mathbf{R}_1 \times m_1\mathbf{q}_1) + \frac{d}{dt} (\mathbf{R}_2 \times m_2\mathbf{q}_2) + \cdots.$$

Since the internal forces $\mathbf{f}_1$, $\mathbf{f}_2$, etc., consist of pairs of equal and opposite co-linear forces, the sum of their moments is zero (see Eq. 9–21). Thus

$$\sum(\mathbf{R} \times \mathbf{F}) = \frac{d\mathbf{H}}{dt}, \tag{9–24}$$

where $\sum(\mathbf{R} \times \mathbf{F})$ is the sum of the moments of the external forces acting on the system, and

$$\mathbf{H} = \sum(\mathbf{R} \times m\mathbf{q}) \tag{9–25}$$

is the *total moment of momentum* of the system of particles about the same moment center.

Equation 9–24 is analogous to Eq. 9–1, which is $\sum\mathbf{F} = d\mathbf{M}/dt$, where $\mathbf{M} = \sum m\mathbf{q}$. For a finite volume V of fluid, consider each infinitesimal volume dV with mass $\rho\,dV$ located at $\mathbf{R}$ from the moment center. Its moment of momentum is $\mathbf{R} \times (\rho\,dV\mathbf{q})$. The total moment of momentum in V is therefore

$$\mathbf{H} = \int_V \mathbf{R} \times (\rho\,dV\mathbf{q}) = \int_V \rho(\mathbf{R} \times \mathbf{q})\,dV,$$

while the total momentum in V is

$$\mathbf{M} = \int_V \rho\mathbf{q}\,dV.$$

Consider the fluid in volume V enclosed by surface S in Fig. 9–1. From a derivation completely analogous to that of Eq. 9–3, we have for Eq. 9–24

$$\sum(\mathbf{R} \times \mathbf{F}) = \frac{d\mathbf{H}}{dt} = \frac{d}{dt}\left[\int_V \rho(\mathbf{R} \times \mathbf{q})\,dV\right] + \int_S \rho(\mathbf{R} \times \mathbf{q})(\mathbf{n} \cdot \mathbf{q})\,dS. \tag{9–26}$$

This equation is analogous to Eq. 9–3. The time rate of change of the moment of momentum of the fluid momentarily occupying the space of volume V is equal to the sum of two items: the local rate of change of the moment of momentum in V, and the rate of flow of moment of momentum out of the enclosing surface S. This is the *theorem of moment of momentum*. For flow systems where the moment of momentum in V does not change with time,

$$\sum(\mathbf{R} \times \mathbf{F}) = \int_S \rho(\mathbf{R} \times \mathbf{q})(\mathbf{n} \cdot \mathbf{q})\,dS. \tag{9–27}$$

In the study of turbomachinery, we are interested particularly in the moment about the axis of rotation of the runner of a turbine or the impeller of a centrifugal pump. Let this axis be the z-axis of a cylindrical coordinate system. For a particle at (r, θ, z) as shown in Fig. 9–28, we have, about any point a on the z-axis,

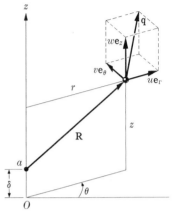

$$\mathbf{R} \times \mathbf{q} = [r\mathbf{e}_r + (z - \delta)\mathbf{e}_z] \times (u\mathbf{e}_r + v\mathbf{e}_\varrho + w\mathbf{e}_z).$$

Since the vector product is distributive (see Eq. 9–19), and according to Eq. 9–15,

$$\mathbf{e}_r \times \mathbf{e}_r = \mathbf{e}_z \times \mathbf{e}_z = 0,$$

$$\mathbf{e}_r \times \mathbf{e}_\theta = \mathbf{e}_z,$$

$$\mathbf{e}_z \times \mathbf{e}_r = -\mathbf{e}_r \times \mathbf{e}_z = \mathbf{e}_\theta,$$

$$\mathbf{e}_z \times \mathbf{e}_\theta = -\mathbf{e}_r,$$

we have

$$\mathbf{R} \times \mathbf{q} = -v(z - \delta)\mathbf{e}_r$$
$$+ [u(z - \delta) - wr]\mathbf{e}_\theta + vr\,\mathbf{e}_z.$$

FIGURE 9–28

Thus the z-component of $\mathbf{R} \times \mathbf{q}$ is vr, which is the moment of the circumferential component v about the axis. Let $\mathbf{T}$ be the total moment $\sum(\mathbf{R} \times \mathbf{F})$ of the forces acting on the control volume V. With $\mathbf{n} \cdot \mathbf{q} = q \cos \gamma$, where γ is the angle between $\mathbf{n}$ and $\mathbf{q}$, the z-component of Eq. 9–27 can be written for steady flows as

$$T_z = \int_S \rho v r q \cos \gamma \, dS. \tag{9–28}$$

Example 9–7. A discharge Q is delivered into the runner of a radial-flow *reaction turbine*, as shown in Fig. 9–29. The fluid is guided by vanes to enter at angle γ_1 with the radial. The design of the runner blades is such that the fluid in this case leaves the runner with speed q_2 at angle γ_2 (see note below). The thickness of the runner is L_1 at the entrance and L_2 at the exit. Find the torque developed by the turbine under this condition.

Take the volume in S between the entrance and the exit of the runner as the control volume. The exact analysis for the velocity distribution at the surface S is difficult. We assume that q_1 is uniform over the entrance area $2\pi r_1 L_1$ and that q_2 is uniform over the exit area $2\pi r_2 L_2$. Since the moment of momentum inside S is constant, we have from Eq. 9–28 the component T_z about the axis of rotation of the torque acting on this volume of fluid:

$$T_z = \int_S \rho v r q \cos \gamma \, dS$$

$$= \rho(q_1 \sin \gamma_1)r_1 q_1 \cos \gamma_1 2\pi r_1 L_1 + \rho(q_2 \sin \gamma_2)r_2 q_2 \cos \gamma_2 2\pi r_2 L_2.$$

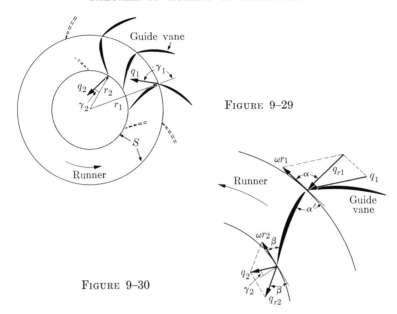

Figure 9–29

Figure 9–30

But the discharge Q is

$$Q = 2\pi r_1 L_1 q_1 \cos (\pi - \gamma_1) = -2\pi r_1 L_1 q_1 \cos \gamma_1$$

and

$$Q = 2\pi r_2 L_2 q_2 \cos \gamma_2.$$

Therefore

$$T_z = -\rho Q r_1 q_1 \sin \gamma_1 + \rho Q r_2 q_2 \sin \gamma_2.$$

This is the torque component T_z acting on the fluid. As the torque due to the shearing forces on the surface S is small, this torque T_z is mainly exerted by the runner on the fluid. Since the torque component T'_z acting on the runner is the reaction of T_z, we have

$$T'_z = -T_z = \rho Q (r_1 q_1 \sin \gamma_1 - r_2 q_2 \sin \gamma_2).$$

[*Note:* Given the angle β of the runner blades (see Fig. 9–30), the runner speed ω, and the discharge Q, the speed q_2 and the angle γ_2 can be determined as follows. Let $\mathbf{q}_{r2}$ be the velocity relative to the blade. According to the principle of relative velocity, the absolute velocity $\mathbf{q}_2$ is the vector sum of the absolute velocity of the blade and $\mathbf{q}_{r2}$. From Fig. 9–30 we have

$$q_2 \cos \gamma_2 = q_{r2} \sin \beta = \frac{Q}{2\pi r_2 L_2}$$

and

$$q_2 \sin \gamma_2 = \omega r_2 - q_{r2} \cos \beta.$$

From these three equations, we can find q_{r2}, q_2, and γ_2.]

PROBLEMS

9-19. In Fig. 9-31 is shown the impeller of a *centrifugal pump* rotating at angular speed ω. The fluid enters from the eye radially and leaves with velocity $\mathbf{q}_2$. Given the angle α of the vanes, and the thickness L_2 of the impeller, show that the torque required to run the pump is

$$T_z = \rho Q r_2 \left(\omega r_2 - \frac{Q \cot \alpha}{2\pi r_2 L_2} \right),$$

where Q is the discharge.

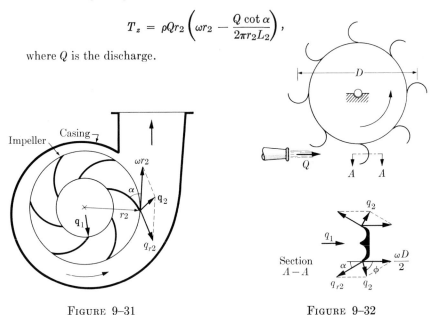

FIGURE 9-31   FIGURE 9-32

9-20. An *impulse turbine* of diameter D is kept at angular speed ω by a jet of liquid of discharge Q and speed q_1, as shown in Fig. 9-32. For an approximate analysis, assume that the speed relative to the bucket remains the same (i.e., $q_{r2} = q_{r1} = q_1 - \frac{1}{2}\omega D$). (a) Show that the torque T exerted on the wheel is $\frac{1}{2}\rho Q D (q_1 - \frac{1}{2}\omega D)(1 + \cos\alpha)$. (b) The power (energy per unit time) in turning the wheel being $T\omega$, show that the maximum power is obtained from the jet when $\omega D = \dot{q}_1$.

10 Flow With Gravity

10-1. Fluid of constant density with prescribed boundaries

In this chapter, the influence of gravity on fluid motion is studied. For simplicity, the effects of internal friction, surface tension, and compressibility are excluded in the discussion. First, consider cases of constant density with prescribed geometrical boundaries. Here the net effect of gravity on the motion can be seen to be nil from the equation of continuity,

$$\frac{\partial u}{\partial x} + \frac{\partial v}{\partial y} + \frac{\partial w}{\partial z} = 0, \tag{10-1}$$

and the momentum equations, such as,

$$\frac{\partial u}{\partial t} + u \frac{\partial u}{\partial x} + v \frac{\partial u}{\partial y} + w \frac{\partial u}{\partial z} = -\frac{1}{\rho} \frac{\partial p}{\partial x} - g \frac{\partial h}{\partial x}. \tag{10-2}$$

For fluids with constant ρ, Eq. 10-2 can be written as

$$\frac{\partial u}{\partial t} + u \frac{\partial u}{\partial x} + v \frac{\partial u}{\partial y} + w \frac{\partial u}{\partial z} = -\frac{\partial}{\partial x} \left(\frac{p}{\rho} + gh \right). \tag{10-3}$$

Instead of p, the quantity $p/\rho + gh$ may be considered as one of the variables to be determined. The magnitude of g can influence only the variable p, and therefore has no effect on the other variables u, v, and w. For example, take the irrotational flow around a circular cylinder shown in Fig. 6–15. The velocity distribution obtained in Example 6–7 for this flow is independent of g.

However, if the shape of a boundary cannot be prescribed but must be determined by dynamical considerations, p and h cannot be grouped together as in Eq. 10–3. For example, for the wavy surface of the sea, the pressure p at the surface is specified to be atmospheric, whatever the elevation h of the surface may be. In such cases, p and h at the surface vary independently, and $p/\rho + gh$ cannot be used as one variable in fitting the boundary conditions. Equation 10–3 is also incorrect if ρ is not uniform (e.g., a system consisting of layers of different fluids). Equation 10–2 must then be used. As a consequence, the velocity components u, v, and w, as well as p, are influenced by the magnitude of the coefficient g in the equations. Such cases are to be discussed in the next section.

10–2. Normalization of equations

In the last section, it has been shown that except for fluids of constant density with prescribed boundaries, the solutions of u, v, w, and p are influenced by the magnitude of gravitational acceleration g. In addition to satisfying the governing equations, these solutions must also satisfy the boundary conditions of the flow under consideration. These conditions will involve linear measurements (e.g., L, the depth of flow in a channel, or the size of an immersed solid body) and speeds (e.g., U, the speed of an approaching stream). For example, take the irrotational flow passing the circular cylinder shown in Fig. 6–15. The solutions must be such that $u = -U$ at large distances from the cylinder, and the radial component of velocity is zero at $r = L$ (or R), the radius of the cylinder. As can be seen from Eq. 6–24, the velocity distribution depends on U and L. In general, the solutions will involve not only the independent variables, such as x, y, z, and t, and the coefficients such as ρ and g in the equations, but also the constants such as U and L in the boundary conditions.

It is to be shown that the number of parameters involved in the solutions can be reduced by using a linear dimension L and a speed U in the boundary conditions as the units of length and speed, respectively. (The unit of time may then be L/U.) Instead of the values of u, v, and w (in ft/sec or cm/sec, etc.) the ratios u/U, v/U, and w/U are considered to be the variables. In place of x, y, z, t, the new independent variables are x/L, y/L, z/L, and tU/L. In other words, the motion is to be described in terms of relative magnitudes of speed at relative positions; e.g., in the flow shown in Fig. 6–15, the speed at two diameters (instead of the distance in ft or cm) in front of the cylinder is described as $\frac{15}{16}$ that of the approaching stream (instead of a speed in ft/sec or cm/sec). To express Eqs. 10–1 and 10–2 in terms of these new variables, one may, for example, multiply the former by the constant L/U, and the latter by the constant L/U^2. For fluids of constant ρ, we obtain

$$\frac{\partial(u/U)}{\partial(x/L)} + \frac{\partial(v/U)}{\partial(y/L)} + \frac{\partial(w/U)}{\partial(z/L)} = 0, \tag{10–4}$$

$$\frac{\partial(u/U)}{\partial(tU/L)} + \left(\frac{u}{U}\right)\frac{\partial(u/U)}{\partial(x/L)} + \left(\frac{v}{U}\right)\frac{\partial(u/U)}{\partial(y/L)} + \left(\frac{w}{U}\right)\frac{\partial(u/U)}{\partial(z/L)}$$
$$= -\frac{\partial[(p-P)/\rho U^2]}{\partial(x/L)} - \frac{1}{(U^2/gL)}\cdot\frac{\partial(h/L)}{\partial(x/L)}, \tag{10–5}$$

where P is a representative pressure (e.g., the pressure in an approaching stream) which can be included here if desired. The momentum equations in the y- and z-directions can be treated in the same manner. With these variables, the boundary conditions are expressed in terms of numbers. For example, we have $u/U = -1$ in the approaching stream in Fig. 6–15. Thus for a given shape of the boundary, the solutions of u/U, v/U, w/U, and $(p-P)/\rho U^2$ will involve, other than the four independent variables x/L, y/L, z/L, and tU/L, only the coefficient U^2/gL in the equations. Note that the number of quantities involved in the solutions has been reduced. This process of simplifying the problem by reducing the boundary conditions to dimensionless numbers is called *normalization of the equations*. This technique will be used again in later chapters.

By normalization of the equations of motion, we have shown that for a fluid of constant density under the influence of gravity, a dimensionless variable, such as $p/\rho U^2$, depends on the geometrical shape of the boundary and the *Fronde number* $U/\sqrt{gL}$. This conclusion has also been arrived at independently in Chapter 2 by dimensional analysis. When there are two fluids with density difference $\Delta\rho$, the net gravitational force per unit mass on a particle of the heavier fluid (of density ρ) in the lighter fluid is $g\,\Delta\rho/\rho$. The effect of gravity is therefore indicated by the *modified Froude number* $U/\sqrt{gL\,\Delta\rho/\rho}$. However, if the upper fluid is a gas and the lower one is a liquid, the density of the gas can be ignored.

PROBLEMS

10–1. The differential equation of the deflection D of a simply supported beam of length L under uniform load w (force per unit length) is

$$\frac{d^2 D}{dx^2} = -\frac{w}{2EI}(L - x)x,$$

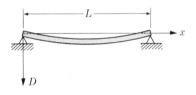

where E is Young's modulus and I is the second moment of the cross-sectional area. The boundary conditions are: $D = 0$ at $x = 0$; and $D = 0$ at $x = L$, as shown in Fig. 10–1. (a) Using L as the unit of length, normalize the equation. How many dimensionless parameters are involved in the solution of D/L? (b) Verify your answer by solving the differential equation.

FIGURE 10–1

10–2. The differential equation for the distribution of temperature through the wall of a circular tube is $(d/dr)[r(dT/dr)] = 0$. The boundary conditions are: $T = 0$ at $r = a$, the radius of the inner surface; and $T = T_1$ at $r = b$, the radius of the outer surface. Show that the quantity T/T_1 depends on two dimensionless quantities, and verify by solving the differential equation.

10–3. The Froude number

For a liquid of constant density with a free surface, the effects of gravity on the fluid motion have been shown to depend on the dimensionless Froude number

$$F = \frac{U}{\sqrt{gL}}.\qquad\qquad (10\text{--}6)$$

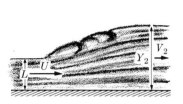

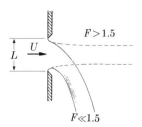

FIGURE 10–2   FIGURE 10–3

That the effects of gravity are indicated by the Froude number can be illustrated with many examples. For example, in the hydraulic jump shown in Fig. 10–2, let the approach speed be U and the upstream depth be L. It has been shown in Example 9–6 that the depth Y_2 downstream of the jump is

$$\frac{Y_2}{L} = \tfrac{1}{2}(\sqrt{1 + 8F^2} - 1). \tag{10–7}$$

From the equation of continuity, the speed V_2 is

$$\frac{V_2}{U} = \frac{L}{Y_2} = \frac{2}{\sqrt{1 + 8F^2} - 1}. \tag{10–8}$$

For another example, consider a jet of liquid with mean speed U from a circular orifice of diameter L as shown in Fig. 10–3. It has been observed that only when the value of the Froude number is larger than 1.5, is the jet essentially horizontal for a distance of several diameters. At lower values of the Froude number, this will not be the case (Fig. 10–3). It has also been observed that the coefficient of discharge in Eq. 5–12 depends on the value of the Froude number, and is equal to about 0.6 only when $U/\sqrt{gL}$ is larger than about 1.5.

The Froude number may be interpreted as a general indication of the ratio of the magnitudes of the *inertial "force"* $m\mathbf{a}$ (where $\mathbf{a}$ is acceleration) to the gravitational force $m\mathbf{g}$ acting on the fluid particles. Since the acceleration is expressed in terms of $u \cdot \partial u/\partial x$, etc., which are of the order of magnitude of U^2/L of the flow, this ratio is

$$\frac{ma}{mg} \sim \frac{mU^2/L}{mg} = \frac{U^2}{gL} = F^2.$$

Here the symbol $\sim$ relates quantities of the same order of magnitude. It should be emphasized that F^2 is not equal to the ratio of the two forces. In the first place, acceleration varies through the flow, and therefore the ratio cannot be represented by a single value. Secondly, any velocity and any linear dimension of the flow may be chosen as U and L in computing F. For example, for the flow in a channel, the maximum velocity or the mean velocity may be used for U. Thus for a given flow, the value of the Froude number depends on the quantities chosen as U and L. However, in flow systems with Froude numbers similarly defined, the influence of gravity is more pronounced in the system with a lower Froude number.

When the Froude number is very large, the gravitational force may become small relative to the inertial and pressure forces acting on the fluid particles. In Eq. 10–5, the term involving gL/U^2 or $1/F^2$ becomes small when F is large. In many cases, this term may become negligible compared with the other terms. Then gravity may be ignored without introducing serious errors. For example, when F is greater than 1.5 for the jet in Fig. 10.3, the shape of the jet near the orifice and the value of the coefficient of discharge remain the same irrespective of the value of F. Only when F is less than 1.5 do they vary with F and therefore are influenced by g. For another example, consider the deflected jet in Fig. 9–3 where the incoming speed and the size of the deflector may be taken as U and L in computing F. When F is large, the effect of gravity is small, and the flow is essentially the same whether the jet is deflected horizontally or vertically.

However, there are many cases where p is more closely related to L than to U (e.g., in the hydraulic jump in Fig. 10–2). The term involving $p/\rho U^2$ in Eq. 10–5 will vary with U^2/gL. This can be seen more clearly by writing Eq. 10–5 as

$$\frac{\partial(u/U)}{\partial(tU/L)} + \left(\frac{u}{U}\right)\frac{\partial(u/U)}{\partial(x/L)} + \left(\frac{v}{U}\right)\frac{\partial(u/U)}{\partial(y/L)} + \left(\frac{w}{U}\right)\frac{\partial(u/U)}{\partial(z/L)}$$
$$= -\frac{1}{(U^2/gL)}\left[\frac{\partial(p/\rho gL)}{\partial(x/L)} + \frac{\partial(h/L)}{\partial(x/L)}\right].$$

So far as the term $p/\rho gL$ is concerned, the gravitational force does not become negligible compared with the pressure force at any value of U^2/gL, and the influence of gravity remains important in all cases. For example, in Eq. 10–7 for the hydraulic jump, $Y_2/L \to \sqrt{2}\,F$ at very high values of the Froude number.

PROBLEM

10–3. In Fig. 9–3 let the jet lie in a vertical plane and let L be the difference in elevation between sections 1 and 2. Show that $R/\rho QU$ depends on a Froude number $F = U/\sqrt{gL}$, where R is the resultant force from the deflector. Also show that, when $F \to \infty$, R becomes independent of gravity. The weight of liquid in the deflector may be assumed to be proportional to $A_1 L\rho g$.

10–4. Speed of small waves in shallow channels

When the free surface of a liquid is disturbed (e.g., by the oscillations of the barrier in Fig. 10–4), the disturbances propagate along the surface with a certain speed called the *wave speed*. This is the speed at which the disturbances are transmitted, and is not the speed of the particles. In this section, the wave speed in shallow channels is to be studied.

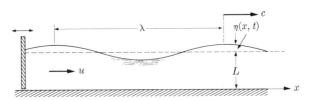

FIGURE 10–4

We consider an open channel of liquid originally at rest with depth L much less than the wavelength λ. (This will be true when the frequency of the disturbances is sufficiently small. See Problem 10–4d.) The amplitude of the wave is so small compared with the depth L that u of the fluid everywhere is much smaller than the wave speed c, and η is much less than L (see Fig. 10–4). For two-dimensional motion, the equations of motion are

$$\frac{\partial u}{\partial x} + \frac{\partial v}{\partial y} = 0, \tag{10–9}$$

$$a_x = \frac{\partial u}{\partial t} + u\frac{\partial u}{\partial x} + v\frac{\partial u}{\partial y} = -\frac{1}{\rho}\frac{\partial p}{\partial x}, \tag{10–10}$$

$$a_y = \frac{\partial v}{\partial t} + u\frac{\partial v}{\partial x} + v\frac{\partial v}{\partial y} = -\frac{1}{\rho}\frac{\partial p}{\partial y} - g. \tag{10–11}$$

Under the conditions $L \ll \lambda$, $\eta \ll L$, and $u \ll c$, many terms in these equations are very small and can be neglected in the analysis. For example, since $\eta \ll L \ll \lambda$, the curvature of the streamlines will be small. The pressure at a cross section is expected to be hydrostatically distributed, in accordance with Eq. 5–10. In other words, $a_y = 0$ in Eq. 10–11. We show this systematically by estimating the order of magnitude of the various terms in Eqs. 10–9 to 10–11, with a technique which will be found useful also in subsequent chapters.

Let $L \sim \delta\lambda$, where the symbol $\sim$ relates quantities of the same order of magnitude. Since $L \ll \lambda$, we have $\delta \ll 1$. Let Δu be the difference in u in a distance Δx of the order of λ. Since $\Delta u \sim u$, we have

$$\frac{\partial u}{\partial x} \sim \frac{u}{\lambda}.$$

Since $y \sim L \sim \delta\lambda$, we have

$$\frac{\partial v}{\partial y} \sim \frac{v}{L} \sim \frac{v}{\delta\lambda}.$$

But the two terms in Eq. 10–9 must be of the same magnitude:

$$\frac{u}{\lambda} \sim \frac{v}{\delta\lambda} \qquad \text{giving} \qquad v \sim \delta u \qquad (v \ll u).$$

We have then the order of magnitude of the terms in Eq. 10–10:

$$\frac{\partial u}{\partial t} \sim \frac{u}{\lambda/c} \sim \frac{uc}{\lambda}, \qquad u\frac{\partial u}{\partial x} \sim \frac{u^2}{\lambda}, \qquad v\frac{\partial u}{\partial y} \sim \delta u\frac{u}{\delta\lambda} \sim \frac{u^2}{\lambda}.$$

Here we have used λ/c as the order of magnitude of Δt during which Δu takes place at a point as a wave of speed c passes by. Similarly, for Eq. 10–11, we have

$$\frac{\partial v}{\partial t} \sim \frac{\delta u}{\lambda/c} \sim \frac{\delta uc}{\lambda}, \qquad u\frac{\partial v}{\partial x} \sim u\frac{\partial u}{\lambda} \sim \frac{\delta u^2}{\lambda}, \qquad v\frac{\partial v}{\partial y} \sim \delta u\frac{\delta u}{\delta\lambda} \sim \frac{\delta u^2}{\lambda}.$$

Since $\delta \ll 1$ and $u \ll c$, the term $\partial u/\partial t$ is much larger than the others. In other words,

$$a_x = \frac{\partial u}{\partial t} \qquad \text{and} \qquad a_y = 0.$$

Equation 10–11 thus becomes

$$0 = -\frac{1}{\rho}\frac{\partial p}{\partial y} - g.$$

With the boundary condition that $p = 0$ at $y = L + \eta(x, t)$ at the surface, this equation gives

$$p = \rho g(L + \eta - y), \tag{10-12}$$

i.e., the pressure at any point is proportional to the depth of liquid above, as in hydrostatics. Thus $\partial p/\partial x = pg \cdot \partial\eta/\partial x$ is not a function of y. In view of Eq. 10–10, we can conclude that a_x is also not a function of y. Thus, if the particles at a cross section have the same velocity at one instant, they will have similar motion and will continue to form a plane.

We shall consider only cases with $u(x, t)$, such as the motion generated by horizontal motion of the barrier in Fig. 10–4. With Eq. 10–12, Eq. 10–10 becomes

$$\frac{\partial u}{\partial t} = -g\frac{\partial\eta}{\partial x}. \tag{10-13}$$

Equation 10–9 can also be expressed in terms of u and η by integrating with respect to y at a constant x:

$$\frac{\partial u}{\partial x}\int_0^{L+\eta} dy + \int_0^{L+\eta}\frac{\partial v}{\partial y}\,dy = 0,$$

where

$$\int_0^{L+\eta}\frac{\partial v}{\partial y}\,dy = \int_{y=0}^{y=L+\eta} dv = v_{(y=L+\eta)} - v_{(y=0)} = \frac{D\eta}{Dt} - 0.$$

The velocity component v at the surface is the particle derivative $D\eta/Dt$, which is equal to

$$\frac{D\eta}{Dt} = \frac{\partial\eta}{\partial t} + u\frac{\partial\eta}{\partial x}.$$

It can be shown that as $u \ll c$, the last term is much smaller than $\partial \eta / \partial t$. Thus

$$\int_0^{L+\eta} \frac{\partial v}{\partial y} \, dy = \frac{\partial \eta}{\partial t} \, .$$

Also, as $\eta \ll L$,

$$\frac{\partial u}{\partial x} \int_0^{L+\eta} dy = L \frac{\partial u}{\partial x} \, .$$

Thus, the equation of continuity for small waves becomes

$$L \frac{\partial u}{\partial x} + \frac{\partial \eta}{\partial t} = 0. \tag{10–14}$$

From Eqs. 10–13 and 10–14, $u(x, t)$ and $\eta(x, t)$ are to be determined. Differentiating these equations with respect to x and t, respectively, and then eliminating u, we have

$$\frac{\partial^2 \eta}{\partial t^2} = gL \frac{\partial^2 \eta}{\partial x^2} \, . \tag{10–15}$$

This is the well-known linear *wave equation*. The general solution can be shown to be

$$\frac{\eta}{L} = F(x - ct) + f(x + ct), \tag{10–16}$$

where F and f are arbitrary functions and

$$c = \sqrt{gL}. \tag{10–17}$$

For example, let $f = 0$, and

$$\frac{\eta}{L} = k \sin \frac{2\pi(x - ct)}{\lambda} \, ,$$

where kL is the amplitude of the sinusoidal wave. At the same $x - ct$ we have the same η. This is therefore a wave propagating in the x-direction at a speed c (see Fig. 10–5). If $F = 0$, $\eta/L = f(x + ct)$ gives the same η at the same $x + ct$. This represents a wave propagating in the negative x-direction. In general, η in Eq. 10–16 is due to the combination of waves in both directions.

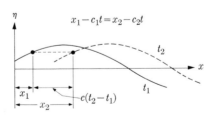

FIGURE 10–5

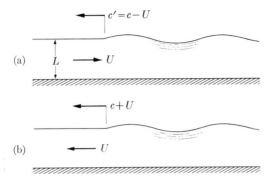

FIGURE 10–6

Similarly, eliminating η from Eqs. 10–13 and 10–14, we have

$$\frac{\partial^2 u}{\partial t^2} = gL \frac{\partial^2 u}{\partial x^2},$$

the general solution of which is

$$\frac{u}{c} = F(x - ct) - f(x + ct). \tag{10–18}$$

The same functions F and f appear in Eqs. 10–16 and 10–18 so as to satisfy Eqs. 10–13 and 10–14.

The value c in Eq. 10–17 is the speed of waves of small amplitude in a shallow channel relative to the liquid in the channel. If there is a current of speed U in the channel, the actual speed of propagation will be $c \pm U$, the sign depending on the direction of the current, as shown in Fig. 10–6.

The analysis of waves of finite amplitude is difficult. Since the wave speed increases with the depth, it may be speculated that different parts of a wave of finite amplitude would propagate at different speeds, with the speed of the crest higher than that of the valley. As a result, the front of the wave would steepen and finally break, as shown in Fig. 10–7. This has been found to be correct in most cases. However, this speculation is based on an analysis for

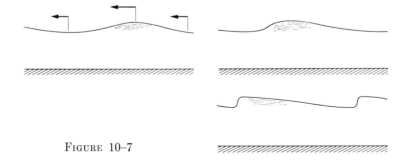

FIGURE 10–7

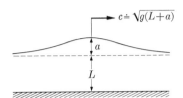

FIGURE 10-8

frictionless fluids with infinitesimal waves. When the effect of internal friction cannot be ignored, as in the case of flood waves in rivers, the waves do not usually break and the wave speed is lower than $\sqrt{gL}$. In an analysis with a closer approximation, it has been found that a certain type of finite waves, called *solitary waves* (see Fig. 10–8) can propagate without change of shape.

PROBLEMS

10-4. (a) Given $\eta = kL \sin [2\pi(x - ct)/\lambda]$, plot η/kL from $x = 0$ to $x = \lambda$ at increments of $\frac{1}{4}\pi\lambda$, for $t = 0$, $\frac{1}{4}\pi\lambda/c$ and $\frac{1}{2}\pi\lambda/c$. (b) Verify that $u = kc \sin [2\pi(x - ct)/\lambda]$ satisfies Eqs. 10–13 and 10–14, in accordance with Eq. 10–18. (c) Find the frequency f of the oscillations of the barrier shown in Fig. 10–4 necessary to produce this wave motion. (d) Show that in order to produce $\lambda \gg L$, f must be much smaller than $\sqrt{g/L}$.

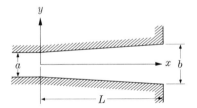

FIGURE 10-9

10-5. Simplify Eqs. 10–9 to 10–11 for the steady flow shown in Fig. 10–9 under the conditions $a \ll L$ and $b \ll L$.

10-5. Characteristics of channel flow

In the previous section, it has been shown that infinitesimal waves propagate in a shallow channel with a wave speed relative to the liquid equal to $\sqrt{gY}$, where Y is the depth of flow. If the speed V and the depth of flow Y at a cross section are used in defining the Froude number $F = V/\sqrt{gY}$ at the section, F can be seen to be the ratio of the channel flow speed to the wave speed of infinitesimal waves. The dependence of the flow on this ratio is immediately evident. If the speed V is smaller than $\sqrt{gY}$, that is, $F < 1$, small waves can propagate upstream with a speed $(\sqrt{gY} - V)$ as shown in Fig. 10–6(a). If V is larger than $\sqrt{gY}$, that is, $F > 1$, small waves cannot propagate upstream but will be pushed downstream. At the critical value $F = 1$, small waves tending to propagate upstream will be brought to a standstill. A flow with a

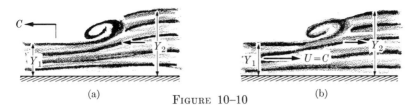

(a)   FIGURE 10–10   (b)

speed V greater than, equal to, or less than $\sqrt{gY}$ is said to be *supercritical, critical,* or *subcritical,* respectively.

A bore of finite amplitude propagates with a speed larger than $\sqrt{gY}$. In Fig. 10–10(a) is shown a bore propagating with a wave speed C into still water. With the coordinate axes moving with the wave, the flow appears steady, as shown in Fig. 10–10(b) as a stationary hydraulic jump. It has been shown in Example 9–6 that

$$Y_2 = \frac{Y_1}{2}\left(\sqrt{1 + \frac{8C^2}{gY_1}} - 1\right),$$

giving

$$C = \sqrt{gY_1} \cdot \sqrt{\frac{1}{2}\left(1 + \frac{Y_2}{Y_1}\right)\frac{Y_2}{Y_1}}. \tag{10–19}$$

For an infinitesimal wave with $Y_2 \to Y_1$, we have again the wave speed $c = \sqrt{gY_1}$. The speed C in Eq. 10–19 is larger than $\sqrt{gY_1}$. Thus, in order to bring a bore to a standstill (i.e., to have a stationary hydraulic jump), the opposing channel flow must have $V_1 = C$, that is, $V_1/\sqrt{gY_1} > 1$. It is important to note that after the jump, the Froude number $V_2/\sqrt{gY_2}$ is less than unity.

The behavior of a channel flow depends greatly on the value of the Froude number $V/\sqrt{gY}$. The following examples will demonstrate further the influence of this number. Consider the steady flow established in a sloping channel by the balance of gravity and internal friction as indicated by the dashed lines in Fig. 10–11. A barrier is then built across the channel. If $V/\sqrt{gY}$ in the

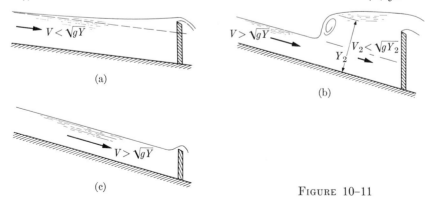

(a)

(b)

(c)

FIGURE 10–11

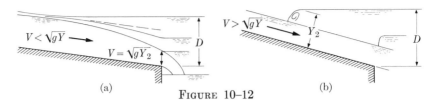

FIGURE 10-12

channel is less than unity, the rise of the free surface at the barrier as a disturbance will propagate upstream indefinitely. The backwater surface will have the shape shown in Fig. 10-11(a). If $V/\sqrt{gY}$ in the channel is originally greater than unity, the disturbance created by the barrier cannot propagate indefinitely upstream, but will be stopped in the form of a hydraulic jump at a section with Y_2 suitable for the jump (see Example 9-6), as shown in Fig. 10-11(b). If the height of the barrier is not sufficient to produce the necessary Y_2 for a hydraulic jump, the disturbance is not large enough to propagate up the stream. The flow will form a standing swell over the barrier, as shown in Fig. 10-11(c).

As another example of the importance of the Froude number $V/\sqrt{gY}$ on channel flows, take the influence of the tailwater on a channel flow. Let the depth of the tailwater D decrease from a value larger than the depth Y of the channel (see Fig. 10-12). If $V/\sqrt{gY}$ in the channel is everywhere less than unity, the change in D as a disturbance can propagate up the channel. As a result we have the two upper surface curves in Fig. 10-12(a) for the corresponding tailwater surfaces. As D decreases further, a value of D will be reached where the value of $V/\sqrt{gY}$ at the end section is raised to unity, as shown by the lowest curve in Fig. 10-12(a). Any further decrease of D, as a disturbance, can no longer propagate across the end section. Thus this depth is the minimum depth at the end section. If $V/\sqrt{gY}$ in the channel is larger than unity, a hydraulic jump will form in the channel when D is large enough to produce the necessary Y_2 for the jump, as shown by the upper curve in Fig. 10-12(b). Otherwise, any change of D cannot propagate into the channel, and the lower curve in Fig. 10-12(b) represents the free surface.

PROBLEMS

10-6. A bore with wave height of 5 ft is formed in a tidal estuary. The estuary is 10 ft in depth with a current of 1 ft/sec. Find the speed of the bore.

10-7. The depth of flow in a long channel is 10 ft. A barrier is built across the channel, raising the depth of flow to 11 ft at the barrier. Sketch the surface profile when the flow was originally (a) 1 ft/sec, and (b) 20 ft/sec.

10-8. A long rectangular flume with a width of 10 ft discharges into an empty tank. Find the depth at the end of the flume when (a) the discharge is 300 cfs and the depth upstream is 3 ft, and (b) the discharge is 200 cfs and the depth upstream is 2.5 ft.

10–9. A rectangular flume with a straight bottom *abc* discharges 100 cfs with a width of 6 ft and a depth of 5 ft as shown in Fig. 10–13. The lower part of the flume is then tilted to a very steep slope along *bc'* so that the depth at *c'* is 1 ft. Find the depth at *b*.

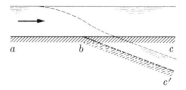

FIGURE 10–13

10–6. Surface profiles of steady channel flows

In the previous section, some characteristics of the surface profiles of steady channel flows have been discussed. In this section, we shall study further the shape of these profiles. In many cases of practical interest, e.g., the backwater in rivers, the influence of internal friction must be included in the analysis (see Eq. 13–50). In this section, however, internal friction is ignored. Such an analysis will indicate the principle involved, and the results are useful in cases where the influence of internal friction is minor, such as the flow in a short flume.

We consider only steady flows with the surface curvature so small that the pressure distribution at a cross section is practically hydrostatic (see Eq. 10–12). Then $p/\rho + gh$ is constant over a cross section, and is equal to $g(Y + h_0)$, where h_0 is the elevation of the bottom of the section (see Fig. 10–14). When frictional effects can be ignored, we have Bernoulli's equation:

$$\frac{p}{\rho} + gh + \frac{V^2}{2} = \text{constant}$$

or

$$\frac{d}{dx}\left(Y + h_0 + \frac{V^2}{2g}\right) = 0. \quad (10\text{–}20)$$

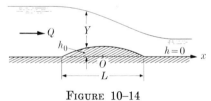

FIGURE 10–14

We have also the equation of continuity for a steady flow:

$$AV = \text{constant } Q, \quad\quad\quad\quad (10\text{–}21)$$

where the area A for a given channel is a function of x and the depth Y. The functions $Y(x)$ and $V(x)$ can be determined from these two equations for a given boundary condition, such as known V_1 and Y_1 of the upstream flow. Eliminating V from Eq. 10–20, we have the differential equation for $Y(x)$:

$$\frac{d}{dx}\left(Y + h_0 + \frac{Q^2}{2gA^2}\right) = 0. \quad\quad\quad (10\text{–}22)$$

Equation 10–22 can be reduced to a more convenient form for particular cases. First consider the case of rectangular cross sections $A = BY$, where the channel width B may vary with x. Then

$$\frac{d}{dx}\left(\frac{1}{A^2}\right) = -\frac{2}{A^3}\frac{dA}{dx} = -\frac{2}{B^3Y^3}\left(B\frac{dY}{dx} + Y\frac{dB}{dx}\right).$$

Equation 10–22 becomes

$$\frac{dY}{dx} = \left(\frac{Q^2}{gB^3Y^2}\cdot\frac{dB}{dx} - \frac{dh_0}{dx}\right)\Big/\left(1 - \frac{Q^2}{gB^2Y^3}\right). \qquad (10\text{–}23)$$

Next take a prismatic channel with the same cross section throughout the course. The area A then depends only on Y. With $dA = T\,dY$, where $T(Y)$ is the width of the surface as shown in Fig. 10–15,

$$\frac{d}{dx}\left(\frac{1}{A^2}\right) = -\frac{2}{A^3}\frac{dA}{dY}\frac{dY}{dx} = -\frac{2T}{A^3}\frac{dY}{dx}.$$

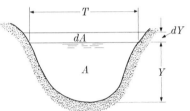

Equation 10–22 becomes

$$\frac{dY}{dx} = -\frac{dh_0/dx}{1 - (Q^2T/gA^3)}. \qquad (10\text{–}24)$$

FIGURE 10–15

Since the functions of x and Y on the right of Eq. 10–23 or 10–24 are known, the surface profile $Y(x)$ can be obtained. When these known functions are complicated, approximate methods of step-by-step integration are usually used.

Example 10–1. Investigate the various possible surface profiles in a symmetrical contraction in a horizontal channel with rectangular cross sections, as shown in Fig. 10–16(a). For $-\frac{1}{2}L < x < \frac{1}{2}L$, the width is $B = W[1 - \frac{1}{2}\cos(\pi x/L)]$, and $L = 8W$. The discharge is numerically equal to $\sqrt{gW^5/32}$.

For rectangular cross sections, Eq. 10–23 is applicable. Since the channel bottom is horizontal, $dh_0/dx = 0$ and Eq. 10–23 becomes

$$\frac{dY}{dx} = \frac{Q^2}{gB^3Y^2}\frac{dB}{dx}\Big/\left(1 - \frac{Q^2}{gB^2Y^3}\right). \qquad (10\text{–}25)$$

The characteristics of the surface profiles can be seen immediately from this equation. The slope dY/dx of the surface at a section depends on whether dB/dx is positive or negative, and whether Q^2/gB^2Y^3 is greater or less than unity. It is convenient to construct a line showing the depth Y at each section for $Q^2/gB^2Y^3 = 1$. This line is called the *critical flow line*, since on this line $V = Q/(BY) = \sqrt{gY}$. For this particular contraction, the equation of this line is

$$Y^3 = \frac{Q^2}{gB^2} = \frac{W^3}{32[1 - \frac{1}{2}\cos(\pi x/8W)]^2},$$

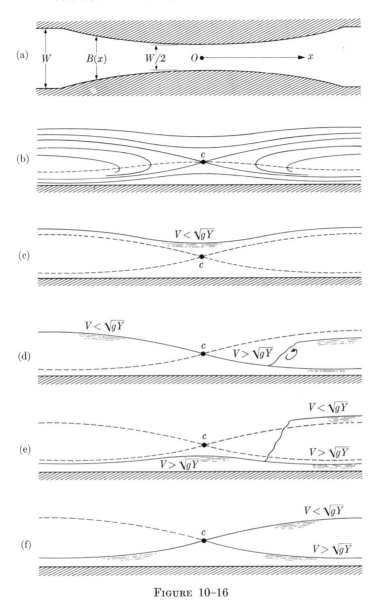

FIGURE 10–16

as shown by the dashed line in Fig. 10–16(b). For a depth above this line, Q^2/gB^2Y^3 is less than unity. Thus, for example, in the converging part of the channel where dB/dx is negative, a profile above the critical flow line must have a negative slope dY/dx, in accordance with Eq. 10–25. Furthermore, except for the point c on the critical flow line, all profiles must have $dY/dx = 0$

at the throat of the contraction where $dB/dx = 0$. All profiles must cross the critical flow line perpendicularly except at point c. At point c, dY/dx becomes indeterminate; i.e., $0/0$.

With this information, all possible profiles can be sketched as shown in Fig. 10–16(b). The coordinates of each of these profiles can be obtained if desired by integrating Eq. 10–25 step by step. For this particular contraction, this equation is reduced to

$$\Delta Y = \frac{\pi (W/Y)^2 \sin (\pi x/8W)}{512[1 - \frac{1}{2}\cos (\pi x/8W)]^3[1 - W^3/\{32 Y^3[1 - \frac{1}{2}\cos (\pi x/8W)]^2\}]} \Delta x.$$

For example, for the profile with $Y = \frac{3}{4}W$ at $x = 2W$, the next point at $\Delta x = W/10$ away is located with

$$\Delta Y = \frac{\pi (\frac{4}{3})^2 \times 0.707}{512(1 - \frac{1}{2} \times 0.707)^3 \left[1 - (\frac{4}{3})^3 \dfrac{1}{32(1 - \frac{1}{2} \times 0.707)^2}\right]} \cdot \frac{W}{10} = 0.00347W,$$

or $Y = 0.75347W$.

An infinite number of possible profiles can be realized in this case, depending on the upstream and the downstream conditions. When the profile is above the critical line (i.e. $V < \sqrt{gY}$ at every cross section), any change of the downstream condition will propagate upstream. The profile is therefore controlled by the downstream condition. A profile above the singular point c, as shown in Fig. 10–16(c), will result if the downstream flow is of sufficient depth. If the downstream depth is decreased, the profile will become lower, and will finally pass through point c, as shown in Fig. 10–16(d). Since $V = \sqrt{gY}$ at c, any further decrease of the downstream depth will not propagate upstream beyond this point. There may be a hydraulic jump joining two separate smooth profiles when suitable depths for a jump are available.

When the profile is below the critical line, disturbances downstream cannot propagate indefinitely upstream. The profile is therefore controlled by the upstream condition. In Fig. 10–16(e) and (f) are shown profiles under this condition. If the downstream flow is sufficiently deep, the whole channel will be drowned and a profile above the critical line will result, as shown in Fig. 10–16(c).

PROBLEM

10–10. In a horizontal rectangular channel with constant width W, there is a barrier with $h_0 = (W/4) \cos (\pi x/L)$, and $L = 4W$, as shown in Fig. 10–14. (a) Given that $Q = \sqrt{gW^5}/8$, find the depth Y at $x = 3L/8$ when the upstream depth is $5W/8$. (b) Sketch all possible smooth profiles over the barrier corresponding to Fig. 10–16(b). (c) Discuss the possible profiles when $V > \sqrt{gY}$ in the upstream channel. (d) Discuss the possible profiles when $V < \sqrt{gY}$ in the upstream channel.

10–7. Surface waves of small amplitude

In Section 10–4, we have discussed small waves in shallow channels. In this section, we shall discuss *surface waves* in channels with a depth not small compared with the wavelength. We limit our discussion to two-dimensional motion due to waves with a small amplitude (see Fig. 10–17). The density of the gas above is considered negligible, and the wavelength is long enough that the effect of surface tension becomes unimportant. To study the wave speed and velocity distribution, the damping effect of internal friction is ignored in the following analysis.

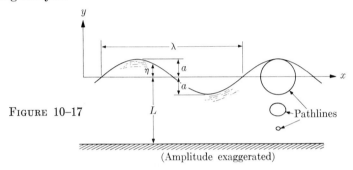

FIGURE 10–17

(Amplitude exaggerated)

The fluid motion being generated from rest, the flow may be considered to be irrotational with a velocity potential. It is convenient to solve for the velocity potential $\phi(x, y, t)$ from which the velocity distribution can then be found. For the incompressible fluid, ϕ must be such that

$$\frac{\partial^2 \phi}{\partial x^2} + \frac{\partial^2 \phi}{\partial y^2} = 0. \tag{10–26}$$

The pressure distribution can be found from the equation

$$-\frac{\partial \phi}{\partial t} + \frac{p}{\rho} + gh + \frac{q^2}{2} = f(t) \tag{10–27}$$

(see Eqs. 6–20 and 6–18). The following boundary conditions are to be satisfied. At the bottom,

$$v = -\frac{\partial \phi}{\partial y} = 0, \quad \text{at } y = -L. \tag{10–28}$$

At the surface where $p = 0$ and $h = y = \eta$, Eq. 10–27 gives

$$g\eta = \frac{\partial \phi}{\partial t} - \frac{q^2}{2}, \quad \text{at } y = \eta,$$

where $f(t)$ has been included in $\partial \phi / \partial t$. To simplify this boundary condition, consider sinusoidal waves with the amplitude much smaller than the wavelength. The speed q of the fluid elements is then expected to be small. It can be verified later that the higher-order term q^2 is much smaller than $g\eta$ (see Problem 10–11). Then

$$g\eta = \frac{\partial \phi}{\partial t}, \qquad \text{at} \quad y = \eta. \tag{10–29}$$

Next, express the variable η in terms of ϕ in this equation as follows. Since at the surface, v is the particle rate of change of η, that is, $D\eta/Dt$,

$$-\frac{\partial \phi}{\partial y} = v = \frac{D\eta}{Dt} = \frac{\partial \eta}{\partial t} + u\frac{\partial \eta}{\partial x} = \frac{\partial \eta}{\partial t},$$

where u and the surface slope $\partial \eta/\partial x$ are both very small and the higher-order term $u \cdot \partial \eta/\partial x$ is dropped. Equation 10–29 becomes

$$-\frac{\partial \phi}{\partial y} = \frac{1}{g}\frac{\partial^2 \phi}{\partial t^2}, \qquad \text{at} \quad y \doteq 0. \tag{10–30}$$

The problem is then to find a solution ϕ of Eq. 10–26 satisfying Eqs. 10–28 and 10–30. For the sinusoidal waves in Fig. 10–17, ϕ at any instant is repeated every wavelength. For a wave propagating with speed c in the $\pm x$-direction, we have the same condition at equal values of $x \mp ct$ (see Fig. 10–5). Therefore we try the solution

$$\phi(x, y, t) = F(y) \sin \frac{2\pi(x \mp ct)}{\lambda},$$

where $F(y)$ and c are to be determined. Substituting this ϕ into Eq. 10–26, we have

$$\frac{d^2 F}{dy^2} - \left(\frac{2\pi}{\lambda}\right)^2 F = 0.$$

The general solution of this equation is

$$F(y) = Ae^{2\pi y/\lambda} + Be^{-2\pi y/\lambda},$$

where the constants A and B are to be determined from the boundary conditions. According to Eq. 10–28, $\partial \phi/\partial y = 0$ and therefore $dF/dy = 0$ at $y = -L$. Thus

$$Ae^{-2\pi L/\lambda} = Be^{2\pi L/\lambda}.$$

Calling this value $C/2$, we have

$$F(y) = \frac{C}{2}\left(e^{2\pi(L+y)/\lambda} + e^{-2\pi(L+y)/\lambda}\right) = C \cosh \frac{2\pi(L+y)}{\lambda},$$

and*

$$\phi = C \cosh \frac{2\pi(L+y)}{\lambda} \sin \frac{2\pi(x \mp ct)}{\lambda}. \tag{10–31}$$

---

* $\cosh x = \frac{1}{2}(e^x + e^{-x})$, $\sinh x = \frac{1}{2}(e^x - e^{-x})$, $\tanh x = \sinh x/\cosh x$, $d(\cosh x)$ = $\sinh x\, dx$, and $d(\sinh x) = \cosh x\, dx$.

Substituting into Eq. 10–30, we have

$$-\sinh \frac{2\pi L}{\lambda} = -\frac{2\pi}{g\lambda} c^2 \cosh \frac{2\pi L}{\lambda},$$

$$c = +\sqrt{\frac{g\lambda}{2\pi} \tanh \frac{2\pi L}{\lambda}}. \tag{10–32}$$

[For shallow-water waves with $L/\lambda \ll 1$, we have $\tanh (2\pi L/\lambda) \doteq 2\pi L/\lambda$ and $c = \sqrt{gL}$, in agreement with Eq. 10–17.] The equation of the free surface is given by Eq. 10–29:

$$\eta = \mp C \sqrt{\frac{2\pi}{g\lambda}} \sinh \frac{2\pi L}{\lambda} \cosh \frac{2\pi L}{\lambda} \cos \frac{2\pi(x \mp ct)}{\lambda}.$$

Since η can be expressed in terms of the wave amplitude a,

$$\eta = a \cos \frac{2\pi(x \mp ct)}{\lambda}, \tag{10–33}$$

the constant C can be expressed in terms of the amplitude a. Therefore the velocity potential is, according to Eq. 10–31,

$$\phi = \mp \frac{a}{\sqrt{\frac{2\pi}{g\lambda}} \sinh \frac{2\pi L}{\lambda} \cosh \frac{2\pi L}{\lambda}} \cosh \frac{2\pi(L + y)}{\lambda} \sin \frac{2\pi(x \mp ct)}{\lambda}. \tag{10–34}$$

This is the solution of $\phi(x, y, t)$.

Several facts of interest may be pointed out. The velocity distribution can be obtained from Eq. 10–34:

$$u = -\frac{\partial \phi}{\partial x} = \pm \frac{a}{\sqrt{\frac{\lambda}{2\pi g}} \sinh \frac{2\pi L}{\lambda} \cosh \frac{2\pi L}{\lambda}} \cosh \frac{2\pi(L + y)}{\lambda} \cos \frac{2\pi(x \mp ct)}{\lambda},$$

$$v = -\frac{\partial \phi}{\partial y} = \pm \frac{a}{\sqrt{\frac{\lambda}{2\pi g}} \sinh \frac{2\pi L}{\lambda} \cosh \frac{2\pi L}{\lambda}} \sinh \frac{2\pi(L + y)}{\lambda} \sin \frac{2\pi(x \mp ct)}{\lambda}.$$

The motion at a point (x, y) can be seen to be oscillatory. For a particle oscillating near the point (x_0, y_0), its velocity can be approximated with $x = x_0$ and $y = y_0$ in these equations. Its displacement components d_x and d_y (which are functions of t) can then be found from

$$d_x = \int u(x_0, y_0, t)\, dt = -\frac{a}{\sinh (2\pi L/\lambda)} \cosh \frac{2\pi(L + y_0)}{\lambda} \sin \frac{2\pi(x_0 \mp ct)}{\lambda},$$

$$d_y = \int v(x_0, y_0, t)\, dt = \frac{a}{\sinh (2\pi L/\lambda)} \sinh \frac{2\pi(L + y_0)}{\lambda} \cos \frac{2\pi(x_0 \mp ct)}{\lambda}.$$

This shows that the path of the particle is an ellipse with the horizontal and vertical semi-axes equal to

$$\frac{a}{\sinh (2\pi L/\lambda)} \cosh \frac{2\pi(L + y_0)}{\lambda} \quad \text{and} \quad \frac{a}{\sinh (2\pi L/\lambda)} \sinh \frac{2\pi(L + y_0)}{\lambda} ,$$

respectively, as shown in Fig. 10–17. The period of the motion is $\lambda/(2\pi c)$.

It is interesting to find the ratio of the horizontal semi-axis for the particles at the surface ($y_0 = 0$) to that for the particles near the bottom ($y_0 \doteq -L$). This ratio is $\cosh (2\pi L/\lambda)/\cosh 0 = \cosh (2\pi L/\lambda)$. For shallow channels with $L \ll \lambda$, this ratio is $\cosh 0 = 1$. However, for deeper channels, the motion is mainly confined to the particles near the surface. For example, for $L = \lambda/2$, this ratio is $\cosh \pi = 12.3$. In fact, for $L \geq \lambda/2$, we have $\cosh (2\pi L/\lambda) \doteq \sinh (2\pi L/\lambda)$. The horizontal axes of the paths are then

$$\frac{a}{\sinh (2\pi L/\lambda)} \cosh \frac{2\pi(L + y_0)}{\lambda}$$

$$= \frac{a}{\sinh (2\pi L/\lambda)} \left(\cosh \frac{2\pi L}{\lambda} \cosh \frac{2\pi y_0}{\lambda} + \sinh \frac{2\pi L}{\lambda} \sinh \frac{2\pi y_0}{\lambda} \right)$$

$$\doteq a \left(\cosh \frac{2\pi y_0}{\lambda} + \sinh \frac{2\pi y_0}{\lambda} \right) = ae^{2\pi y_0/\lambda}.$$

Similarly, the vertical axis can also be shown to be $ae^{2\pi y_0/\lambda}$. Thus for $L \geq \lambda/2$, the paths are circles with the radius decreasing rapidly with depth. With $\tanh (2\pi L/\lambda) \doteq 1$, the wave speed in Eq. 10–32 becomes

$$c = \sqrt{g\lambda/2\pi}. \qquad (10\text{--}35)$$

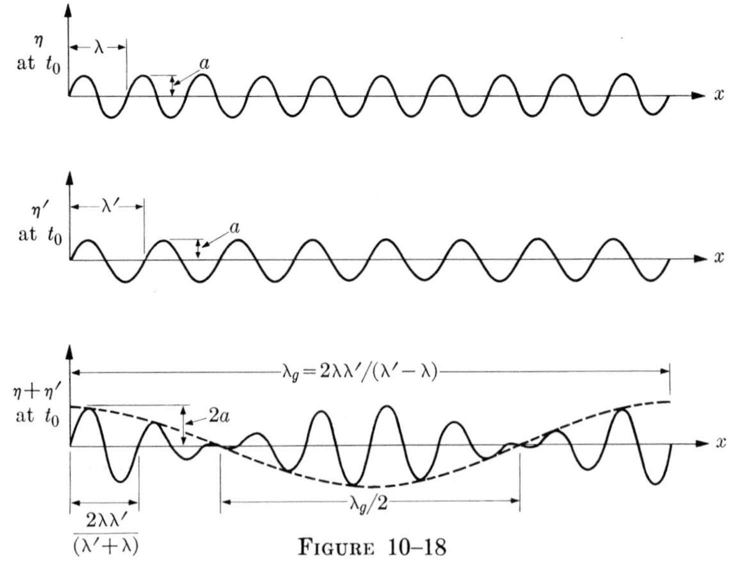

FIGURE 10–18

Another interesting fact about surface waves is that the wave speed varies with the wavelength. Consider two small waves of different wavelengths λ and λ'. Since the equations involved are linear (see Eqs. 10–26, 10–28, and 10–30), the resultant motion may be obtained by superposition, as indicated in Fig. 10–18, for a certain instant t_0. Since the two wave systems travel at different speeds, the shape of the resultant wave will change with time. Of particular interest is the case with the two wavelengths nearly the same. Consider two wave systems of equal amplitude:

$$\eta = a \sin \left[(x - ct)2\pi/\lambda \right] \quad \text{and} \quad \eta' = a \sin \left[(x - c't)2\pi/\lambda' \right].$$

The resultant wave is

$$\eta + \eta' = a \sin \left[\frac{2\pi}{\lambda} (x - ct) \right] + a \sin \left[\frac{2\pi}{\lambda'} (x - c't) \right]$$

$$= 2a \cos \left[\frac{\pi(\lambda' - \lambda)}{\lambda\lambda'} \left(x - \frac{c\lambda' - c'\lambda}{\lambda' - \lambda} t \right) \right]$$

$$\times \sin \left[\frac{\pi(\lambda' + \lambda)}{\lambda\lambda'} \left(x - \frac{c\lambda' + c'\lambda}{\lambda' + \lambda} t \right) \right]. \quad (10\text{–}36)$$

In Fig. 10–18 is shown the waveform at a certain instant t_0. Since $\lambda' - \lambda$ is small compared with λ, this waveform may be considered as one with a wavelength $2\lambda\lambda'/(\lambda' + \lambda)$, a wave speed $(c\lambda' + c'\lambda)/(\lambda' + \lambda)$, and a gradually varying amplitude

$$2a \cos \left[\frac{\pi(\lambda' - \lambda)}{\lambda\lambda'} \left(x - \frac{c\lambda' - c'\lambda}{\lambda' - \lambda} t \right) \right],$$

which varies from $2a$ to $-2a$ and back to $2a$ in a distance equal to $\lambda_g = 2\lambda\lambda'/(\lambda' - \lambda)$. The waves appear to form groups with a spacing of $\frac{1}{2}\lambda_g$, as shown in Fig. 10–18. These groups travel with a *group speed*

$$c_g = \frac{c\lambda' - c'\lambda}{\lambda' - \lambda}. \quad (10\text{–}37)$$

To illustrate the magnitude of c_g, consider the limiting case with $\lambda' = \lambda + d\lambda$, and $c' = c + dc$:

$$c_g = \frac{c(\lambda + d\lambda) - (c + dc)\lambda}{d\lambda} = c - \lambda \frac{dc}{d\lambda}.$$

With $c = \sqrt{g\lambda/2\pi}$ from Eq. 10–35 for deep-water waves ($L \geq \lambda/2$), this equation yields $c_g = c/2$. Thus the waves travel twice as fast as the groups. Each wave will pass from one group into another while its amplitude varies in the process.

PROBLEMS

10–11. In deriving Eq. 10–34 for waves with the amplitude a much less than the wavelength λ, the term $q^2/2$ is considered very small compared with $g\eta$ at the surface (see Eq. 10–29). Verify with Eq. 10–34 that $u^2 \ll g\eta$ and $v^2 \ll g\eta$ at the surface.

10–12. Find the stream function $\psi(x,\, y,\, t)$ for small surface waves propagating in the x-direction. Simplify your answer for the case of deep-water wave $(L \geq \lambda/2)$. Plot a few streamlines for the instant $t = 0$, such as $\psi = 0$, $\pm\frac{1}{2}a\sqrt{g\lambda/2\pi}$ and $\pm a\sqrt{g\lambda/2\pi}$. [*Hint:* $\sinh(A + B) = \sinh A \cosh B + \cosh A \sinh B$.]

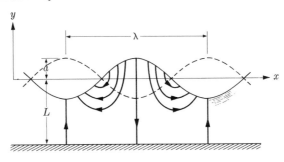

FIGURE 10–19

10–13. Under suitable conditions, *standing waves*, as shown in Fig. 10–19, can be generated. The free surface oscillates between the full line and the dashed line. Solve Eq. 10–26 for the velocity potential $\phi(x,\, y,\, t)$ of this motion in terms of a, λ, and L. [*Hint:* Since the motion is repeated every wavelength λ and every period τ of the oscillations, you may assume that $\phi = f(y) \cos(2\pi x/\lambda) \cos(2\pi t/\tau)$, where $f(y)$ and τ are to be determined.]

11 Flow of Viscous Fluids

11-1. Newtonian fluids

All real fluids exhibit internal friction. At solid boundaries, there is practically no relative velocity between the solid and the contacting fluid particles. As a result of these properties, mechanical energy is dissipated into heat, and there is skin friction on solid surfaces. Due to the presence of the slow-moving fluid particles near solid surfaces, there is a tendency for separation of the flow from the boundary in zones of deceleration (see Fig. 5–11). These phenomena occur no matter how small the actual internal friction may be. Since they cannot occur in a frictionless fluid, internal friction must be considered when dissipation, skin friction, separation, and other phenomena related to internal friction are being studied.

It has been found experimentally that most common fluids, including air and water, when tested as shown in Fig. 1–1, offer a shearing resistance proportional to the rate of deformation:

$$\tau = \frac{F}{A} = \mu \frac{U}{d},\tag{11-1}$$

where τ is the shearing stress, and μ is the *absolute*, or *dynamic*, *viscosity* of the fluid. At ordinary temperature and pressure, μ of a *Newtonian* fluid depends on temperature only. However, Eq. 11–1 has been written for the particular flow system shown in Fig. 1–1, usually referred to as plane *Couette flow*. The definition of viscosity must be generalized to be useful for flow systems in general. This will be discussed in Section 11–7.

It will be seen presently that it is the quantity $\nu = \mu/\rho$ that appears in the equations of motion. Since the dimensional formula of ν is L^2T^{-1}, it is called the *kinematic viscosity*. Values of μ and ν of air and water are listed in the Appendix. Note that while the viscosity of a gas generally increases with temperature, that of a liquid generally decreases with temperature.

11-2. Simple laminar flow systems of viscous fluids

For laminar flows where the velocity varies only in a direction perpendicular to the direction of flow, as shown in Fig. 11–1, with $u(y)$, and $v = w = 0$, Eq. 11–1 can be generalized immediately. Since U/d is the rate of change of

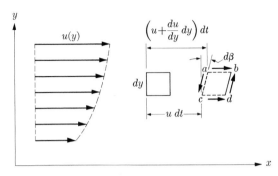

FIGURE 11-1

speed through the depth in Fig. 1-1, Eq. 11-1 can be written as

$$\tau = \mu \frac{du}{dy}.$$ (11-2)

This equation is assumed to be applicable to flow systems such as that shown in Fig. 11-1, where the shearing stress varies from layer to layer depending on the local value of du/dy. In Fig. 11-1 are also indicated the directions of the shearing stresses on a fluid element for a positive value of local du/dy. Although the normal stresses at a point in a viscous flow may vary with direction, they are independent of direction in these simple flow systems, as will be seen in Problem 11-20(b).

Several simple laminar flow systems will be studied with Eq. 11-2 in the following examples. For incompressible flows in general, the momentum equations are given in Eqs. 11-30 to 11-32 for rectangular coordinates, and in Eqs. 11-34 to 11-36 for cylindrical coordinates. Having studied Examples 11-1 and 11-2, the reader may proceed to study Examples 11-5 and 11-6 with these equations. However, to fully understand their derivation, he should study Sections 11-3 to 11-8.

Example 11-1. A viscous liquid flows laminarly down a large inclined plane at a slope S. The uniform depth of flow is L. Find the velocity distribution.

Since a laminar flow is one without turbulence, and the flow is steady, the flow variables are independent of time. For this flow, we have $v = 0$, $w = 0$, and u is a function of y only. The equation of continuity is satisfied.

To derive the momentum equations, consider the fluid element shown in Fig. 11-2. Let the shearing stress at y_0 be τ_0. At $y_0 + dy$, the shearing stress is $\tau_0 + (d\tau/dy)_0 \, dy$. (Compare the assumed directions of the shearing forces in Fig. 11-2 with those in Fig. 11-1.) The net viscous force in the x-direction is therefore

$$\left[\tau_0 + \left(\frac{d\tau}{dy}\right)_0 dy\right] dx \, dz - \tau_0 \, dx \, dz = \left(\frac{d\tau}{dy}\right)_0 dx \, dy \, dz.$$

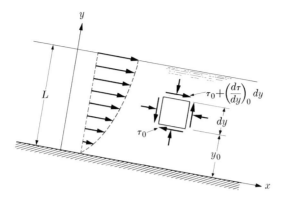

FIGURE 11–2

Since $\tau = \mu(du/dy)$, and the mass of the fluid element is $\rho\, dx\, dy\, dz$, the viscous force per unit mass is therefore

$$\frac{(d/dy)[\mu(du/dy)]\, dx\, dy\, dz}{\rho\, dx\, dy\, dz} = \frac{\mu}{\rho}\frac{d^2 u}{dy^2} = \nu\frac{d^2 u}{dy^2}.$$

Since the value y_0 is arbitrary, the subscript 0 has been omitted. Adding this force per unit mass to Eq. 5–1, we have the momentum equation for the x-direction:

$$-\frac{\partial}{\partial x}\left(\frac{p}{\rho} + gh\right) + \nu\frac{d^2 u}{dy^2} = a_x = 0.$$

Since there is no net viscous force in the y- and z-directions, we have from Eqs. 5–2 and 5–3

$$-\frac{\partial}{\partial y}\left(\frac{p}{\rho} + gh\right) = 0, \qquad -\frac{\partial}{\partial z}\left(\frac{p}{\rho} + gh\right) = 0.$$

From the last two equations, it can be seen that $p/\rho + gh$ is a function of x only, and can therefore be evaluated at any y. At the surface where $p = 0$, we have

$$\frac{d}{dx}\left(\frac{p}{\rho} + gh\right) = g\frac{dh}{dx} = -gS.$$

Therefore

$$\nu\frac{d^2 u}{dy^2} = -gS.$$

Integrating, we obtain

$$\frac{du}{dy} = -\frac{gS}{\nu}y + c_1.$$

From the boundary condition that $\tau = \mu \cdot du/dy = 0$ at $y = L$ at the surface, we have $c_1 = gSL/\nu$:

$$\frac{du}{dy} = \frac{gS}{\nu}(L - y).$$

Integrating again and using the boundary condition that $u = 0$ at $y = 0$, we have

$$u = \frac{gS}{\nu}\, y \left(L - \frac{y}{2} \right).$$

The velocity at any point is therefore proportional to the slope and inversely proportional to viscosity in this laminar flow.

Due to the presence of shearing forces, work may be done by one layer upon another. Energy is thereby transferred among the layers. There is also dissipation of mechanical energy into heat. As a result, Bernoulli's equation is not applicable. There is a decrease of the quantity $p/\rho + gh + q^2/2$ along a streamline. In this case, we have the rate of decrease

$$\frac{\partial}{\partial x}\left(\frac{p}{\rho} + gh + \frac{q^2}{2} \right) = \frac{d}{dx}\left(\frac{p}{\rho} + gh \right) = -gS,$$

which is the same for all the streamlines. In other words, the sum of net energy transferred and dissipation into heat is the same for all particles. Since there is no energy transfer out of the fluid system, the total energy transferred is zero. Therefore $-gS$ is also the average dissipation of mechanical energy, usually referred to as the loss of mechanical energy per unit mass per unit distance.

Example 11-2. In an incompressible flow in a straight circular pipe, the velocity distribution may be considered to be the same at each cross section except for the stretch near the entrance. Study this flow when it is laminar and steady. (This flow is usually referred to as *Poiseuille flow*.)

Let the axis of the pipe be the z-axis. Assuming that the flow is axially symmetric and is directed along the axis, we have $u = 0$, $v = 0$, and w is a function of r only. The equation of continuity is satisfied by this flow system.

To derive the momentum equations, consider a fluid element as shown in Fig. 11-3. The shearing stress τ_0 at r_0 acts on a surface of area $r_0\, d\theta\, dz$. At $r_0 + dr$, the shearing stress $\tau_0 + (d\tau/dr)_0\, dr$ acts on an area $(r_0 + dr)\, d\theta\, dz$. The net viscous force in the z-direction is therefore

$$\left[\tau_0 + \left(\frac{d\tau}{dr}\right)_0 dr \right](r_0 + dr)\, d\theta\, dz - \tau_0 r_0\, d\theta\, dz = \left[\left(\frac{d\tau}{dr}\right)_0 + \frac{\tau_0}{r_0} \right] r_0\, d\theta\, dr\, dz.$$

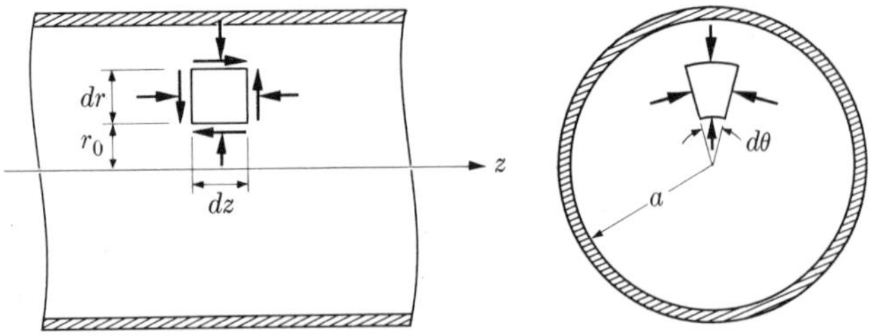

FIGURE 11-3

Since $\tau = \mu \cdot dw/dr$, and the mass of the fluid element is $\rho r_0 \, d\theta \, dr \, dz$, the viscous force per unit mass in the z-direction is

$$\frac{\left[\dfrac{d}{dr}\left(\mu \dfrac{dw}{dr}\right) + \dfrac{\mu}{r}\dfrac{dw}{dr}\right] r \, d\theta \, dr \, dz}{\rho r \, d\theta \, dr \, dz} = \nu\left(\frac{d^2w}{dr^2} + \frac{1}{r}\frac{dw}{dr}\right).$$

(The subscript 0 has been omitted.) Adding this to Eq. 5–4, we have the momentum equation for the z-direction:

$$-\frac{\partial}{\partial z}\left(\frac{p}{\rho} + gh\right) + \nu\left(\frac{d^2w}{dr^2} + \frac{1}{r}\frac{dw}{dr}\right) = a_z = 0. \tag{11–3}$$

Since there is no net viscous force in the r- and θ-directions, we have from Eqs. 5–5 and 5–6,

$$-\frac{\partial}{\partial r}\left(\frac{p}{\rho} + gh\right) = a_r = 0,$$

$$-\frac{\partial}{r\,\partial\theta}\left(\frac{p}{\rho} + gh\right) = a_\theta = 0.$$

From the last two equations, it can be seen that $p/\rho + gh$ must be independent of r and θ. Since w depends on r only, the quantity in the parentheses with w in Eq. 11–3 can only be a function of r. Now that $(d/dz)(p/\rho + gh)$ cannot be a function of r, we must have

$$\nu\left(\frac{d^2w}{dr^2} + \frac{1}{r}\frac{dw}{dr}\right) = \frac{d}{dz}\left(\frac{p}{\rho} + gh\right) = \text{constant}.$$

Call this constant $-gs$. The velocity distribution is then given by

$$\frac{d^2w}{dr^2} + \frac{1}{r}\frac{dw}{dr} = \frac{1}{r}\frac{d}{dr}\left(r\frac{dw}{dr}\right) = -\frac{gs}{\nu}.$$

Integrating and using the condition $dw/dr = 0$ at $r = 0$, we have

$$r\frac{dw}{dr} = -\frac{gsr^2}{2\nu}.$$

Integrating again and using the condition $w = 0$ at $r = a$ at the wall, we have

$$w = \frac{gs}{4\nu}(a^2 - r^2). \tag{11–4}$$

The velocity distribution is therefore parabolic.

The quantity gs can be seen to be the average dissipation of mechanical energy per unit mass per unit distance by comparing the definition of gs with the last equation in Example 11–1. For this reason, s is called the *frictional slope* of pipe flow. It is interesting to find the relationship between s and the mean velocity V:

$$V = \frac{Q}{A} = \frac{1}{A}\int_A w\, dA = \frac{1}{\pi a^2}\int_0^a \frac{gs}{4\nu}(a^2 - r^2)2\pi r\, dr = \frac{ga^2 s}{8\nu}. \quad (11\text{--}5)$$

Thus the frictional slope s is proportional to V in laminar pipe flow. In technical literature, the quantity gs is often expressed as

$$gs = f\frac{V^2}{2D}, \quad (11\text{--}6)$$

where $D = 2a$ is the diameter of the pipe. The dimensionless coefficient f is called the *resistance coefficient* of pipe flows. For laminar flow, this coefficient is then

$$f = \frac{2gDs}{V^2} = \frac{2gD}{V^2}\left(\frac{32\nu V}{gD^2}\right) = \frac{64}{VD/\nu}, \quad (11\text{--}7)$$

where VD/ν is the Reynolds number. It has been found that the flow is laminar and this solution is valid when VD/ν is less than 2000 (see Fig. 2–4).

PROBLEMS

11–1. Show that in Fig. 1–1 with $\partial p/\partial x = 0$ in the laminar flow, the velocity distribution is linear.

11–2. A viscous fluid flows laminarly in the x-direction between two large parallel walls fixed at $y = \pm b$. Show that the velocity distribution is parabolic.

11–3. A viscous fluid flows laminarly in the annular space between two long concentric cylinders as shown in Fig. 11–4. Given that the average dissipation of energy per unit mass per unit distance $(d/dz)(p/\rho + gh) = -gs$, find the discharge through the annulus and the shearing stresses at the walls.

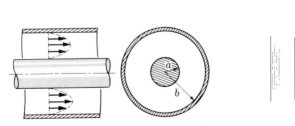

FIGURE 11–4   FIGURE 11–5

11–4. A viscous liquid flows laminarly down a vertical rod as shown in Fig. 11–5. Given the radii a and b, find the velocity distribution. From this solution, verify that the weight of the nonaccelerating liquid is balanced by the shearing force at the surface of the rod.

11–3. Cartesian tensors

The definition of viscosity in Eq. 11–2 must be further generalized in order to be useful for flow systems in general. Since du/dy is the rate of shearing strain $d\beta/dt$ of a fluid element in Fig. 11–1, it can be seen from this equation that μ relates the rate of shearing strain to the shearing stresses of a fluid element. A more general definition of μ is to be found in the relationship between the rate of strain and the stress of the fluid elements. These two quantities will be shown to be tensors in the following sections. We shall first study some general properties of tensors.

A vector has been seen to be a quantity which can be represented by three components along a set of rectangular coordinate axes. For the same vector, the values of its three components will be different if a different set of axes is used. However, these different sets of components are not independent. From any one set, the other sets can be determined. Let A_1, A_2, A_3 be the components of vector $\mathbf{A}$ along three rectangular coordinate axes. We call these axes x_1-, x_2-, and x_3-axes instead of x-, y-, and z-axes for a reason which will become apparent presently. Let x_1'-, x_2'-, and x_3'-axes be another set of rectangular coordinate axes. The component A_1' of $\mathbf{A}$ along the x_1'-axis can be found as the sum of the components along this axis of the vectors representing A_1, A_2, and A_3:

$$A_1' = a_{11}A_1 + a_{21}A_2 + a_{31}A_3,$$

where a_{ij} is used to designate the cosine of the angle between the x_i- and the x_j'-axes. (We shall use the convention that the first and the second subscripts in the directional cosines a_{ij} always refer to the unprimed and primed systems respectively. In Fig. 11–6 is shown a case

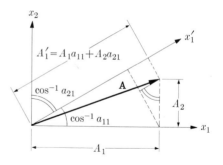

FIGURE 11–6

with $\mathbf{A}$ and the x_1'-axis in the plane of the x_1- and x_2-axes ($A_3 = 0$).) The last equation can be written as

$$A_1' = \sum_{i=1}^{3} a_{i1}A_i.$$

Similarly, the components of $\mathbf{A}$ along the x_2'- and x_3'-axes are

$$A_2' = a_{12}A_1 + a_{22}A_2 + a_{32}A_3 = \sum_{i=1}^{3} a_{i2}A_i,$$

$$A_3' = a_{13}A_1 + a_{23}A_2 + a_{33}A_3 = \sum_{i=1}^{3} a_{i3}A_i.$$

These three equations can be written as

$$A'_m = \sum_i a_{im} A_i. \tag{11-8}$$

The index m which takes one value (1, 2, or 3) in each of the three equations is called the *free index*. The index i which is to be summed from 1 to 3 in each equation is called the *dummy index*. It is so called because $\sum a_{im} A_i$ and $\sum a_{jm} A_j$ mean exactly the same thing. Equation 11–8 is the *transformation rule* of vectors. Since any set can be regarded as the primed or unprimed set, we also have

$$A_j = \sum_m a_{jm} A'_m. \tag{11-9}$$

In this equation, j is the free index taking one value in each of the three equations represented, and m is the dummy index to be summed from 1 to 3 in each equation. As defined above, a_{jm} is the cosine of the angle between the x_j- and the x'_m-axes.

A second-rank *cartesian tensor*, usually referred to simply as a *tensor*, is defined as a quantity with nine components $T_{ij}(i = 1, 2,$ or $3; j = 1, 2,$ or $3)$ which obey the following *transformation rule* between rectangular coordinate systems:

$$T'_{mn} = \sum_i \sum_j a_{im} a_{jn} T_{ij}. \tag{11-10}$$

Here m and n are the free indices, each taking one value (1, 2, or 3) in each of the nine equations represented by Eq. 11–10. The indices i and j are the dummy indices to be summed from 1 to 3 in each equation. In each equation, there are nine terms at the right of the equation. Equation 11–10 gives a purely mathematical definition of a cartesian tensor, but many physical quantities are found to fit this definition.

It is visually helpful to write out a set of components of a tensor in an array. With respect to an unprimed set of coordinate axes,

$$T_{ij} = \begin{bmatrix} T_{11} & T_{12} & T_{13} \\ T_{21} & T_{22} & T_{23} \\ T_{31} & T_{32} & T_{33} \end{bmatrix}.$$

The same tensor has the components T'_{mn} with respect to another set of axes:

$$T'_{mn} = \begin{bmatrix} T'_{11} & T'_{12} & T'_{13} \\ T'_{21} & T'_{22} & T'_{23} \\ T'_{31} & T'_{32} & T'_{33} \end{bmatrix}.$$

According to the transformation rule, the component T'_{12} in the primed system, for example, is related to the components T_{ij} as follows:

$$T'_{12} = a_{11}a_{12}T_{11} + a_{11}a_{22}T_{12} + a_{11}a_{32}T_{13}$$
$$+ a_{21}a_{12}T_{21} + a_{21}a_{22}T_{22} + a_{21}a_{32}T_{23}$$
$$+ a_{31}a_{12}T_{31} + a_{31}a_{22}T_{32} + a_{31}a_{32}T_{33}.$$

In this case, we have $m = 1$ and $n = 2$ in Eq. 11–10. In each row, i is kept at one value while j is being summed from 1 to 3. It can be easily verified that the order of summation is immaterial:

$$\sum_i a_{im}\left(\sum_j a_{jn}T_{ij}\right) = \sum_j a_{jn}\left(\sum_i a_{im}T_{ij}\right).$$

Since any set can be regarded as the primed or unprimed set, the transformation rule of tensors can also be written as

$$T_{ij} = \sum_m \sum_n a_{im}a_{jn}T'_{mn}. \tag{11–11}$$

Tensors of other *ranks* are defined in a similar manner. For example, a fourth-rank tensor has 3^4 or 81 components which obey the transformation rule:

$$G'_{mnop} = \sum_i \sum_j \sum_k \sum_l a_{im}a_{jn}a_{ko}a_{lp}G_{ijkl}.$$

Thus a vector as defined by Eq. 11–8 is a first-rank tensor with 3^1 components, and a scalar is a zero-rank tensor with 3^0 or one component. A tensor without specified rank refers to a second-rank tensor.

Before proceeding to study the properties of tensors, we first study the relationship among the nine directional cosines between two sets of rectangular coordinate axes. Take a vector **A**. According to Eq. 11–9, any component A_j in the unprimed system can be expressed in terms of the three components in the primed system:

$$A_j = \sum_m a_{jm}A'_m.$$

For any value of j, there are three terms on the right of the equation, each with A'_1, A'_2, and A'_3, respectively. Each of these primed components can in turn be expressed in terms of the three unprimed components:

$$A'_m = \sum_i a_{im}A_i.$$

Thus A_j can be expressed in nine terms as

$$A_j = \sum_m a_{jm}\sum_i a_{im}A_i = \sum_i\left(\sum_m a_{im}a_{jm}\right)A_i.$$

But A_j cannot be dependent on the other unprimed components, and the right-hand side must be equal to A_j. The directional cosines must therefore be such that

$$\sum_m a_{im}a_{jm} \begin{cases} = 1, & \text{for } i = j, \\ = 0, & \text{for } i \neq j. \end{cases}$$

With six combinations of i and j, this expression contains six equations relating the nine directional cosines. Therefore only three of them can be independent. (To become familiar with the new notation, the reader is advised to write out in full the last four equations.)

Introducing the quantity δ_{ij}, known as the *Kronecker delta*, which has the values

$$\delta_{ij} \begin{cases} = 1, & \text{if } i = j, \\ = 0, & \text{if } i \neq j, \end{cases} \tag{11-12}$$

we can write

$$\sum_m a_{im}a_{jm} = \delta_{ij}. \tag{11-13}$$

Since any coordinate system can be considered as the primed or unprimed system, we also have

$$\sum_m a_{mi}a_{mj} = \delta'_{ij}. \tag{11-14}$$

It can be shown that the six equations represented by Eq. 11-14 are equivalent to those represented by Eq. 11-13.

In this section, a new notation has been introduced which facilitates a concise expression of lengthy equations. It is further noticed that the dummy indices (such as i and j in Eq. 11-10), but not the free indices, always appear twice under a summation sign. If an index appearing twice in a term is taken to mean also a summation, the summation sign can then be omitted. We shall adopt this *summation convention*. Thus Eq. 11-10 is to be written as

$$T'_{mn} = \sum_i \sum_j a_{im}a_{jn}T_{ij} = a_{im}a_{jn}T_{ij}.$$

The summation signs have been omitted, since the dummy indices i and j appear twice in the same term.

Example 11-3. Show that δ_{ij} is a tensor.

There are nine values of δ_{ij}: $\delta_{11} = \delta_{22} = \delta_{33} = 1$, and the other six equal to zero. There are also nine values of δ'_{mn} in the primed system: $\delta'_{11} = \delta'_{22} = \delta'_{33} = 1$, and the other six equal to zero. It is to be shown that any δ'_{mn} can be computed from $\delta'_{mn} = a_{im}a_{jn}\,\delta_{ij}$ in accordance with the transformation rule of tensors:

$$a_{im}a_{jn}\,\delta_{ij} = a_{im}(a_{jn}\,\delta_{ij}) = a_{im}a_{in}.$$

Of the three terms in the summation $(a_{jn}\delta_{ij})$, only the term a_{in} with $j = i$ is nonzero. Since from Eq. 11–14, $a_{im}a_{in} = \delta'_{mn}$, we have

$$\delta'_{mn} = a_{im}a_{jn}\,\delta_{ij}.$$

The Kronecker delta is therefore a tensor which appears as

$$\delta_{ij} = \begin{bmatrix} 1 & 0 & 0 \\ 0 & 1 & 0 \\ 0 & 0 & 1 \end{bmatrix} \quad \text{and} \quad \delta'_{mn} = \begin{bmatrix} 1 & 0 & 0 \\ 0 & 1 & 0 \\ 0 & 0 & 1 \end{bmatrix}.$$

It is also called a *unit tensor*.

Example 11–4. $\mathbf{u}(x_1, x_2, x_3)$ is a vector field with components u_1, u_2, and u_3 which vary over space (e.g., the velocity in a flow). Show that the nine partial derivatives $\partial u_i/\partial x_j$ at any point are components of a tensor.

Since $\mathbf{u}$ is a vector, its components in another (primed) coordinate system are, according to Eq. 11–8,

$$u'_m = a_{im}u_i.$$

It is to be shown that each of the nine partial derivatives $\partial u'_m/\partial x'_n$ in the primed system is related to $\partial u_i/\partial x_j$ by the transformation rule of tensors:

$$\frac{\partial u'_m}{\partial x'_n} = a_{im}a_{jn}\,\frac{\partial u_i}{\partial x_j}.$$

Consider two points a distance dx'_1 apart at constant x'_2 and x'_3. The difference in u'_m ($m = 1, 2,$ or 3) between these two points is

$$du'_m = \frac{\partial u'_m}{\partial x'_1}\,dx'_1.$$

(We do not use a dummy index here with dx'_1 because no summation is involved.) This vector of magnitude dx'_1 has the following components in the unprimed system:

$$dx_1 = a_{11}dx'_1, \qquad dx_2 = a_{21}dx'_1, \qquad \text{and} \qquad dx_3 = a_{31}dx'_1.$$

But in the unprimed system, this difference of du'_m between these two points (at dx_1, dx_2, and dx_3 apart) is expressed as

$$du'_m = \frac{\partial u'_m}{\partial x_1}\,dx_1 + \frac{\partial u'_m}{\partial x_2}\,dx_2 + \frac{\partial u'_m}{\partial x_3}\,dx_3.$$

Equating the last two equations, we obtain

$$\frac{\partial u'_m}{\partial x'_1}\,dx'_1 = \left(a_{11}\frac{\partial u'_m}{\partial x_1} + a_{21}\frac{\partial u'_m}{\partial x_2} + a_{31}\frac{\partial u'_m}{\partial x_3} \right) dx'_1.$$

Since $\mathbf{u}$ is a vector, $u'_m = a_{im}u_i$,

$$\frac{\partial u'_m}{\partial x'_1} = a_{11}\frac{\partial}{\partial x_1}(a_{im}u_i) + a_{21}\frac{\partial}{\partial x_2}(a_{im}u_i) + a_{31}\frac{\partial}{\partial x_3}(a_{im}u_i) = a_{im}a_{j1}\frac{\partial u_i}{\partial x_j}.$$

In general,

$$\frac{\partial u'_m}{\partial x'_n} = a_{im}a_{jn}\frac{\partial u_i}{\partial x_j}.$$

PROBLEMS

11–5. Given the components of a force $\mathbf{F}$: $F_1 = 3$, $F_2 = 4$, and $F_3 = 0$, find by using Eq. 11–8 its components when the coordinate system rotates through an angle of 30° about the x_3-axis.

11–6. In general, the order of summation is immaterial. Verify by expansion that for $m = 1$ and $n = 2$,

$$T'_{mn} = \sum_i a_{im}\left(\sum_j a_{jn}T_{ij}\right) = \sum_j a_{jn}\left(\sum_i a_{im}T_{ij}\right).$$

11–7. Write out in full the six equations relating the directional cosines in Eq. 11–14.

11–8. According to the summation convention, what are T_{ii} and δ_{jj}?

11–9. Verify that $\delta_{ij}A_j = A_i$ by writing out this equation in full.

11–10. Given that A_i and B_j are the components of vectors $\mathbf{A}$ and $\mathbf{B}$, respectively, with respect to a coordinate system (i.e., they obey the transformation rule in Eq. 11–8), show that the nine values A_iB_j are the components of a tensor; i.e., $A'_mB'_n = a_{im}a_{jn}A_iB_j$, in accordance with Eq. 11–10. (A_iB_j are called the components of the *dyadic product* of $\mathbf{A}$ and $\mathbf{B}$.)

11–11. Given a scalar function ϕ which varies with the coordinates x_1, x_2, and x_3, show that the three derivatives $\partial\phi/\partial x_1$ are components of a vector obeying Eq. 11–8.

11–12. Given a scalar function ϕ, show that the nine second partial derivatives $\partial^2\phi/\partial x_i\partial x_j$ are components of a tensor.

11–4. Properties of cartesian tensors

We now proceed to study some properties of tensors. While the components of a tensor vary with the coordinate system used for reference, the tensor is a definite entity with some properties which do not vary with the coordinate system used. Such properties are called *invariants*. For our purpose, we are going to show that the sum T_{ii} of the diagonal terms in the array of a tensor is an invariant, that is, $T_{ii} = T'_{mm}$:

$$T_{ii} = \delta_{ij}T_{ij} = \delta_{ij}a_{im}a_{jn}T'_{mn} = a_{im}(\delta_{ij}a_{jn})T'_{mn} = a_{im}a_{in}T'_{mn} = \delta'_{mn}T'_{mn} = T'_{mm},$$

that is,

$$T_{11} + T_{22} + T_{33} = T'_{11} + T'_{22} + T'_{33}. \tag{11–15}$$

We now consider two special types of tensors, namely symmetric and anti-symmetric tensors. If $T_{ij} = T_{ji}$ (i.e., $T_{12} = T_{21}$, etc.), the tensor is said to be *symmetric*. The last tensor in Eq. 11–18 below is such a tensor. It can be shown that being symmetric is a property of this tensor, and is independent of the coordinate system used for reference. With $T_{ij} = T_{ji}$, we have in any primed coordinate system

$$T'_{mn} = a_{im}a_{jn}T_{ij} = a_{jn}a_{im}T_{ji} = T'_{nm}.$$

Thus a symmetric tensor remains symmetric after transformation. It can be proved that by a proper choice of the set of axes of reference, a symmetric tensor can always be reduced to a *diagonal tensor* with diagonal terms only:

$$\begin{bmatrix} T_{11} & 0 & 0 \\ 0 & T_{22} & 0 \\ 0 & 0 & T_{33} \end{bmatrix}.$$

These axes are called *principal axes* of the symmetric tensor. (For proof, see textbooks on tensor analysis.)

A tensor is said to be *antisymmetric* if $T_{ij} = -T_{ji}$ (i.e., $T_{12} = -T_{21}$, etc.). The middle tensor in Eq. 11–18 below is such a tensor. It can be shown that an antisymmetric tensor remains antisymmetric after transformation (see Problem 11–14). Obviously, all the diagonal terms of such a tensor must be zero. An antisymmetric tensor, like a vector, can be completely specified with three independent components, such as T_{12}, T_{23}, and T_{31}.

We now define the addition of two tensors. Since the sum **A** of two vectors **B** and **C** may be defined to have the three components.

$$A_i = B_i + C_i, \tag{11–16}$$

the *sum S of two tensors* U and T is defined to have the nine components

$$S_{ij} = U_{ij} + T_{ij}. \tag{11–17}$$

A numerical example is

$$\begin{bmatrix} 3 & 9 & 3 \\ -1 & 0 & 1 \\ 5 & -3 & 1 \end{bmatrix} = \begin{bmatrix} 0 & 5 & -1 \\ -5 & 0 & 2 \\ 1 & -2 & 0 \end{bmatrix} + \begin{bmatrix} 3 & 4 & 4 \\ 4 & 0 & -1 \\ 4 & -1 & 1 \end{bmatrix}, \tag{11–18}$$

where $S_{12} = U_{12} + T_{12} = 5 + 4 = 9$, and so on. It is to be shown that the nine values of S_{ij} as defined are the components of a tensor; i.e., the nine values of $S'_{mn} = U'_{mn} + T'_{mn}$ in a primed system are related to S_{ij} by Eq. 11–10:

$$S'_{mn} = U'_{mn} + T'_{mn} = a_{im}a_{jn}U_{ij} + a_{im}a_{jn}T_{ij} = a_{im}a_{jn}(U_{ij} + T_{ij}) = a_{im}a_{jn}S_{ij}.$$

According to this definition of addition, any tensor can be expressed as the sum of an antisymmetric tensor and a symmetric tensor. Any given tensor component S_{ij} can be written as $S_{ij} = U_{ij} + T_{ij}$, where

$$U_{ij} = \tfrac{1}{2}(S_{ij} - S_{ji}) \qquad \text{and} \qquad T_{ij} = \tfrac{1}{2}(S_{ij} + S_{ji}). \qquad (11\text{--}19)$$

Since $U_{ji} = \tfrac{1}{2}(S_{ji} - S_{ij}) = -U_{ij}$, tensor U is antisymmetric. It can also be seen that T is symmetric. In Eq. 11–18, the first tensor has been written as the sum of two tensors, one antisymmetric and the other symmetric. For example, $U_{13} = \tfrac{1}{2}(3 - 5) = -1$, and $T_{13} = \tfrac{1}{2}(3 + 5) = 4$.

PROBLEMS

11–13. Show that while the components of vector **A** vary with the coordinate system used, $\delta_{ij}A_iA_j$ is an invariant. If **A** is velocity, what is $\delta_{ij}A_iA_j$?

11–14. Show that an antisymmetric tensor remains antisymmetric after transformation.

11–15. Given vectors **B** and **C**, show that the three values $A_i = B_i + C_i$ are components of a vector obeying Eq. 11–8.

11–16. Given a set of components of a tensor

$$S_{ij} = \begin{bmatrix} 2 & -1 & 0 \\ 1 & 3 & 2 \\ 0 & -2 & 4 \end{bmatrix},$$

write out the full arrays of the symmetric tensor and the antisymmetric tensor which combine to form tensor S.

11–5. Stress at a point

The internal force per unit area at a point of a continuous body depends on the orientation of the imaginary surface upon which the internal force acts. The internal force on an area may be resolved into normal and tangential components. For an area with its outward normal along the x_1-axis, we have the normal stress S_{11} and the shearing stresses S_{12} and S_{13} in the x_2- and x_3-directions, respectively, as shown in Fig. 11–7(a). (By the law of actions and reactions, these same stresses act on the surface of the adjacent particle with its outward normal in the direction opposite the x_1-axis as shown in Fig. 11–8. Thus we can use the symbol S_{ij} to designate the stress component on a surface with its outward normal in the $\pm x_1$-direction acting in the $\pm x_j$-direction.) At the same point, for an area with an outward normal in the x_2-direction, we have a different total force and S_{21}, S_{22}, and S_{33}, as shown in Fig. 11–7(b). Similarly, we have S_{31}, S_{32}, and S_{33} as shown in Fig. 11–7(c). With a surface inclined to these axes, we have other stress components at this point.

It is to be shown that the nine stress components at a point, as shown in Fig. 11–7, are the components of a tensor which describes the state of stress completely at this point. To show this, consider an infinitesimal tetrahedron of the material around the point in question, as shown in Fig. 11–9. It is to be

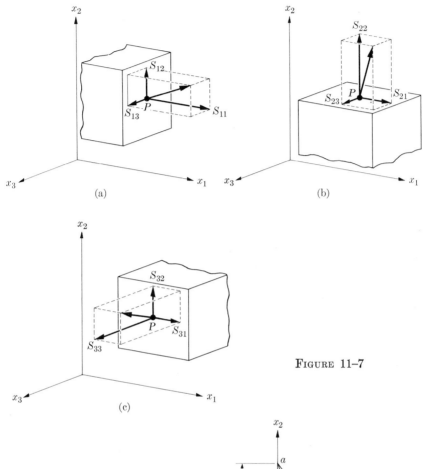

(a)

(b)

(c)

FIGURE 11–7

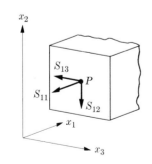

FIGURE 11–8

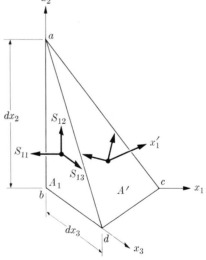

FIGURE 11–9

shown that S'_{1n} ($n = 1$, 2, or 3) acting on the inclined surface is related to S_{ij} on the three perpendicular surfaces by $S'_{1n} = a_{i1}a_{jn}S_{ij}$. Apply Newton's law of motion $\mathbf{F} = d(m\mathbf{q})/dt$ to the mass. Here $\mathbf{F}$ includes the surface forces and the body forces acting on the element. The surface forces are proportional to the surface area and are therefore of the order of $dx_1\,dx_2$. The body force and the mass m are proportional to the volume and are therefore of the order of $dx_1\,dx_2\,dx_3$, which vanishes in comparison with $dx_1\,dx_2$. Therefore, for the infinitesimal volume, Newton's law becomes $\mathbf{F} = 0$, where $\mathbf{F}$ is the sum of the surface forces. First find the sum $\mathbf{f}$ of the forces on the three perpendicular surfaces. In the x_1-direction, we have

$$f_1 = -A_1S_{11} - A_2S_{21} - A_3S_{31} = -A_iS_{i1},$$

where S_{21} acts on area A_2 of surface abc, and S_{31} acts on area A_3 of surface abd. In general, component f_j of the sum $\mathbf{f}$ is

$$f_j = -A_iS_{ij}.$$

Let the outward normal of the inclined surface A' be the x'_1-axis. The component of the vector $\mathbf{f}$ along any of the primed axes x'_n ($n = 1$, 2, or 3) is, according to Eq. 11–8,

$$f'_n = a_{jn}f_j = -a_{jn}A_iS_{ij}.$$

Since the total force $\mathbf{F}$ on all four surfaces must be zero, $\mathbf{f}$ must be balanced by the force on A'. Therefore f'_n must be balanced by S'_{1n} on A':

$$S'_{1n}A' = -f'_n = a_{jn}A_iS_{ij}.$$

Since A_2 is the projection of A' along the x_2-axis, we have $A_2 = A'a_{21}$. In general, $A_i = A'a_{i1}$. Therefore $S'_{1n} = a_{i1}a_{jn}S_{ij}$. Since the normal of A' could have been called the x'_m-axis,

$$S'_{mn} = a_{im}a_{jn}S_{ij},$$

which is the transformation rule of tensors. Thus, the stress components S_{ij} at a point are the components of a tensor. It is called the *stress tensor* at this point. As scalar and vector quantities may vary from point to point, so may this tensor quantity.

Next, we show that the stress tensor is symmetric, that is, $S_{ij} = S_{ji}$. To do this, apply to the infinitesimal element shown in Fig. 11–10 Newton's law of motion in the form of Eq. 9–23:

$$\sum \mathbf{R} \times \mathbf{F} = \frac{d}{dt}\left(\sum \mathbf{R} \times m\mathbf{q}\right),$$

which states that the rate of change of total moment of momentum of a system of particles is equal to the moment of the external forces acting on the system. It can be shown that this equation is valid with either a fixed point or the center of mass of the element as the moment center (for proof, see textbooks on mechanics of particles). Here $\sum \mathbf{R} \times \mathbf{F}$ includes the moments due to the body force and the surface forces. Again, the body force and the mass, being of higher order, vanish in comparison with the surface forces. This equation becomes $\sum \mathbf{R} \times \mathbf{F} = 0$, in which $\mathbf{F}$ now includes only the surface forces, i.e., the total moment of the surface forces is zero. Take the moment about the center of the mass. The component of the moment along the x_1-axis is $(S_{23}\, dx_3\, dx_1)\, dx_2 - (S_{32}\, dx_1\, dx_2)\, dx_3$. Since the moment is zero, we have $S_{23} = S_{32}$. Simi-

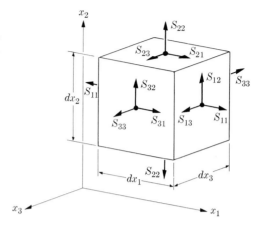

FIGURE 11–10

larly, we can show that $S_{31} = S_{13}$ and $S_{12} = S_{21}$. Therefore the stress tensor is symmetric, with $S_{ij} = S_{ji}$.

We have shown that because of Newton's law of motion, the stress at a point in a continuous material is a symmetric tensor. The state of stress at a point is completely specified by any one set of nine components of the tensor. Its components with respect to another set of axes can be computed according to the transformation rule in Eq. 11–10. Since the tensor is symmetric, there are three pairs of equal shearing stress components, namely, $S_{12} = S_{21}$, $S_{23} = S_{32}$, and $S_{31} = S_{13}$. With respect to a set of principal axes, all these shearing stress components are zero, as shown in Fig. 11–11(b). The three normal stress components are in general not equal. However, while the stress tensor has a different set of components with respect to another coordinate system, the sum of the three normal stresses $(S_{11} + S_{22} + S_{33})$ in any set is the same (see Eq. 11–15). Since the difference between two tensors $(S_{ij} - k\,\delta_{ij})$ is also a tensor, any arbitrary datum k, such as the atmospheric pressure, can be used for the normal components of the stress tensor.

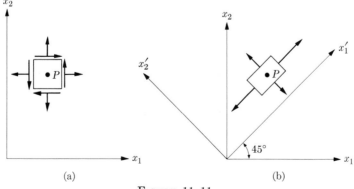

FIGURE 11–11

PROBLEM

11–17. The stress at a point has the following components: $S_{11} = S_{22} = S_{33} = 2$, $S_{12} = 1$, $S_{23} = S_{31} = 0$. (a) Verify that the primed axes shown in Fig. 11–11 are principal axes of the tensor. (b) Compute the normal stress components in the primed system, and verify that their sum is an invariant.

11–6. The rate of strain at a point

To study the rate of strain of an element at a point O in Fig. 11–12, consider the motion at point P relative to O. Let the velocity components be designated by u_1, u_2, and u_3, which are functions of x_1, x_2, x_3, and t. At the same instant, the difference of velocity component u_1 between these two points is

$$du_1 = \frac{\partial u_1}{\partial x_1}\,dx_1 + \frac{\partial u_1}{\partial x_2}\,dx_2 + \frac{\partial u_1}{\partial x_3}\,dx_3 = \frac{\partial u_1}{\partial x_j}\,dx_j.$$

In general,

$$du_i = \frac{\partial u_i}{\partial x_j}\,dx_j.$$

The behavior of the element relative to O depends therefore on the nine partial derivatives $\partial u_i/\partial x_j$ at this point. These derivatives of the components of a vector have been shown in Example 11–4 to be the components of a tensor. The physical significance of this tensor can be revealed by writing it as the sum of two tensors, one antisymmetric and the other symmetric, in accordance with Eq. 11–19:

$$\begin{bmatrix} \dfrac{\partial u_1}{\partial x_1} & \dfrac{\partial u_1}{\partial x_2} & \dfrac{\partial u_1}{\partial x_3} \\[2mm] \dfrac{\partial u_2}{\partial x_1} & \dfrac{\partial u_2}{\partial x_2} & \dfrac{\partial u_2}{\partial x_3} \\[2mm] \dfrac{\partial u_3}{\partial x_1} & \dfrac{\partial u_3}{\partial x_2} & \dfrac{\partial u_3}{\partial x_3} \end{bmatrix} = \begin{bmatrix} 0 & \dfrac{1}{2}\left(\dfrac{\partial u_1}{\partial x_2} - \dfrac{\partial u_2}{\partial x_1}\right) & -\dfrac{1}{2}\left(\dfrac{\partial u_3}{\partial x_1} - \dfrac{\partial u_1}{\partial x_3}\right) \\[2mm] -\dfrac{1}{2}\left(\dfrac{\partial u_1}{\partial x_2} - \dfrac{\partial u_2}{\partial x_1}\right) & 0 & \dfrac{1}{2}\left(\dfrac{\partial u_2}{\partial x_3} - \dfrac{\partial u_3}{\partial x_2}\right) \\[2mm] \dfrac{1}{2}\left(\dfrac{\partial u_3}{\partial x_1} - \dfrac{\partial u_1}{\partial x_3}\right) & -\dfrac{1}{2}\left(\dfrac{\partial u_2}{\partial x_3} - \dfrac{\partial u_3}{\partial x_2}\right) & 0 \end{bmatrix}$$

$$+ \begin{bmatrix} \dfrac{\partial u_1}{\partial x_1} & \dfrac{1}{2}\left(\dfrac{\partial u_1}{\partial x_2} + \dfrac{\partial u_2}{\partial x_1}\right) & \dfrac{1}{2}\left(\dfrac{\partial u_3}{\partial x_1} + \dfrac{\partial u_1}{\partial x_3}\right) \\[2mm] \dfrac{1}{2}\left(\dfrac{\partial u_1}{\partial x_2} + \dfrac{\partial u_2}{\partial x_1}\right) & \dfrac{\partial u_2}{\partial x_2} & \dfrac{1}{2}\left(\dfrac{\partial u_2}{\partial x_3} + \dfrac{\partial u_3}{\partial x_2}\right) \\[2mm] \dfrac{1}{2}\left(\dfrac{\partial u_3}{\partial x_1} + \dfrac{\partial u_1}{\partial x_3}\right) & \dfrac{1}{2}\left(\dfrac{\partial u_2}{\partial x_3} + \dfrac{\partial u_3}{\partial x_2}\right) & \dfrac{\partial u_3}{\partial x_3} \end{bmatrix}$$

or in brief

$$\frac{\partial u_i}{dx_j} = R_{ij} + e_{ij},$$

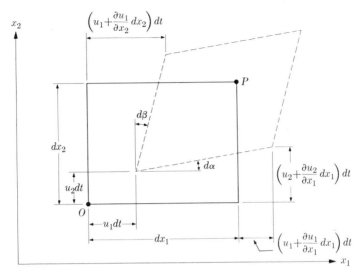

FIGURE 11–12

where

$$R_{ij} = \frac{1}{2}\left(\frac{\partial u_i}{\partial x_j} - \frac{\partial u_j}{\partial x_i}\right) \quad \text{and} \quad e_{ij} = \frac{1}{2}\left(\frac{\partial u_i}{\partial x_j} + \frac{\partial u_j}{\partial x_i}\right). \quad (11\text{–}20)$$

The three independent components of the antisymmetric tensor R_{ij} have been shown to be the three components of the angular velocity of the fluid element (see Eqs. 6–1 to 6–3). The tensor R_{ij} therefore is responsible for the rotating motion of the element. The motion of P relative to O is then due to a rotation and the tensor e_{ij}, which is to be shown presently to be the rate of strain.

The nature of the symmetric tensor e_{ij} can be studied as follows. Consider the position of the element after a time dt in Fig. 11–12. Let u_1, u_2, and u_3 be the velocity components at point O. During dt, this element increases in length in the x_1-direction by the amount

$$\left[dx_1 + \left(u_1 + \frac{\partial u_1}{\partial x_1}\, dx_1\right) dt - u_1\, dt\right] - dx_1 = \frac{\partial u_1}{\partial x_1}\, dx_1\, dt.$$

The rate of longitudinal strain in the x_1-direction is therefore $\partial u_1/\partial x_1$ which is e_{11}. Similarly, the other diagonal terms e_{22} and e_{33} are the rates of longitudinal strain in the x_2- and x_3-directions, respectively. It can also be seen from Fig. 11–12 that two sides of the element have rotated through angles $d\alpha$ and $d\beta$, respectively, during dt. Since

$$d\alpha = (\partial u_2/\partial x_1)\, dt \quad \text{and} \quad d\beta = (\partial u_1/\partial x_2)\, dt$$

(see derivation of Eq. 6–1), and the shearing strain in dt is $d\alpha + d\beta$, the rate of shearing strain is therefore $\partial u_2/\partial x_1 + \partial u_1/\partial x_2$, which is $2e_{12}$ or $2e_{21}$. Thus, in general, $2e_{ij}$, when $i \neq j$, represents a rate of shearing strain. The symmetric tensor e_{ij} is therefore called the *rate of strain tensor* at point O.

Since e_{ij} is a tensor, the sum $\partial u_i / \partial x_i$ of the three diagonal terms (the three rates of longitudinal strain) is an invariant through transformation. It can be shown to be the time rate of fractional increase of volume of the element. Since e_{11} is a rate of longitudinal strain, the length of the element in Fig. 11–12 after dt is $dx_1(1 + e_{11} \, dt)$. Its volume becomes

$$dx_1(1 + e_{11} \, dt) \cdot dx_2(1 + e_{22} \, dt) \cdot dx_3(1 + e_{33} \, dt) =$$
$$dx_1 \, dx_2 \, dx_3[1 + (e_{11} + e_{22} + e_{33}) \, dt].$$

The rate of fractional increase of volume is therefore $e_{11} + e_{22} + e_{33}$. This invariant is the divergence of $\mathbf{u}$ (see Eq. 4–29):

$$e_{11} + e_{22} + e_{33} = \frac{\partial u_1}{\partial x_1} + \frac{\partial u_2}{\partial x_2} + \frac{\partial u_3}{\partial x_3}$$

$$= \nabla \cdot \mathbf{u}. \qquad (11\text{–}21)$$

For an incompressible fluid, the volume of an element is constant. Therefore $\nabla \cdot \mathbf{u} = 0$, which is the equation of continuity for incompressible fluids in Eq. 4–19.

Since e_{ij} is a symmetric tensor, there are principal axes with respect to which there is no rate of shearing strain. For example, tensor e_{ij} at point O in Fig. 11–13 may be described with the rates of strain of the element $abcd$. This ele-

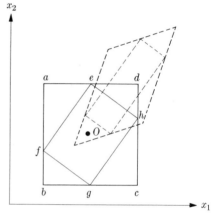

FIGURE 11–13

ment can be seen to be undergoing shearing strain as well as longitudinal strain. However, for the element with sides ef and fg along the principal axes of the tensor at point O, there is no rate of shearing strain.

PROBLEMS

11–18. In the flow in Fig. 1–1, (a) find the components of the rate of strain with respect to the coordinate axes shown. Draw a diagram similar to Fig. 11–12 for a cubic fluid element in this case. (b) Verify that the set of axes obtained by rotating through 45° about the x_3- or z-axis are principal axes of the rate of strain tensor. [Note that the rates of longitudinal strain with respect to these principal axes are indicated by those of the diagonals of the square seen in the diagram constructed in (a).] (c) Compute the divergence of the velocity in both systems. The answers should be the same. Why?

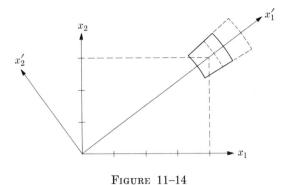

FIGURE 11–14

11–19. For a line source of strength $Q = 2\pi$, the velocity components of an incompressible fluid are

$$u_1 = x_1/(x_1^2 + x_2^2), \qquad u_2 = x_2/(x_1^2 + x_2^2), \qquad u_3 = 0.$$

(a) Find the components of the rate of strain tensor at the point $(4, 3, 0)$ with respect to the unprimed system shown in Fig. 11–14. (b) Verify that the radial and the other primed axes are the principal axes of this tensor.

11–7. Navier-Stokes equations

To derive the momentum equation for a continuous material, we consider the element shown in Fig. 11–15. The net surface force per unit mass in the x_1-direction is

$$\frac{1}{\rho\, dx_1\, dx_2\, dx_3}$$

$$\times \left[\left(\frac{\partial S_{11}}{\partial x_1}\, dx_1\right) dx_2\, dx_3 + \left(\frac{\partial S_{21}}{\partial x_2}\, dx_2\right) dx_3\, dx_1 + \left(\frac{\partial S_{31}}{\partial x_3}\, dx_3\right) dx_1\, dx_2 \right]$$

$$= \frac{1}{\rho}\, \frac{\partial S_{i1}}{\partial x_i}.$$

Since the component in this direction of the weight per unit mass is $-g(\partial h/\partial x_1)$, the momentum equation for this direction is

$$a_1 = -g\,\frac{\partial h}{\partial x_1} + \frac{1}{\rho}\,\frac{\partial S_{i1}}{\partial x_i},$$

where a_1 is the component in the x_1-direction of the particle acceleration **a**. In general, the momentum equation for

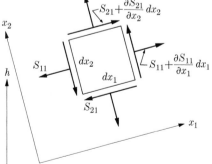

FIGURE 11–15

the x_j-direction is

$$a_j = -g \frac{\partial h}{\partial x_j} + \frac{1}{\rho} \frac{\partial S_{ij}}{\partial x_i}. \tag{11-22}$$

This equation is valid for all continuous materials with the weight as the only body force.

To study fluid motion with Eq. 11–22, it is necessary to propose a relationship between the stress and the rate of strain of a fluid particle. In proposing this relationship, one is guided by the following experimental observations. First, elements of common fluids are believed to be *isotropic*, with mechanical properties independent of directions. Secondly, when a fluid is at rest with zero rate of strain, there are normal stress components $(-p)$ which are the same in all directions; i.e., $S_{ij} = -p\delta_{ij}$. Thirdly, in the flow shown in Fig. 1–1 with rate of shearing strain $2e_{12} = 2e_{21} = U/d$, it is found that for common fluids, the shearing stress component S_{21} is linearly related to the rate of shearing strain

$$S_{21} = \frac{F}{A} = \mu \frac{U}{d} = 2\mu e_{21}.$$

Therefore, for motion in general, the simplest hypothesis one can frame is that a stress component is linearly related to the components of the rate of strain.

Let the unprimed coordinate axes shown in Fig. 11–16 be principal axes of the stress tensor of the fluid element under consideration. For isotropic materials, the principal axes of the rate of strain tensor of the element coincide with those of its stress tensor. For a Newtonian fluid, it is postulated that with respect to the principal axes,

$$S_{11} = -p + Ce_{11} + \lambda e_{22} + \lambda e_{33},$$
$$S_{22} = -p + \lambda e_{11} + Ce_{22} + \lambda e_{33},$$
$$S_{33} = -p + \lambda e_{11} + \lambda e_{22} + Ce_{33}.$$

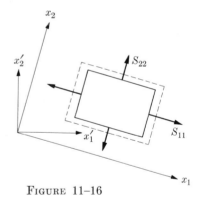

FIGURE 11–16

A stress component with respect to principal axes is related to the component of the rate of strain in the same direction by a coefficient C, and to the components of the rate of strain in the other two directions by a coefficient λ. These two coefficients are assumed to be material properties that are independent of the velocity distribution of the flow. The scalar quantity $-p$ would be equal to the normal stresses if e_{11}, e_{22}, and e_{33} are zero (e.g., in fluid statics). The meaning of p in a flow will be discussed subsequently. The equations above can be rewritten as

$$S_{11} = -p + \lambda\theta + 2\mu e_{11},$$
$$S_{22} = -p + \lambda\theta + 2\mu e_{22},$$
$$S_{33} = -p + \lambda\theta + 2\mu e_{33},$$

or, in general,

$$S_{mn} = -p\delta_{mn} + \lambda\theta\delta_{mn} + 2\mu e_{mn}, \tag{11–23}$$

where $2\mu = C - \lambda$, and θ is the divergence of the velocity:

$$\theta = \frac{\partial u_1}{\partial x_1} + \frac{\partial u_2}{\partial x_2} + \frac{\partial u_3}{\partial x_3} = e_{11} + e_{22} + e_{33}. \tag{11–24}$$

For (primed) axes other than these principal axes, we have by transformation

$$S'_{ij} = a_{mi}a_{nj}S_{mn} = a_{mi}a_{nj}(-p\delta_{mn} + \lambda\theta\delta_{mn} + 2\mu e_{mn})$$

$$= (-p + \lambda\theta)a_{mi}a_{mj} + 2\mu a_{mi}a_{nj}e_{mn} = -p\delta'_{ij} + \lambda\theta\delta'_{ij} + 2\mu e'_{ij}.$$

Omitting the primes to obtain an equation for any set of axes, we have

$$S_{ij} = -p\delta_{ij} + \lambda\theta\delta_{ij} + 2\mu e_{ij}.$$

This equation relates the stress tensor and the rate of strain tensor of Newtonian fluids. The coefficient μ is the dynamic viscosity of the fluid. Usually, another coefficient $\mu' = \lambda + \frac{2}{3}\mu$ is used instead of λ; μ' is called the *second coefficient of viscosity*. Then

$$S_{ij} = -p\delta_{ij} + (\mu' - \tfrac{2}{3}\mu)\theta\delta_{ij} + 2\mu e_{ij}. \tag{11–25}$$

The momentum equations can now be written for Newtonian fluids: Since

$$\frac{\partial S_{ij}}{\partial x_i} = \frac{\partial}{\partial x_i}[-p\delta_{ij} + (\mu' - \tfrac{2}{3}\mu)\theta\delta_{ij} + 2\mu e_{ij}]$$

$$= -\frac{\partial p}{\partial x_j} + \frac{\partial}{\partial x_j}[(\mu' - \tfrac{2}{3}\mu)\theta] + \frac{\partial}{\partial x_i}\left[\mu\left(\frac{\partial u_i}{\partial x_j} + \frac{\partial u_j}{\partial x_i}\right)\right].$$

Equation 11–22 becomes

$$\rho a_j = -\rho g\frac{\partial h}{\partial x_j} - \frac{\partial p}{\partial x_j} + \frac{\partial}{\partial x_j}[(\mu' - \tfrac{2}{3}\mu)\theta] + \frac{\partial}{\partial x_i}\left[\mu\left(\frac{\partial u_i}{\partial x_j} + \frac{\partial u_j}{\partial x_i}\right)\right]. \tag{11–26}$$

This is called the *Navier-Stokes equation* for Newtonian fluids.

This equation can be simplified when compressibility of the fluid can be ignored so that the divergence θ is negligible, and when μ can be considered constant. Then

$$a_j = -\frac{1}{\rho}\frac{\partial p}{\partial x_j} - g\frac{\partial h}{\partial x_j} + \frac{\mu}{\rho}\frac{\partial^2 u_j}{\partial x_i\,\partial x_i}. \tag{11–27}$$

Here we have used the relation $\partial^2 u_i/\partial x_i\partial x_j = \partial\theta/\partial x_j = 0$. Note that only the coefficient of viscosity μ is significant under this condition. The coefficient μ' is effective only when the element can expand or contract. The meaning of the quantity p in this case can be seen from Eqs. 11–23 and 11–24:

$$S_{11} + S_{22} + S_{33} = -3p + (3\lambda + 2\mu)\theta = -3p$$

or

$$-p = \tfrac{1}{3}(S_{11} + S_{22} + S_{33}). \tag{11–28}$$

Thus $-p$ in this case is the mean of the normal stresses in three orthogonal directions. This quantity is invariant through transformation. When there is no rate of strain, as in hydrostatics, we have, from Eq. 11–23, $S_{11} = S_{22} = S_{33} = -p$.

In flow systems where compressibility cannot be ignored, two coefficients of viscosity are involved. It is difficult to measure μ' experimentally, but available evidence shows that it is negligibly small. From Eqs. 11–23 and 11–24, one can show that

$$S_{11} + S_{22} + S_{33} = -3p + 3(\mu' - \tfrac{2}{3}\mu)\theta + 2\mu\theta = -3p + 3\mu'\theta.$$

Stokes assumed $\mu' = 0$. Then $-p$ is again the mean of the normal stresses, and Eq. 11–26 becomes

$$\rho a_j = -\rho g \frac{\partial h}{\partial x_j} - \frac{\partial p}{\partial x_j} - \frac{2}{3}\frac{\partial}{\partial x_j}\left(\mu \frac{\partial u_i}{\partial x_i}\right) + \frac{\partial}{\partial x_i}\left[\mu\left(\frac{\partial u_i}{\partial x_j} + \frac{\partial u_j}{\partial x_i}\right)\right]. \tag{11–29}$$

Equation 11–27 is the Navier-Stokes equation for Newtonian fluids when μ is constant and the effect of compressibility is unimportant. For convenience of reference, this equation is written out in full for use with x-, y-, and z-axes:

$$\frac{\partial u}{\partial t} + u\frac{\partial u}{\partial x} + v\frac{\partial u}{\partial y} + w\frac{\partial u}{\partial z} = -\frac{\partial}{\partial x}\left(\frac{p}{\rho} + gh\right) + \nu\left(\frac{\partial^2 u}{\partial x^2} + \frac{\partial^2 u}{\partial y^2} + \frac{\partial^2 u}{\partial z^2}\right),$$
$$\tag{11–30}$$

$$\frac{\partial v}{\partial t} + u\frac{\partial v}{\partial x} + v\frac{\partial v}{\partial y} + w\frac{\partial v}{\partial z} = -\frac{\partial}{\partial y}\left(\frac{p}{\rho} + gh\right) + \nu\left(\frac{\partial^2 v}{\partial x^2} + \frac{\partial^2 v}{\partial y^2} + \frac{\partial^2 v}{\partial z^2}\right),$$
$$\tag{11–31}$$

$$\frac{\partial w}{\partial t} + u\frac{\partial w}{\partial x} + v\frac{\partial w}{\partial y} + w\frac{\partial w}{\partial z} = -\frac{\partial}{\partial z}\left(\frac{p}{\rho} + gh\right) + \nu\left(\frac{\partial^2 w}{\partial x^2} + \frac{\partial^2 w}{\partial y^2} + \frac{\partial^2 w}{\partial z^2}\right).$$
$$\tag{11–32}$$

With the equation of continuity in Eq. 4–19, we have four equations for the four functions u, v, w, and p. At a solid boundary, the boundary condition is that there is no relative motion between the solid surface and the contacting fluid particles. These equations are so complex that only a few exact solutions have been found for simple flow systems. However, these solutions give an insight into the general behavior of laminar viscous flow. Examples of solutions for laminar flows can be found in Sections 11–2, 11–7, and 11–8. Approximate solutions will be discussed in subsequent sections.

Example 11–5. The infinite plane surface shown in Fig. 11–17 oscillates in its own plane with $u_0 = U \cos (2\pi t/T)$, where U is the amplitude of its speed and T is the period, or duration of each cycle, of the oscillations. Find the velocity distribution in the Newtonian fluid in steady oscillations.

Assume that the motion is in layers with u and $p/\rho + gh$ as functions of y and t only, and $v = w = 0$. For such a flow, the equation of continuity and Eq. 11–32 are satisfied. Equations 11–30 and 11–31 become

$$\frac{\partial u}{\partial t} = \nu \frac{\partial^2 u}{\partial y^2} \qquad (11\text{–}33)$$

and

$$0 = \frac{\partial}{\partial y} \left(\frac{p}{\rho} + gh \right) .$$

The last equation indicates that $p/\rho + gh$ is independent of y.

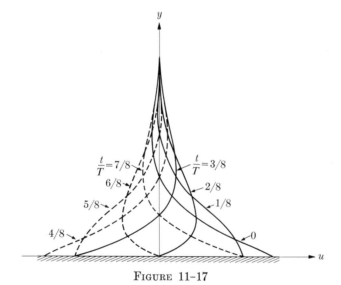

FIGURE 11–17

While standard methods of solution are available for Eq. 11–33, the details are too involved to be presented here. Because of the simplicity of the boundary condition in this case, we can proceed as follows. Expecting that in steady oscillation, each layer oscillates with the same frequency with decreasing amplitude with the distance from the plate, try the solution

$$u(y, t) = U e^{-ky} \cos \frac{2\pi(t - \tau)}{T} ,$$

where $\tau(y)$ is the time lag of the oscillations of the layer at y, with $\tau = 0$ at $y = 0$, and k is a positive constant to be determined. The factor e^{-ky} decreases with increase of y. This solution satisfies the boundary condition that $u = u_0$ at $y = 0$. Substituting this $u(y, t)$ into Eq. 11–33, we have

$$\left[k^2 - \left(\frac{2\pi}{T}\frac{d\tau}{dy}\right)^2\right]\cos\frac{2\pi(t - \tau)}{T} + \frac{2\pi}{T}\left[\frac{d^2\tau}{dy^2} - 2k\frac{d\tau}{dy} + \frac{1}{\nu}\right]\sin\frac{2\pi(t - \tau)}{T} = 0.$$

Since this equation is to hold for all time, we must have

$$k^2 - \left(\frac{2\pi}{T}\frac{d\tau}{dy}\right)^2 = 0$$

and

$$\frac{d^2\tau}{dy^2} - 2k\frac{d\tau}{dy} + \frac{1}{\nu} = 0.$$

From the first equation with the condition that $\tau = 0$ at $y = 0$, we obtain $\tau = Tky/2\pi$. The second equation yields $k = \sqrt{\pi/(\nu T)}$. Thus

$$u = Ue^{-\sqrt{\pi}y/\sqrt{\nu T}}\cos 2\pi\left(\frac{t}{T} - \frac{y}{2\sqrt{\pi\nu T}}\right).$$

The velocity profiles for several values of t are shown in Fig. 11–17.

Note that $\sqrt{\nu T}$ is a characteristic length in this flow. At a distance y equal to this length, the amplitude of u is only $0.17U$. It can be seen that the influence of viscosity is mainly confined to a depth of the order of $\sqrt{\nu T}$.

The displacement of the plane surface is $\int u_0 dt = (UT/2\pi)\sin(2\pi t/T)$. Let L be the amplitude $UT/2\pi$ of its displacement, and let δ be the thickness $\sqrt{\nu T}$. It can be easily verified that $\delta/L = \sqrt{2\pi/(UL/\nu)}$. Thus the relative thickness of the layer of viscous action is inversely proportional to the square root of the Reynolds number UL/ν.

PROBLEMS

11–20. (a) Write out Eq. 11–25 in full for each of the components S_{ij}. (b) Compute S_{11}, S_{22}, S_{33}, and $-p$ in the flow of Example 11–1. (c) With the stress components in (a), derive Eq. 11–30 independently, starting from Eq. 11–22.

11–21. (a) Simplify the Navier-Stokes equations for the flow system in Example 11–1. (b) Show that if an incompressible flow is irrotational (therefore with a velocity potential), the net viscous force acting on a fluid element is zero.

11–8. Navier-Stokes equations in cylindrical coordinates

When cylindrical coordinates are used, the Navier-Stokes equations must be rewritten due to the variation of the r- and θ-directions. First consider the components $e_{\theta\theta}$ and $e_{r\theta}$ of the rate of strain; $e_{\theta\theta}$ is the rate of longitudinal strain in the θ-direction, and $2e_{r\theta}$ is the rate of shearing strain shown in Fig. 11–18.

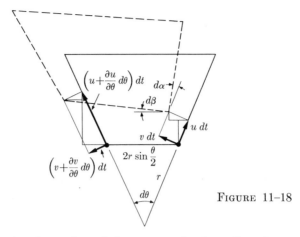

FIGURE 11–18

During a time dt, the elongation of the element in the θ-direction can be seen to be

$$\left(v + \frac{\partial v}{\partial \theta}\,d\theta\right) dt \cos \frac{d\theta}{2} + \left(u + \frac{\partial u}{\partial \theta}\,d\theta\right) dt \sin \frac{d\theta}{2} + u\,dt \sin \frac{d\theta}{2} - v\,dt \cos \frac{d\theta}{2}.$$

Dividing this by the product of dt and the initial length $2r \sin (d\theta/2)$, and using $\cos (d\theta/2) = 1$ and $\sin (d\theta/2) = d\theta/2$ for infinitesimal $d\theta$, we have the rate of longitudinal strain

$$e_{\theta\theta} = \frac{1}{r}\frac{\partial v}{\partial \theta} + \frac{u}{r}.$$

From Fig. 11–18, it can also be seen that

$$d\alpha = \frac{\partial v}{\partial r}\,dt \quad \text{and} \quad d\beta = \left(\frac{1}{r}\frac{\partial u}{\partial \theta} - \frac{v}{r}\right) dt.$$

(See also Problem 6–2.) Thus the rate of shearing strain $2e_{r\theta}$ is

$$2e_{r\theta} = \frac{d\alpha + d\beta}{dt} = \frac{\partial v}{\partial r} + \frac{1}{r}\frac{\partial u}{\partial \theta} - \frac{v}{r}.$$

It is not difficult to see that

$$e_{rr} = \frac{\partial u}{\partial r}, \qquad\qquad e_{zz} = \frac{\partial w}{\partial z},$$

$$e_{\theta z} = \frac{1}{2}\left(\frac{\partial v}{\partial z} + \frac{\partial w}{r\,\partial \theta}\right), \qquad e_{zr} = \frac{1}{2}\left(\frac{\partial w}{\partial r} + \frac{\partial u}{\partial z}\right).$$

For Newtonian fluids, the stress components of this element can be obtained from Eq. 11–25. When compressibility can be ignored, this equation becomes

$$S_{ij} = -p\,\delta_{ij} + 2\mu e_{ij}.$$

This equation relates the stress and the rate of strain at a point and is therefore not invalidated by the variation of the r- and θ-directions from point to point. Using the subscripts r, θ, and z for 1, 2, and 3, respectively, we have

$$S_{rr} = -p + 2\mu e_{rr} = -p + 2\mu \frac{\partial u}{\partial r},$$

$$S_{\theta\theta} = -p + 2\mu \left(\frac{1}{r}\frac{\partial v}{\partial \theta} + \frac{u}{r}\right),$$

$$S_{zz} = -p + 2\mu \frac{\partial w}{\partial z},$$

$$S_{r\theta} = 2\mu e_{r\theta} = \mu \left(\frac{\partial v}{\partial r} + \frac{1}{r}\frac{\partial u}{\partial \theta} - \frac{v}{r}\right),$$

$$S_{\theta z} = \mu \left(\frac{\partial v}{\partial z} + \frac{1}{r}\frac{\partial w}{\partial \theta}\right),$$

$$S_{zr} = \mu \left(\frac{\partial w}{\partial r} + \frac{\partial u}{\partial z}\right).$$

To derive the momentum equations, we first consider the surface forces on the element shown in Fig. 11–19. For the r-direction, we have

$$\left(S_{rr} + \frac{\partial S_{rr}}{\partial r}\, dr\right)(r + dr)\, d\theta\, dz - S_{rr}r\, d\theta\, dz$$

$$- \left(S_{\theta\theta} + \frac{\partial S_{\theta\theta}}{\partial \theta}\, d\theta\right) dr\, dz \sin \frac{d\theta}{2} - S_{\theta\theta}\, dr\, dz \sin \frac{d\theta}{2}$$

$$+ \left(S_{\theta r} + \frac{\partial S_{\theta r}}{\partial \theta}\, d\theta\right) dr\, dz \cos \frac{d\theta}{2} - S_{\theta r}\, dr\, dz \cos \frac{d\theta}{2}$$

$$+ \left(S_{zr} + \frac{\partial S_{zr}}{\partial z}\, dz\right) r\, d\theta\, dr - S_{zr}r\, d\theta\, dr.$$

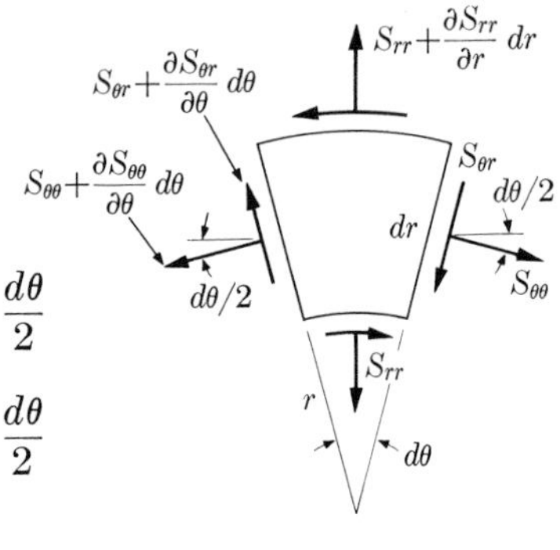

FIGURE 11–19

Dividing this by the mass $\rho r\, d\theta\, dr\, dz$, using $\cos (d\theta/2) = 1$, and $\sin (d\theta/2) = d\theta/2$ for infinitesimal $d\theta$, assuming μ to be constant, and neglecting higher-order terms, we have the surface forces per unit mass in the r-direction:

$$\frac{1}{\rho}\left(\frac{\partial S_{rr}}{\partial r} + \frac{\partial S_{\theta r}}{r\, \partial \theta} + \frac{\partial S_{zr}}{\partial z} + \frac{S_{rr} - S_{\theta\theta}}{r}\right)$$

$$= -\frac{1}{\rho}\frac{\partial p}{\partial r} + \frac{\mu}{\rho}\left[\left(\frac{\partial^2 u}{\partial r^2} + \frac{1}{r}\frac{\partial u}{\partial r} + \frac{1}{r^2}\frac{\partial^2 u}{\partial \theta^2} + \frac{\partial^2 u}{\partial z^2}\right)\right.$$

$$\left. - \frac{u}{r^2} - \frac{2}{r^2}\frac{\partial v}{\partial \theta} + \frac{\partial}{\partial r}\left(\frac{\partial u}{\partial r} + \frac{u}{r} + \frac{1}{r}\frac{\partial v}{\partial \theta} + \frac{\partial w}{\partial z}\right)\right].$$

According to the equation of continuity in Eq. 4–21, the last parenthesis is zero when compressibility is negligible. With these surface forces in Eq. 5–4, we have the momentum equation for the r-direction:

$$\frac{\partial u}{\partial t} + u \frac{\partial u}{\partial r} + \frac{v}{r} \frac{\partial u}{\partial \theta} + w \frac{\partial u}{\partial z} - \frac{v^2}{r} = -\frac{\partial}{\partial r}\left(\frac{p}{\rho} + gh\right) + \nu \left(\nabla^2 u - \frac{u}{r^2} - \frac{2}{r^2} \frac{\partial v}{\partial \theta}\right),$$

where
$$\nabla^2 = \frac{\partial^2}{\partial r^2} + \frac{1}{r} \frac{\partial}{\partial r} + \frac{1}{r^2} \frac{\partial^2}{\partial \theta^2} + \frac{\partial^2}{\partial z^2}. \tag{11–34}$$

In a similar manner, we obtain for the θ- and z-directions

$$\frac{\partial v}{\partial t} + u \frac{\partial v}{\partial r} + \frac{v}{r} \frac{\partial v}{\partial \theta} + w \frac{\partial v}{\partial z} + \frac{uv}{r} = -\frac{\partial}{r \partial \theta}\left(\frac{p}{\rho} + gh\right) + \nu \left(\nabla^2 v + \frac{2}{r^2} \frac{\partial u}{\partial \theta} - \frac{v}{r^2}\right),$$
$$\tag{11–35}$$

$$\frac{\partial w}{\partial t} + u \frac{\partial w}{\partial r} + \frac{v}{r} \frac{\partial w}{\partial \theta} + w \frac{\partial w}{\partial z} = -\frac{\partial}{\partial z}\left(\frac{p}{\rho} + gh\right) + \nu \nabla^2 w. \tag{11–36}$$

Example 11–6. When a steady laminar flow is produced by a long circular cylinder rotating in an infinite body of viscous fluid, the motion can be shown to be irrotational with $v = \Gamma_0/2\pi r$, where Γ_0 is the common circulation along any of the circular paths (see Problem 11–24). Find the unsteady motion when the cylinder stops suddenly. For simplicity, the radius of the cylinder may be considered to be infinitesimal.

To find the unsteady motion with $v(r, t)$ and $u = w = 0$, the momentum equation is, according to Eq. 11–35,

$$\frac{1}{\nu} \frac{\partial v}{\partial t} = \frac{\partial^2 v}{\partial r^2} + \frac{1}{r} \frac{\partial v}{\partial r} - \frac{v}{r^2},$$

with the boundary conditions that $v = \Gamma_0/2\pi r$ at $t = 0$, and $v = 0$ at $r = 0$ for $t > 0$. It can be seen that the solution of v must depend on Γ_0, r, and (νt). These four quantities can be expressed dimensionally in terms of length and time. Therefore there are two independent dimensionless products, say rv/Γ_0 and $r^2/\nu t$, and rv/Γ_0 must be a function of $r^2/\nu t$ (see Eqs. 2–3 and 2–4); i.e.,

$$v = \frac{\Gamma_0}{r} f(\eta), \qquad \eta = \frac{r^2}{\nu t}.$$

To solve for $f(\eta)$, substitute the following into the momentum equation:

$$\frac{\partial v}{\partial t} = \frac{\Gamma_0}{r} \frac{df}{d\eta} \frac{\partial \eta}{\partial t} = -\frac{\Gamma_0 r}{\nu t^2} \frac{df}{d\eta}, \qquad \frac{\partial v}{\partial r} = \frac{\Gamma_0}{r} \frac{df}{d\eta} \frac{\partial \eta}{\partial r} - \frac{\Gamma_0}{r^2} f = \frac{2\Gamma_0}{\nu t} \frac{df}{d\eta} - \frac{\Gamma_0}{r^2} f,$$

$$\frac{\partial^2 v}{\partial r^2} = \frac{2\Gamma_0}{\nu t} \frac{d^2 f}{d\eta^2} \frac{\partial \eta}{\partial r} - \left(\frac{\Gamma_0}{r^2} \frac{df}{d\eta} \frac{\partial \eta}{\partial r} - \frac{2\Gamma_0}{r^3} f\right) = \frac{4r\Gamma_0}{\nu^2 t^2} \frac{d^2 f}{d\eta^2} - \frac{2\Gamma_0}{r\nu t} \frac{df}{d\eta} + \frac{2\Gamma_0}{r^3} f,$$

to obtain
$$\frac{d^2 f}{d\eta^2} + \frac{1}{4} \frac{df}{d\eta} = 0.$$

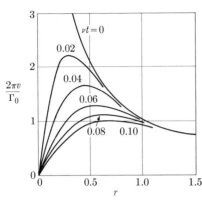

FIGURE 11–20

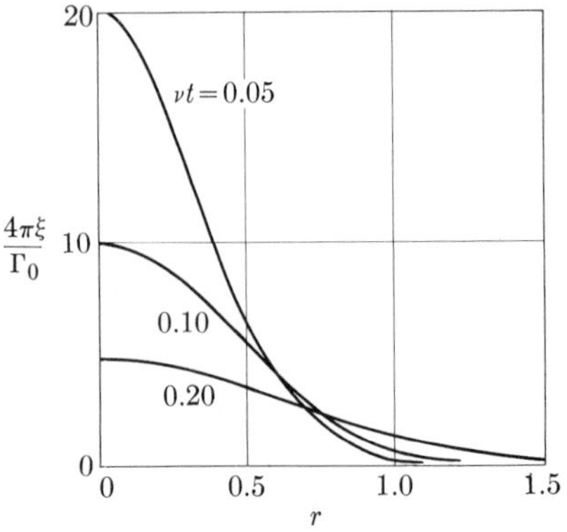

FIGURE 11–21

Integrating, we have

$$\frac{df}{d\eta} + \frac{f}{4} = c_1.$$

The general solution of this linear equation is

$$f = 4c_1 + c_2 e^{-\eta/4} \qquad \text{or} \qquad v = \frac{\Gamma_0}{r}\left(4c_1 + c_2 e^{-r^2/4\nu t}\right).$$

When the constants of integration are determined from the boundary conditions, we have

$$\frac{2\pi r v}{\Gamma_0} = 1 - e^{-r^2/4\nu t}.$$

This solution is shown graphically in Fig. 11–20.

Since $\Gamma_0/2\pi r$ is the initial speed at a point, $2\pi r v/\Gamma_0$ is therefore the ratio of v to the initial speed at a point. At a given time t, this ratio is reduced to one half at $r = \frac{5}{3}\sqrt{\nu t}$. Thus the viscous effect is confined mainly to a distance of the order of $\sqrt{\nu t}$.

Another interesting fact in this case is related to the rotational motion of the fluid elements. The magnitude of the vorticity of a fluid element in this case is (see Eq. 6–7)

$$\xi = \frac{1}{r}\frac{d(rv)}{dr} = \frac{\Gamma_0}{4\pi\nu t} e^{-r^2/4\nu t}.$$

This equation is plotted as Fig. 11–21. Originally, ξ is zero everywhere in the steady irrotational motion. When the cylinder stops, fluid elements in contact with the cylinder are given rotational motion. Vorticity is then spread by viscous action. The spread of vorticity in a viscous fluid can be shown to be analogous to the spread of heat by conduction.

PROBLEMS

11–22. Derive Eq. 11–35.

11–23. The annular space between two long concentric cylinders is filled with a viscous fluid. The inner cylinder is stationary while the outer cylinder rotates steadily. Assuming that the flow is axially symmetric as shown in Fig. 11–22, (a) simplify Eqs. 11–34 to 11–36 for this flow; (b) show that the velocity distribution is

$$v = \frac{r^2 - a^2}{b^2 - a^2} \cdot \frac{b^2 \omega}{r} \; ;$$

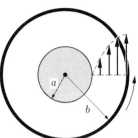

and (c) show that the torque per unit length acting on the outer or inner cylinder is

FIGURE 11–22

$$4\pi\mu\omega a^2 b^2/(b^2 - a^2).$$

11–24. A long cylinder of radius a rotates with a constant angular speed ω in an infinite body of viscous fluid. Show that the steady motion induced is irrotational with $v = a^2\omega/r$.

11–9. The Reynolds number

In the study of the effects of gravity on fluid motion in Section 10–2, the method of normalization of equations was used. This same method is to be employed here to study the effects of viscosity. The compressibility of the fluid is ignored in the following discussion.

Let L be a linear measurement (e.g., the size of a body) and U be a reference velocity of flow (e.g., the approach velocity of a stream). To change the dependent variables in the Navier-Stokes equations to u/U, v/U, w/U, and $p/\rho U^2$, and the independent variables to x/L, y/L, z/L, and tU/L, multiply each term of the equations by L/U^2. Equation 11–30 then becomes

$$\frac{\partial(u/U)}{\partial(tU/L)} + \left(\frac{u}{U}\right)\frac{\partial(u/U)}{\partial(x/L)} + \left(\frac{v}{U}\right)\frac{\partial(u/U)}{\partial(y/L)} + \left(\frac{w}{U}\right)\frac{\partial(u/U)}{\partial(z/L)}$$

$$= -\frac{\partial(p/\rho U^2)}{\partial(x/L)} - \frac{1}{(U^2/gL)}\frac{\partial(h/L)}{\partial(x/L)}$$

$$+ \frac{1}{(UL/\nu)}\left[\frac{\partial^2(u/U)}{\partial(x/L)^2} + \frac{\partial^2(u/U)}{\partial(y/L)^2} + \frac{\partial^2(u/U)}{\partial(z/L)^2}\right]. \quad (11\text{–}37)$$

The equations of motion for the y- and z-directions can be treated in the same manner. The equation of continuity has been written in terms of these new variables in Eq. 10–4. These equations are to be solved to fit the boundary conditions which are expressed as numbers; e.g., in the approach stream,

$u/U = 1$. Thus for a given shape of the boundary, the solutions of u/U, v/U, w/U, and $p/\rho U^2$ will involve, other than the independent variables x/L, y/L, z/L, and tU/L, only the dimensionless coefficients U^2/gL and UL/ν. The Froude number $U/\sqrt{gL}$, which represents the effects of gravity, has been discussed in Section 10–3. The quantity UL/ν is called the *Reynolds number* of the flow. It represents the effects of viscosity on the fluid motion. In other words, the effects of viscosity are the same if UL/ν is the same in geometrically similar systems, irrespective of the magnitude of ν itself.

That the effects of viscosity are indicated by the Reynolds number can be demonstrated with many examples. Take Poiseuille flow in a circular pipe in Example 11–2. The resistance coefficient f for laminar flow can be seen from Eq. 11–7 to be dependent on the Reynolds number VD/ν, where V is the mean velocity, and D is the diameter of the pipe. It has also been observed that the stability of the laminar flow depends on VD/ν. The flow becomes unstable when VD/ν exceeds about 2000, and will become turbulent if disturbed.

The Reynolds number may be interpreted as a general indication of the ratio of the inertial "force" to the viscous force acting on the fluid particles. Since the inertial "force" per unit mass (i.e., acceleration), expressed in terms such as $u \cdot \partial u/\partial x$, is of the order of magnitude of U^2/L, and the viscous force per unit mass, expressed in such terms as $\nu \cdot \partial^2 u/\partial x^2$, is of the order of magnitude of $\nu U/L^2$, this ratio of the forces is of the order of magnitude of

$$\frac{U^2/L}{\nu U/L^2} = \frac{UL}{\nu}.$$

It should be emphasized, however, that the Reynolds number indicates only the order of magnitude of the ratio of these two forces. In a given flow, different representative lengths and velocities of the flow may be chosen for L and U in defining a Reynolds number, thus giving different values for the number. However, in comparing flow systems with Reynolds numbers similarly defined, one can say that the system with a lower value has more pronounced viscous effects.

While the equations of motion of viscous fluid have been derived, the solution of these equations is generally very difficult. The main mathematical difficulty lies with the nonlinear convective terms $u \cdot \partial u/\partial x$, etc. Most solutions of the equations have been obtained for simple flow systems where these nonlinear terms are equal to zero (see the examples in Sections 11–2, 11–7, and 11–8). So far, only a handful of solutions have been obtained involving nonzero nonlinear terms. Most solutions of practical interest have been obtained for cases where, with approximations, the equations can be reduced to linear equations. One scheme of approximation is used for flows with an exceedingly low Reynolds number. The nonlinear acceleration terms everywhere are assumed to be negligible compared with the viscous terms. Several approximate solutions obtained in this manner will be presented in the next section. Another scheme of approximation is employed when the Reynolds number is exceedingly high. In this case, the viscous terms are assumed to be negligible (i.e., the fluid is assumed to be frictionless) everywhere except in a boundary layer. Because of the high rate of strain in a boundary layer, the viscous force is not small no matter how small the viscosity may be. This method of approximation will be discussed in Chapter 12.

11–10. Fluid flow with low Reynolds number

In dealing with flows with a very low Reynolds number, such as the slow fall of small particles in a fluid or the slow motion of a very viscous fluid through narrow passages, approximate solutions of the Navier-Stokes equations may be obtained by neglecting the nonlinear convective terms $u \cdot \partial u/\partial x$, etc. The possibility of this scheme of approximation is suggested by the fact that at a low Reynolds number, the inertial "force" is small compared with the viscous force acting on most, though not necessarily all, of the fluid particles. In cases where the velocity is low, the derivatives $\partial u/\partial x$, etc., may be expected to be small. The nonlinear terms $u \cdot \partial u/\partial x$, etc., being the product of two small quantities, may probably be neglected in comparison with the other terms in the equations. In some cases, the viscous forces are predominant mainly because the passage is narrow (see Example 11–8). For low Reynolds numbers, the effects of compressibility are always negligible. The approximate momentum equations are then assumed to be

$$\frac{\partial u}{\partial t} = -\frac{\partial}{\partial x}\left(\frac{p}{\rho} + gh\right) + \nu \left(\frac{\partial^2 u}{\partial x^2} + \frac{\partial^2 u}{\partial y^2} + \frac{\partial^2 u}{\partial z^2}\right). \tag{11–38}$$

The equations for the y- and z-directions are similar.

In this manner, Stokes obtained the solution for the slow motion of a sphere in a viscous fluid. He found that the drag D of the sphere of diameter d at a speed U is

$$D = 3\pi\mu U d.$$

With the *drag coefficient* C_D of a body defined by

$$D = C_D \tfrac{1}{2}\rho U^2 A,$$

where A is a projected area of the body, the drag coefficient of a sphere in slow motion is

$$C_D = \frac{D}{\tfrac{1}{2}\rho U^2 (\pi/4)\, d^2} = \frac{24}{Ud/\nu}. \tag{11–39}$$

The approximate solution has been found experimentally to be valid for $Ud/\nu < 1$, as shown in Fig. 2–9.

Several examples of solutions obtained with this method of approximation are presented below.

Example 11–7. The *Hele-Shaw apparatus* consists of a thin layer of fluid of uniform thickness in laminar motion around an obstacle, as shown in Fig. 11–23. Show that, when inertial "forces" are negligible, the flow pattern in each plane of the flow is similar to that of a two-dimensional irrotational flow.

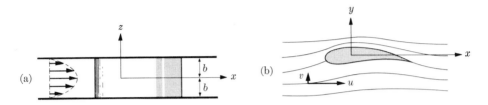

FIGURE 11–23

When the inertial "forces" are negligible, Eq. 11–38 and similar equations for the other directions are applicable. Since the thickness of the fluid is small, we further assume that the derivatives of u and v with respect to x and y are much smaller than those with respect to z (see Problem 11–28). For steady flow with $w = 0$, the equations of motion become

$$\frac{\partial}{\partial x}\left(\frac{p}{\rho} + gh\right) = \nu\,\frac{\partial^2 u}{\partial z^2},$$

$$\frac{\partial}{\partial y}\left(\frac{p}{\rho} + gh\right) = \nu\,\frac{\partial^2 v}{\partial z^2}, \tag{11–40}$$

$$\frac{\partial}{\partial z}\left(\frac{p}{\rho} + gh\right) = 0.$$

The last equation indicates that $p/\rho + gh$ is independent of z. With the derivatives of $p/\rho + gh$ independent of z, the first two equations yield upon integration with respect to z,

$$\frac{\partial u}{\partial z} = \frac{z}{\nu}\frac{\partial}{\partial x}\left(\frac{p}{\rho} + gh\right) + f_1(x, y) \qquad \text{and} \qquad \frac{\partial v}{\partial z} = \frac{z}{\nu}\frac{\partial}{\partial y}\left(\frac{p}{\rho} + gh\right) + f_2(x, y).$$

The functions f_1 and f_2 are equal to zero because $\partial u/\partial z = 0$ and $\partial v/\partial z = 0$ at $z = 0$ due to symmetry. With the boundary conditions that $u = v = 0$ at $z = \pm b$, these two equations give

$$u = -\frac{b^2 - z^2}{2\nu}\frac{\partial}{\partial x}\left(\frac{p}{\rho} + gh\right) \qquad \text{and} \qquad v = -\frac{b^2 - z^2}{2\nu}\frac{\partial}{\partial y}\left(\frac{p}{\rho} + gh\right).$$

It can be seen that the velocity distribution is parabolic along the z-direction. For any plane at $z = z_0$, we have

$$u = -\frac{\partial}{\partial x}\left[\frac{b^2 - z_0^2}{2\nu}\left(\frac{p}{\rho} + gh\right)\right], \qquad v = -\frac{\partial}{\partial y}\left[\frac{b^2 - z_0^2}{2\nu}\left(\frac{p}{\rho} + gh\right)\right].$$

This velocity distribution is the same as a two-dimensional irrotational flow with the velocity potential

$$\phi(x, y) = \frac{b^2 - z_0^2}{2\nu}\left(\frac{p}{\rho} + gh\right),$$

where $p/\rho + gh$ is a function of x and y. Unlike an irrotational flow, however, the velocity at the solid surface is zero. The flow in the apparatus is rotational, with components of angular velocity in the x- and y-directions, as witnessed from the parabolic velocity distribution in the z-direction. However, the viscous forces on each particle are such that the component of angular velocity in the z-direction remains practically zero except at the solid surface.

It has been found that ink injected into the Hele-Shaw apparatus makes the apparatus useful in the visualization of irrotational flow around two-dimensional bodies of an arbitrary shape. The velocity should be kept very low and the curvature of the body should be gentle such that acceleration is negligible everywhere, as assumed in the above analysis, and separation of the flow from the body is avoided.

Example 11–8. In Fig. 11–24 is shown a slipper bearing in which two inclined surfaces move relative to each other. Show that with a thin film of viscous fluid in between, there is very little frictional resistance even under large normal load

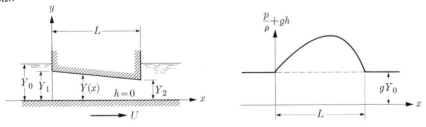

FIGURE 11–24

The relative speed U in this case is not necessarily small. It is to be shown presently that if the thickness Y of the film is sufficiently small compared with the width L of the bearing, Eq. 11–38 can be used in this case. We show this by estimating the order of magnitude of the various terms in the equations of motion. (This method of simplification has been used in Section 10–4 in the study of shallow-water waves.)

For the assumed steady two-dimensional motion, the equations of motion are

$$\frac{\partial u}{\partial x} + \frac{\partial v}{\partial y} = 0, \tag{11-41}$$

$$u\frac{\partial u}{\partial x} + v\frac{\partial u}{\partial y} = -\frac{\partial}{\partial x}\left(\frac{p}{\rho} + gh\right) + \nu\left(\frac{\partial^2 u}{\partial x^2} + \frac{\partial^2 u}{\partial y^2}\right), \tag{11-42}$$

$$u\frac{\partial v}{\partial x} + v\frac{\partial v}{\partial y} = -\frac{\partial}{\partial y}\left(\frac{p}{\rho} + gh\right) + \nu\left(\frac{\partial^2 v}{\partial x^2} + \frac{\partial^2 v}{\partial y^2}\right). \tag{11-43}$$

Let $Y \sim \delta L$, where the symbol $\sim$ relates quantities of the same order of magnitude. As $Y \ll L$, we have $\delta \ll 1$. Since the variation Δu in a distance L is of the same order as U, we have for Eq. 11–41 $\partial u/\partial x \sim U/L$. Let V be the maximum v. Then $\partial v/\partial y \sim V/Y \sim V/\delta L$. Since the two terms in Eq. 11–41 must be of the same magnitude,

$$\frac{U}{L} \sim \frac{V}{\delta L}, \quad \text{giving} \quad V \sim \delta U \quad \text{(i.e., } V \ll U\text{)}.$$

For Eq. 11–42, we have

$$u\frac{\partial u}{\partial x} \sim \frac{U^2}{L}, \qquad v\frac{\partial u}{\partial y} \sim \delta U\frac{U}{\delta L} \sim \frac{U^2}{L},$$

$$\nu\frac{\partial^2 u}{\partial x^2} \sim \left(\frac{\nu}{UL}\right)\frac{U^2}{L}, \qquad \nu\frac{\partial^2 u}{\partial y^2} \sim \left(\frac{\nu}{UL}\right)\frac{U^2}{\delta^2 L} \sim \left(\frac{L}{Y}\right)^2\left(\frac{\nu}{UL}\right)\frac{U^2}{L}.$$

So long as

$$\frac{UL}{\nu}\left(\frac{Y}{L}\right)^2 \ll 1 \quad \text{and} \quad \frac{Y}{L} \ll 1,$$

the other terms are negligible compared with $\nu \cdot \partial^2 u/\partial y^2$. Similarly, for Eq. 11–43,

$$u\frac{\partial v}{\partial x} \sim \frac{\delta U^2}{L}, \qquad v\frac{\partial v}{\partial y} \sim \frac{\delta U^2}{L},$$

$$\nu\frac{\partial^2 v}{\partial x^2} \sim \left(\frac{\nu}{UL}\right)\frac{\delta U^2}{L}, \qquad \nu\frac{\partial^2 v}{\partial y^2} \sim \frac{L}{Y}\left(\frac{\nu}{UL}\right)\frac{U^2}{L}.$$

All these terms are negligible compared with $\nu \cdot \partial^2 u/\partial y^2$ in Eq. 11–42. Thus the momentum equations become

$$0 = -\frac{\partial}{\partial x}\left(\frac{p}{\rho} + gh\right) + \nu\frac{\partial^2 u}{\partial y^2}, \qquad 0 = -\frac{\partial}{\partial y}\left(\frac{p}{\rho} + gh\right).$$

The last equation indicates that $p/\rho + gh$ is a function of x only. Integrating the first equation successively with respect to y, we have

$$\frac{\partial u}{\partial y} = \frac{y}{\nu} \frac{d}{dx} \left(\frac{p}{\rho} + gh \right) + f_1(x),$$

$$u = \frac{y^2}{2\nu} \frac{d}{dx} \left(\frac{p}{\rho} + gh \right) + f_1(x)y + f_2(x).$$

The functions f_1 and f_2 are to be determined from the boundary conditions that $u = U$ at $y = 0$, and that $u = 0$ at $y = Y(x)$, where $Y(x)$ is the known varying thickness of the fluid. Thus

$$u = U \left(1 - \frac{y}{Y} \right) - \frac{Y}{2\nu} y \left(1 - \frac{y}{Y} \right) \frac{d}{dx} \left(\frac{p}{\rho} + gh \right). \qquad (11\text{–}44)$$

For a bearing with a given shape $Y(x)$, the velocity distribution $u(x, y)$ depends on $(d/dx)(p/\rho + gh)$. Another available equation is the equation of continuity from which $p/\rho + gh$ can be determined. Using the continuity equation in the form $Q = $ constant, where Q is the discharge per unit length, we have

$$Q = \int_0^{Y(x)} u \, dy = \text{constant}.$$

Remembering that Y and $(d/dx)(p/\rho + gh)$ are functions of x only, we have by using Eq. 11–44

$$Q = \frac{UY}{2} - \frac{Y^3}{12\nu} \frac{d}{dx} \left(\frac{p}{\rho} + gh \right),$$

$$\frac{d}{dx} \left(\frac{p}{\rho} + gh \right) = \frac{6\nu U}{Y^2} - \frac{12\nu Q}{Y^3},$$

$$\frac{p}{\rho} + gh = 6\nu U \int \frac{dx}{Y^2} - 12\nu Q \int \frac{dx}{Y^3} + C. \qquad (11\text{–}45)$$

The constants Q and C can be determined from the condition that $p/\rho + gh$ is the same at $x = 0$ and at $x = L$. Note that $p/\rho + gh$ is not constant along the x-direction.

For the plane bearing shown in Fig. 11–24 with $Y(x) = Y_1 - mx$, where $m = (Y_1 - Y_2)/L$, Eq. 11–45 gives

$$\frac{p}{\rho} + gh = \frac{6\nu U}{m(Y_1 - mx)} - \frac{6\nu Q}{m(Y_1 - mx)^2} + C.$$

To determine Q and C, use $h = y$, so that $p/\rho = gh = gY_0$ at $x = 0$ and at

$x = L$. With these boundary conditions, we find

$$Q = \frac{U Y_1 Y_2}{Y_1 + Y_2}$$

and

$$C = g Y_0 - \frac{6\nu U}{m(Y_1 + Y_2)}.$$

Thus

$$\frac{p}{\rho} + gh = g Y_0 + \frac{6\nu U(Y_1 - mx - Y_2)x}{(Y_1 + Y_2)(Y_1 - mx)^2}.$$

It can be seen that with $Y = Y_1 - mx > Y_2$, the pressure is higher in between the two surfaces, as shown in Fig. 11–24(b). The velocity $u(x, y)$ can now be obtained from Eq. 11–44, if desired.

The normal force F per unit length that the bearing can support is

$$F = \int_0^L p \, dx = \int_0^L g(Y_0 - y) \, dx + \frac{6\mu U}{Y_1 + Y_2} \int_0^L \frac{(Y_1 - mx - Y_2)x}{(Y_1 - mx)^2} \, dx.$$

The first integral is the hydrostatic force. The force due to dynamic effects is

$$F_d = \frac{6\mu U}{Y_1 + Y_2} \int_0^L \frac{(Y_1 - mx - Y_2)x}{(Y_1 - mx)^2} \, dx$$

$$= \frac{6\mu U L^2}{(Y_1 - Y_2)^2} \left(\ln \frac{Y_1}{Y_1} - 2 \frac{Y_1 - Y_2}{Y_1 + Y_2} \right).$$

Other things being equal, this force is maximum when $Y_1 = 2.2Y_2$. Conversely, given F_d per unit length, we can determine the film thickness from this equation. For $Y_1 = 2.2Y_2$, we obtain $Y_1/L = 0.88\sqrt{\mu U/F_d}$.

It is left to the reader to show that the drag force D per unit length of the lower surface is

$$D = \frac{2\mu U L}{Y_1 - Y_2} \left(2 \ln \frac{Y_1}{Y_2} - 3 \frac{Y_1 - Y_2}{Y_1 + Y_2} \right).$$

For the case with $Y_1 = 2.2Y_2$, the ratio D/F_d is only $2.1 Y_1/L$ or $1.8\sqrt{\mu U/F_d}$.

PROBLEMS

11–25. Derive the expression for the drag of the bearing in Example 11–8.

11–26. Show that for very slow motion, the quantity $p/\rho + gh$ satisfies the Laplace equation

$$\frac{\partial^2}{\partial x^2} \left(\frac{p}{\rho} + gh \right) + \frac{\partial^2}{\partial y^2} \left(\frac{p}{\rho} + gh \right) + \frac{\partial}{\partial z^2} \left(\frac{p}{\rho} + gh \right) = 0.$$

11–27. Show that for a very slow two-dimensional viscous motion, the stream function $\psi(x, y)$ defined in Section 7–2 must satisfy the biharmonic equation

$$\frac{\partial^4 \psi}{\partial x^2} + 2 \frac{\partial^4 \psi}{\partial x^2 \partial y^2} + \frac{\partial^4 \psi}{\partial y^4} = 0,$$

in accordance with the momentum equations.

11–28. Show that Eq. 11–40 can be used for the flow in the Hele-Shaw apparatus in Fig. 11–23 if $(UL/\nu)(b/L)^2 \ll 1$.

12 Two-Dimensional Laminar Boundary Layers

12-1. Introduction

In Section 11–9, it was found that a dimensionless parameter called the Reynolds number, UL/ν, enters into consideration when viscous forces are included in the momentum equation. For water, ν is about 10^{-5} sec/ft^2, and for air, ν is about 6×10^{-3} sec/ft^2 (at room temperature). Thus we see that unless we are dealing with flows of very low velocity or very small characteristic length, the value of the Reynolds number will in general be rather large. The theory of boundary layers is a study of certain phenomena associated with flows with large Reynolds numbers.

At this point, it is appropriate to reconsider the assumption of frictionless flow. In reality, there is no such thing as a frictionless fluid. All fluids are viscous to some extent. However, when we consider a fluid-flow problem with very large Reynolds numbers, from the simple interpretation given in Section 11–9, it is intuitively clear that the effects of viscosity should be small. In fact, an idealized frictionless fluid (with $\nu = 0$) will yield an infinite Reynolds number in any problem. Hence, we may properly consider the theory of frictionless flows as a study of flows with extremely large Reynolds numbers. In other words, when the Reynolds number is very large, the flow pattern, the pressure distribution, etc., may be expected to be approximated by the theory of frictionless flow. This is really the basic justification for the study of frictionless flows.

There are certain phenomena, however, which are intrinsically viscous and cannot be dealt with by the theory of frictionless flow alone. For example, consider the large Reynolds number flow over a streamlined body. The theory of frictionless flow can describe quite adequately the flow pattern and pressure distribution (see Section 6–7). But suppose it is desired to know how much heat should be supplied to the body in order to maintain the surface temperature of this body at a certain specified level above the free-stream temperature. This is a convective "heat transfer" problem which is entirely out of reach by frictionless flow theory alone. As a second example, consider the problem of a submarine cruising under water at constant speed. If one applies the friction-

232

less flow theory, the flow pattern and pressure distribution can again be predicted. But if one proceeds to calculate the drag force experienced by the submarine based on the irrotational frictionless flow theory, the answer will be zero (D'Alembert's paradox, Section 8–4). The point here is that a submarine cruising at constant speed under water experiences drag due to viscous friction. Calculation of skin friction is another problem which cannot be treated by frictionless flow theory alone. As a last example, consider the large Reynolds number flow over a circular cylinder. A detailed calculation based on frictionless flow theory is available (see Examples 6–7 to 6–9). Problem 6–18 indicated that according to this theory the circular cylinder also experiences no drag. Also, it is readily shown from the solutions (such as Eq. 6–24) that the flow pattern predicted by the frictionless theory is symmetric fore and aft, which from common experience we know to be generally not true (compare Figs. 6–14 and 6–23). While the flow pattern near the front of the circular cylinder is reasonably approximated by the frictionless flow theory, near the rear of the body a "dead water" region with unsteady eddies is generally found. The flow appears to "separate" from the body shortly after it passes the mid-point of the body. We have here an example that the frictionless flow theory fails to predict correctly even the flow pattern, especially at the rear of the body.

It is an experimental fact that at solid boundaries the relative velocity between the fluid and the solid is zero. In other words, the fluid does not slip on a solid surface. According to our formulation of the frictionless flow theory, no attention was given to this phenomenon. Consequently, the frictionless flow theory requires only that the fluid velocity be tangential to the solid surface, and in general the theory predicts that the fluid will slip on a solid surface. Thus we see that while the frictionless flow pattern and pressure distribution may be quite accurate, the description of the tangential component of fluid velocity on and near the body surface is entirely incorrect. What actually happens is that immediately adjacent to the solid surface there is a thin layer, called the *boundary layer*, for which the frictionless flow theory ceases to be valid. Within this thin layer, frictional or viscous forces are of magnitudes comparable with inertial "force." The tangential component of the fluid velocity varies smoothly across this layer, going from the slip velocity predicted by the frictionless flow to zero on the surface of the body. Outside this thin layer, the frictionless description is generally adequate, but inside this layer, proper consideration must be given to the viscous forces. The boundary-layer theory in essence accepts the description given by the frictionless flow theory on the flow field as a whole, and confines its attention to the study of the flow *inside* these thin layers.

As shown in Fig. 12–1, the thickness of a boundary layer generally increases downstream. The flow in the upstream portion of the layer is usually laminar. This portion is called a *laminar boundary layer*. The flow in the layer may become turbulent at some downstream station. In this chapter, we shall confine

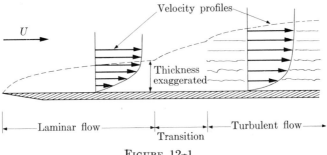

FIGURE 12-1

our attention to laminar boundary layers. Turbulent boundary layers are treated separately in Section 13-7. We shall study the problems of skin friction and the separation of the boundary layer. The problem of convective heat transfer in the boundary layer will not be considered due to the limited scope of this book.

12-2. The boundary-layer approximations

Let us first confine our attention to problems for which the frictionless flow theory gives accurate flow patterns and pressure distributions. We can generally expect this to be true whenever the body in question is streamlined. We shall indicate in this section what approximations we shall use and their physical meanings as a preliminary to the following sections.

One proceeds from the basic assumption that for large Reynolds number flows, the boundary layers are thin compared with the characteristic dimension of the problem. A number of simplifying approximations are immediately available, and collectively they are called the *boundary-layer approximations*. We itemize them as follows.

(1) The pressure within the boundary layer is approximately constant in the direction normal to the surface. Hence the value of the pressure at any point inside the boundary layer can be approximately identified with the pressure prevailing at the outer edge of the boundary layer where the frictionless flow theory is valid.

(2) The flow within this thin boundary layer is essentially parallel, so that the shearing stresses acting on a fluid element can be approximated by $\mu(\partial u/\partial y)$ only, where u is the tangential component of the fluid velocity and y is measured normal to the body surface.

(3) The thickness of this thin layer is small, not only compared with the characteristic length of the body, but is also small compared with the local radius of curvature of the body surface. This simplifying assumption allows us to write down the governing equations later in the so-called "boundary-layer coordinates."

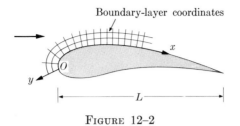

Boundary-layer coordinates

FIGURE 12–2

For simplicity we shall deal only with two-dimensional flows in this chapter. The physical derivation of the governing equations proceeds as follows. We first set up a coordinate system (x, y) on the body as shown in Fig. 12–2. The curvilinear x-axis lies along the body surface and coordinate y is measured in a direction perpendicular to it. Such a coordinate system is called a *boundary-layer coordinate system*. The basic assumption that the boundary layer is thin implies that in the (x, y) coordinate system our region of interest is confined to

$$\frac{x}{L} = O(1), \qquad \frac{y}{L} \ll 1. \tag{12–1}$$

Here the symbol $O(\ )$ is read "order of" and $O(1)$ indicates a magnitude which is estimated to be not much different from unity. Although the boundary-layer coordinate system is curvilinear, it is clear that, provided our attention is confined to this thin layer, *locally* it must *appear* approximately as a rectangular coordinate system. We shall demonstrate with a circular cylindrical surface that the equation of continuity for the boundary layer coordinates is the same equation as for rectangular coordinates (see Eq. 4–18); i.e.,

$$\frac{\partial \rho}{\partial t} + \frac{\partial (\rho u)}{\partial x} + \frac{\partial (\rho v)}{\partial y} = 0, \tag{12–2}$$

where u is the velocity component parallel to the body surface (in the x-direction), and v is the velocity component perpendicular to the body surface (in the y-direction).

For the cylindrical coordinates shown in Fig. 12–3, the equation of continuity is, from Eq. 4–20,

$$\frac{\partial \rho}{\partial t} + \frac{1}{r}\frac{\partial (\rho v r)}{\partial r} + \frac{1}{r}\frac{\partial (\rho u)}{\partial \theta} = 0,$$

where u and v are again the velocity components along and normal to the surface, respectively. With the boundary-layer coordinates

$$x = r_0 \theta,$$

$$y = r - r_0,$$

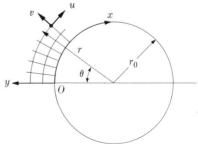

FIGURE 12–3

where r_0 is the radius of the cylinder, this equation becomes

$$r_0\left(1 + \frac{y}{r_0}\right)\frac{\partial\rho}{\partial t} + \frac{\partial}{\partial y}\left[\rho v r_0\left(1 + \frac{y}{r_0}\right)\right] + r_0\frac{\partial}{\partial x}(\rho u) = 0. \qquad (12\text{--}3)$$

Since our interest is confined to the thin boundary layer with $y \ll r_0$, we have

$$\frac{y}{r_0} \ll 1 \qquad \text{and} \qquad \frac{\rho v}{r_0} \ll \frac{\partial(\rho v)}{\partial y}.$$

Thus, approximately,

$$\frac{\partial\rho}{\partial t} + \frac{\partial(\rho u)}{\partial x} + \frac{\partial(\rho v)}{\partial y} = 0,$$

which is Eq. 12–2. Our demonstration is then complete.

We proceed in an analogous manner to derive the momentum equation. The particle acceleration a_x is, from the first equation in Eq. 4–15,

$$a_x = \frac{\partial u}{\partial t} + u\frac{\partial u}{\partial x} + v\frac{\partial u}{\partial y}. \qquad (12\text{--}4)$$

The net force per unit mass acting on the particle is simply

$$-\frac{1}{\rho}\frac{dp}{dx} + \frac{1}{\rho}\frac{\partial}{\partial y}\left(\mu\frac{\partial u}{\partial y}\right), \qquad (12\text{--}5)$$

where for simplicity we have neglected body forces such as gravity. Thus the x-momentum equation for the boundary layer is

$$\frac{\partial u}{\partial t} + u\frac{\partial u}{\partial x} + v\frac{\partial u}{\partial y} = -\frac{1}{\rho}\frac{dp}{dx} + \frac{1}{\rho}\frac{\partial}{\partial y}\left(\mu\frac{\partial u}{\partial y}\right). \qquad (12\text{--}6)$$

It is instructive to compare this equation with Eq. 11–29. There are two important differences. First, $\partial p/\partial x$ in Eq. 11–29 is written as dp/dx in Eq. 12–6. The total derivative indicates that we have taken advantage of the first boundary-layer approximation to assume that p is a known function of x only. Secondly, the viscous terms in Eq. 11–29, when written out in full, are

$$-\frac{\partial}{\partial x}\left[\tfrac{2}{3}\mu\left(\frac{\partial u}{\partial x} + \frac{\partial v}{\partial y}\right)\right] + \frac{\partial}{\partial x}\left[\mu\left(\frac{\partial u}{\partial x} + \frac{\partial u}{\partial x}\right)\right] + \frac{\partial}{\partial y}\left[\mu\left(\frac{\partial v}{\partial x} + \frac{\partial u}{\partial y}\right)\right].$$

In accordance with the second boundary-layer approximation, only the last term $\partial(\mu\,\partial u/\partial y)/\partial y$ needs to be retained.

We shall treat only problems in which it is possible to assume that ρ and μ are constants. Therefore, the following developments are valid only for relatively low Mach number flows *and* for surface temperature not too different from the free-stream temperature. The continuity and momentum equations,

Eqs. 12–2 and 12–6, therefore reduce to

$$\frac{\partial u}{\partial x} + \frac{\partial v}{\partial y} = 0, \tag{12–7a}$$

$$\frac{\partial u}{\partial t} + u\frac{\partial u}{\partial x} + v\frac{\partial u}{\partial y} = -\frac{1}{\rho}\frac{dp}{dx} + \nu\frac{\partial^2 u}{\partial y^2}, \tag{12–7b}$$

which are called the *boundary-layer equations*. Their derivations were due to L. Prandtl. Formally, the statement of a boundary-layer problem is as follows. Given a solid body in a flow such that the characteristic Reynolds number is large, find solutions $u(x, y)$ and $v(x, y)$ for Eqs. 12–7, using $p(x)$ calculated from the frictionless flow theory. The solutions must satisfy the following boundary conditions.

(1) On the body surface, $y = 0$, we require $u = v = 0$ to satisfy the *no-slip condition*.

(2) At the outer edge of the boundary layer, we require the tangential velocity component u to approach $U(x)$, the slip velocity calculated from frictionless flow theory. The values of $p(x)$ and $U(x)$ are related by Bernoulli's equation $p + \frac{1}{2}\rho U^2 = $ constant.

(3) At some upstream station in the boundary layer, the velocity profile must agree with some initial condition dictated by the physical problem. In some problems, this condition is automatically met. For example, for the body shown in Fig. 12–2, the velocity profile at the front stagnation point $(x = 0)$ of the body is $u = U = 0$.

With Eqs. 12–7a and 12–7b, we can proceed with the study of boundary-layer problems, as shall be done in Section 12–4. The derivation of these equations is based on physical arguments. The following section illustrates that it is also possible to arrive at the same results by purely mathematical means.

12–3. The boundary-layer equations

To demonstrate the derivation of the boundary-layer equations, we shall consider the flow over a circular cylinder as shown in Fig. 12–3. The equation of continuity has been derived in the previous section as Eq. 12–2. The Navier-Stokes momentum equations in cylindrical coordinates are given by Eqs. 11–34 to 11–36. Since the flow is assumed two-dimensional, we need not consider Eq. 11–36. Let us first nondimensionalize our variables:

$$\frac{u}{U} = u^*, \qquad \frac{v}{U} = v^*,$$

$$\frac{r}{r_0} = r^*, \qquad \frac{Ut}{r_0} = t^*, \qquad \frac{p}{\rho U^2} = p^*, \tag{12–8}$$

where u and v are the tangential and normal velocity components, respectively, and U is a characteristic speed. Then Eqs. 11–34 and 11–35 become

$$\frac{\partial u^*}{\partial t^*} + v^* \frac{\partial u^*}{\partial r^*} + \frac{u^*}{r^*} \frac{\partial u^*}{\partial \theta} + \frac{u^* v^*}{r^*} = -\frac{1}{r^*} \frac{\partial p^*}{\partial \theta} + \frac{1}{R} \left[\nabla^{*2} u^* + \frac{2}{r^{*2}} \frac{\partial v^*}{\partial \theta} - \frac{u^*}{r^{*2}} \right],$$

(12–9)

$$\frac{\partial v^*}{\partial t^*} + v^* \frac{\partial v^*}{\partial r^*} + \frac{u^*}{r^*} \frac{\partial v^*}{\partial \theta} - \frac{u^{*2}}{r^*} = -\frac{\partial p^*}{\partial r^*} + \frac{1}{R} \left[\nabla^{*2} v^* - \frac{v^*}{r^{*2}} - \frac{2}{r^{*2}} \frac{\partial v^*}{\partial \theta} \right],$$

(12–10)

where

$$R = \frac{U r_0}{\nu}, \qquad \nabla^{*2} = \frac{\partial^2}{\partial r^{*2}} + \frac{1}{r^*} \frac{\partial}{\partial r^*} + \frac{1}{r^{*2}} \frac{\partial^2}{\partial \theta^2}.$$

It is clear from Eqs. 12–9 and 12–10 that for very large R, the bracket term in each of the equations will become negligible, and the equations then reduce to that of frictionless flow theory.

In order to study the boundary layer, it is clear that some viscous terms must be kept. Physically, we know that for large R, the magnitudes of v^* and $r^* - 1$ are very small in the boundary layer. In fact, we suspect that they may be of the order of $1/\sqrt{R}$ (see Example 11–5). In order to magnify them so as to study them more easily, we introduce now the following dimensionless boundary-layer variables and coordinates, each identified by a subscript b:

$$u_b = u^*, \qquad v_b = \sqrt{R}\, v^*, \qquad p_b = p^*,$$

(12–11)

$$x_b = \theta, \qquad y_b = \sqrt{R}\,(r^* - 1), \qquad t_b = t^*.$$

With a large value of R, the magnitudes of v_b and y_b are now comparable to u_b and x_b, respectively, and are of the order of unity. Then Eqs. 11–9 and 11–10 become

$$\frac{\partial u_b}{\partial t_b} + v_b \frac{\partial u_b}{\partial y_b} + \frac{1}{1 + (y_b/\sqrt{R})} u_b \frac{\partial u_b}{\partial x_b} + \frac{u_b v_b}{\sqrt{R} + y_b}$$

$$= -\frac{1}{1 + (y_b/\sqrt{R})} \frac{\partial p_b}{\partial x_b} + \frac{1}{R} \left[R \frac{\partial^2 u_b}{\partial y_b^2} + \frac{\sqrt{R}}{1 + (y_b/\sqrt{R})} \frac{\partial u_b}{\partial y_b} \right.$$

$$\left. + \frac{1}{[1 + (y_b/\sqrt{R})]^2} \left(\frac{\partial^2 u_b}{\partial x_b^2} + 2 \frac{\partial v_b}{\partial x_b} - u_b \right) \right],$$

(12–12a)

$$\frac{1}{R} \left\{ \frac{\partial v_b}{\partial t_b} + v_b \frac{\partial v_b}{\partial y_b} + \frac{u_b}{1 + (y_b/\sqrt{R})} \frac{\partial v_b}{\partial x_b} \right\} - \frac{u_b^2}{\sqrt{R} + y_b} = -\frac{\partial p_b}{\partial y_b} + \frac{1}{R^2}$$

$$\times \left[R \frac{\partial^2 v_b}{\partial y_b^2} + \frac{\sqrt{R}}{1 + (y_b/\sqrt{R})} \frac{\partial v_b}{\partial y_b} + \frac{1}{[1 + (y_b/\sqrt{R})]^2} \left(\frac{\partial^2 v_b}{\partial x_b^2} - 2 \frac{\partial v_b}{\partial x_b} - v_b \right) \right].$$

(12–12b)

When Reynolds number R is very large, Eqs. 12–12a and 12–12b reduce to

$$\frac{\partial u_b}{\partial t_b} + u_b \frac{\partial u_b}{\partial x_b} + v_b \frac{\partial u_b}{\partial y_b} = -\frac{\partial p_b}{\partial x_b} + \frac{\partial^2 u_b}{\partial y_b^2}, \qquad (12\text{–}13)$$

$$\frac{\partial p_b}{\partial y_b} = 0, \qquad (12\text{–}14)$$

where terms of order $R^{-1/2}$ have been neglected. From Eq. 12–14, we conclude that p_b is independent of y_b, and depends on x_b only. Hence the term $\partial p_b/\partial x_b$ in Eq. 12–13 can be replaced by dp_b/dx_b. By inspection, Eq. 12–13 is identical with 12–7b if the comparison is made in terms of the same variables.

From Eqs. 12–11, 12–13, and 12–14, two significant results emerge. First, we see that the thickness of the boundary layer is proportional to $R^{-1/2}$, since $y_b = 0(1)$ and

$$r^* - 1 = \frac{r - r_0}{r_0} = \frac{y_b}{\sqrt{R}}.$$

This result is in agreement with our original assumption that the boundary layer is thin when R is large. Secondly, we see that R does not appear explicitly in Eq. 12–13. The boundary conditions also turn out to be independent of R (see Section 12–4). Thus, after a solution is found for a laminar boundary-layer problem in these variables, the result can be used in any large Reynolds number laminar flow.

PROBLEMS

12–1. Using the transformation Eqs. 12–8 and 12–11, repeat the derivation of Eq. 12–2 for a circular cylinder. Show that the resultant continuity equation is accurate to order $R^{-1/2}$.

12–2. Show that the skin friction coefficient defined as

$$C_f = \frac{[\mu(\partial u/\partial y)]_{y=0}}{\frac{1}{2}\rho U^2}$$

is proportional to $R^{-1/2}$ for laminar boundary layers.

12–4. The flat plate

The simplest problem in boundary-layer theory is that of the two-dimensional steady laminar flow over a semi-infinite flat plate at zero angle of attack. This problem was first solved by Blasius in 1913.

We consider here a rectangular cartesian coordinate system (x, y) with its origin at the leading edge of the flat plate (see Fig. 12–1). The flat plate itself lies on the positive x-axis, and there is a uniform flow parallel to the x-axis from left to right with constant velocity U. Note that this coordinate system is identically the boundary-layer coordinate system without modification. The frictionless flow theory will predict that no disturbance will be created by the presence of the plate; hence its solution is simply

$$u = U = \text{constant},$$
$$v = 0, \qquad \qquad (12\text{--}15)$$
$$p = p_\infty = \text{constant}.$$

Using the concept of a two-dimensional stream function discussed in Section 7–2, we write

$$u(x, y) = -\frac{\partial \psi}{\partial y}, \qquad v(x, y) = \frac{\partial \psi}{\partial x}, \qquad (12\text{--}16)$$

where $\psi(x, y)$ is a stream function. Substituting Eq. 12–16 into Eq. 12–7b, we obtain

$$\frac{\partial \psi}{\partial y} \frac{\partial^2 \psi}{\partial x\, \partial y} - \frac{\partial \psi}{\partial x} \frac{\partial^2 \psi}{\partial y^2} = \nu \frac{\partial^3 \psi}{\partial y^3}. \qquad (12\text{--}17)$$

The appropriate boundary conditions are as follows:
   (a) No-slip condition at the plate, i.e.:

$$\frac{\partial \psi}{\partial y} = -u = 0 \qquad \text{at } y = 0,$$

and

$$\frac{\partial \psi}{\partial x} = v = 0 \qquad \text{at } y = 0. \qquad (12\text{--}18)$$

   (b) Free-stream condition: Away from the plate (at some finite y), the velocity component parallel to the plate should approach U. This boundary condition is difficult to write down mathematically at this moment, but we shall see later how it is resolved.
   (c) Upstream condition: We expect that at the leading edge, the flow is essentially undisturbed except for the no-slip condition. We may write

$$u = -\frac{\partial \psi}{\partial y} = U \qquad \text{at } x = 0, \quad y > 0,$$

$$u = -\frac{\partial \psi}{\partial y} = 0 \qquad \text{at } x = 0, \quad y = 0. \qquad (12\text{--}19)$$

Blasius found that it is possible to reduce Eq. 12–17 to an ordinary differential equation by using the following dimensionless variables, generally known as the Blasius variables:

$$\xi = x/L, \qquad \eta = y \sqrt{\frac{U}{\nu x}}, \qquad f(\xi, \eta) = -\frac{1}{\sqrt{\nu x U}} \psi(x, y), \qquad (12\text{--}20)$$

where L is a constant with the unit of length. Note that the normal coordinate

y has been defined in a manner similar to y_b in Eq. 12–11. In terms of these variables, we have

$$u = -\frac{\partial \psi}{\partial y} = U \frac{\partial f}{\partial \eta}, \qquad v = \frac{\partial \psi}{\partial x} = -\frac{1}{2} \sqrt{\frac{\nu U}{x}} \left(f - \eta \frac{\partial f}{\partial \eta} + 2\xi \frac{\partial f}{\partial \xi} \right). \quad (12\text{--}21)$$

Then Eq. 12–17 becomes

$$\frac{\partial^3 f}{\partial \eta^3} + \frac{1}{2} f \frac{\partial^2 f}{\partial \eta^2} = \xi \left(\frac{\partial f}{\partial \eta} \frac{\partial^2 f}{\partial \xi \partial \eta} - \frac{\partial f}{\partial \xi} \frac{\partial^2 f}{\partial \eta^2} \right). \quad (12\text{--}22)$$

The boundary conditions are now:

(a) No-slip condition: From Eq. 12–18, we have directly

$$\frac{\partial f}{\partial \eta} = 0 \qquad \text{at } \eta = 0, \quad (12\text{--}23a)$$

$$f = 0 \qquad \text{at } \eta = 0. \quad (12\text{--}23b)$$

(b) Free-stream condition: From the definition of η, it is seen that if the "edge" of the boundary layer is located at some small but finite value of y, the corresponding value of η will in general depend on ν, and will become infinitely large for vanishingly small ν. To put it another way, we recall that η is defined as

$$\eta = y \sqrt{\frac{U}{\nu x}} = \frac{y}{x} \sqrt{\frac{Ux}{\nu}},$$

where the quantity inside the square root has the form of a Reynolds number. Hence, in the limit of very large Ux/ν, for any finite value of y/x, the value of η will approach infinity. The free-stream condition is written accordingly as

$$\frac{u}{U} = \frac{\partial f}{\partial \eta} = 1 \qquad \text{at } \eta = \infty. \quad (12\text{--}24)$$

(c) Upstream condition: Conditions in Eqs. 12–23a and 12–24 have already satisfied this condition as given by Eq. 12–19.

Note that the boundary conditions are independent of the value of the Reynolds number UL/ν. (This is generally *not* true whenever the wall is porous and there is a flow from or into the wall.) At this point, the physical problem has been formulated in mathematical terms, and the remaining task is to solve for $f(\xi, \eta)$. We observe from Eq. 12–22 and conditions in Eqs. 12–23 and 12–24 that $f(\xi, \eta)$ need not depend on ξ. As a matter of fact, $f(\xi, \eta)$ must be independent of ξ, since the definition of ξ involves an arbitrary length L. Therefore, Eq. 12–22 can be written simply as an ordinary differential equation:

$$\frac{d^3 f}{d\eta^3} + \frac{1}{2} f \frac{d^2 f}{d\eta^2} = 0. \quad (12\text{--}25)$$

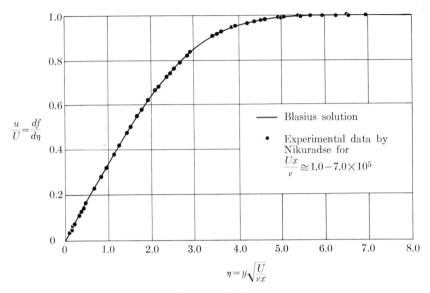

$$\frac{u}{U} = \frac{df}{d\eta}$$

Blasius solution

• Experimental data by Nikuradse for $\frac{Ux}{\nu} \cong 1.0 - 7.0 \times 10^5$

$$\eta = y\sqrt{\frac{U}{\nu x}}$$

FIGURE 12–4

This is generally known as the Blasius equation and $f(\eta)$ is called the *Blasius function*. It has been accurately computed, and Fig. 12–4 shows a plot of the velocity profile. Data from several experiments are also plotted, and it is seen that the agreement is excellent.

The skin friction distribution on the plate can be obtained as follows. Defining the skin friction coefficient as

$$c_f = \frac{\tau_0}{\frac{1}{2}\rho U^2},$$

where $\tau_0 = \mu(\partial u/\partial y)$ at $y = 0$, we obtain

$$c_f = \frac{\mu U}{\frac{1}{2}\rho U^2} \frac{d^2 f}{d\eta^2} \frac{\partial \eta}{\partial y} = 2\sqrt{\frac{\nu}{Ux}} \frac{d^2 f}{d\eta^2} \qquad \text{at } \eta = 0.$$

The numerical value of $d^2 f/d\eta^2$ at $\eta = 0$ is found from this solution and is 0.332. Thus, the laminar skin friction coefficient distribution is given by the simple formula

$$c_f = 0.664\sqrt{\nu/Ux}. \tag{12–26}$$

If the thickness $\delta(x)$ of the boundary layer is arbitrarily defined as the value of y, where $u = 0.99U$, it can be seen from Fig. 12–4 that, approximately,

$$\delta(x) \doteq 5.0\sqrt{\nu x/U}. \tag{12–27}$$

Thus, the thickness of a laminar boundary layer on a flat plate grows parabolically.

PROBLEM

12–3. Find the drag coefficient C_D (per unit width) as defined in Eq. 2–13 for skin friction of a rectangular flat plate of length L.

12–5. Application of momentum theorem

In spite of the importance of boundary-layer problems, there are only a limited number of known exact solutions. At this moment there is no general method available for the solution of a boundary-layer problem. However, there are many approximate methods, and the simplest and most popular is called the Kármán-Pohlhausen momentum integral technique. The basis for this technique is the application of the momentum theorem presented in Chapter 9. The theme of the arguments is as follows. For most engineering applications, it is seldom necessary to find the detailed velocity distributions inside the boundary layers. The important engineering results are usually the skin friction distribution and sometimes the thickness estimation of the boundary layer. If we begin the analysis by assuming some reasonable shape for the velocity profiles, it is possible to obtain the desired results by requiring that the flow in the boundary layer as a whole satisfies the laws of conservation of momentum and mass. As an introduction, we study again the laminar boundary layer of a flat plate by this approximate method. The general form of the Kármán-Pohlhausen momentum integral equation will be derived in the next section.

To apply this approximate method, the shape of the velocity distribution in the boundary layer must be assumed. The usefulness of this method lies in the fact that a usable result can be obtained with any reasonable assumed velocity profile.

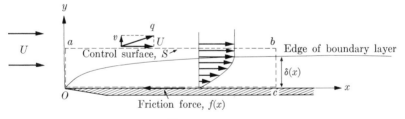

FIGURE 12–5

Example 12–1. Find the skin friction and the thickness $\delta(x)$ of the laminar boundary layer on a flat plate in a steady flow by using the momentum theorem.

Take a control surface S of unit width as shown in Fig. 12–5. From the momentum theorem, we have from Eq. 9–5:

$$\Sigma F_x = \int_S \rho u q \cos \gamma \, dS.$$

Since the pressure in this flow does not vary in the x-direction, we have

$$\Sigma F_x = -f(x),$$

where $f(x)$ is the total frictional force exerted by the plate over the length x. To evaluate the flux of momentum, assume the velocity distribution in the layer to be, say,

$$\frac{u}{U} = \sin\left(\frac{\pi}{2}\frac{y}{\delta}\right),$$

such that $u = 0$ at $y = 0$, and $u = U$ at $y = \delta(x)$. Then we have for the surface bc,

$$\int_{bc} \rho u q \cos \gamma \, dS = \rho U^2 \int_0^\delta \sin^2\left(\frac{\pi}{2}\frac{y}{\delta}\right) dy = \frac{\rho U^2 \delta}{2}.$$

For the surface oa, where $u = U$ and $q \cos \gamma = -U$, we have

$$\int_{oa} \rho u q \cos \gamma \, dS = -\rho U^2 \delta.$$

For the surface ab, where $u = U$ and $q \cos \gamma = v$, we have

$$\int_{ab} \rho u q \cos \gamma \, dS = \rho U \int_{ab} v \, dS.$$

But $\rho \int_{ab} v \, dx$ is the mass flux out of the surface ab. By continuity, it is equal to the difference between the mass flux over surfaces oa and bc. Thus

$$\rho \int_{ab} v \, dS = \rho \int_{oa} u \, dS - \rho \int_{bc} u \, dS = \rho \int_0^\delta U \, dy - \rho \int_0^\delta U \sin\left(\frac{\pi}{2}\frac{y}{\delta}\right) dy$$

$$= \left(1 - \frac{2}{\pi}\right) \rho U \delta.$$

Hence we have

$$-f(x) = \int_S \rho u q \cos \gamma \, dS = \tfrac{1}{2}\rho U^2 \delta - \rho U^2 \delta + \left(1 - \frac{2}{\pi}\right)\rho U^2 \delta$$

$$= -\frac{4 - \pi}{2\pi} \rho U^2 \delta. \tag{12–28}$$

The viscous shearing stress τ_0 on the wall can be computed from the assumed velocity profile

$$\tau_0 = \left(\mu \frac{\partial u}{\partial y}\right)_{y=0} = \frac{\pi}{2} \mu \frac{U}{\delta}. \tag{12–29}$$

In Eqs. 12–28 and 12–29, we have three unknown functions, namely, $f(x)$, $\delta(x)$, and $\tau_0(x)$. However, the total frictional force f is related to τ_0 by

$$f = \int_0^x \tau_0 \, dx.$$

Upon differentiation, we have

$$\tau_0 = \frac{df}{dx},$$ (12–30)

which provides the third equation.

A single equation for f can be obtained by eliminating τ_0 between Eqs. 12–29 and 12–30, and then using Eq. 12–28 to eliminate δ. We obtain

$$f\frac{df}{dx} = \frac{4 - \pi}{4}\,\mu\rho U^3,$$

which can readily be integrated to give

$$f = \sqrt{\frac{4 - \pi}{2}}\,\sqrt{\mu\rho U^3 x}.$$ (12–31)

From Eq. 12–28, the value of δ is then

$$\delta = \frac{2\pi}{4 - \pi}\,\frac{f}{U^2} = 4.80\,\sqrt{\frac{\mu x}{\rho U}}.$$ (12–32)

The value of τ_0 is obtained from Eq. 12–29:

$$\tau_0 = 0.328\rho U^2\,\sqrt{\frac{\mu}{\rho U x}}.$$ (12–33)

Thus the coefficient of skin friction c_f is

$$c_f = \frac{\tau_0}{\frac{1}{2}\rho U^2} = 0.656\,\sqrt{\frac{\mu}{\rho U x}}.$$ (12–34)

Note that although the velocity profile has been arbitrarily assumed, Eqs. 12–32 and 12–34 are good approximations of Blasius' results in Eqs. 12–27 and 12–26, respectively.

PROBLEM

12–4. Repeat the problem in Example 12–1 by assuming that

$$u = Uy/\delta \qquad (0 \le y \le \delta(x)),$$
$$u = U \qquad\quad (y \ge \delta(x)).$$

12–6. The Kármán-Pohlhausen momentum integral equation

The simple example worked out in the previous section illustrates the basic features of the momentum integral method. In this section, we shall derive formally the Kármán-Pohlhausen momentum integral equation starting from the two-dimensional steady laminar boundary-layer equations. The pressure gradient dp/dx can be arbitrary and boundary-layer coordinates are used so that the solid surface may be curved.

The boundary-layer equations are from Eq. 12–7

$$\frac{\partial u}{\partial x} + \frac{\partial v}{\partial y} = 0,$$

$$u \frac{\partial u}{\partial x} + v \frac{\partial u}{\partial y} = U \frac{dU}{dx} + \nu \frac{\partial^2 u}{\partial y^2},$$

where $U(x)$ is the velocity of the flow just outside of the boundary layer. The term $-(dp/dx)/\rho$ in Eq. 12–7b has been replaced by $U(dU/dx)$, since p and U outside the boundary layer are related by Bernoulli's equation, Eq. 5–11. Multiplying the continuity equation by $U - u$ and subtracting from the product the momentum equation, we obtain

$$\frac{\partial}{\partial x} [u(U - u)] + \frac{\partial}{\partial y} [v(U - u)] + (U - u) \frac{dU}{dx} = -\nu \frac{\partial^2 u}{\partial y^2}. \qquad (12\text{–}35)$$

This equation is then integrated with respect to y from $y = 0$ to a point outside the boundary layer $y = h$, $u = U$ where $\partial u/\partial y$ can be considered to be zero. After some manipulations, we obtain the *Kármán-Pohlhausen momentum integral equation*

$$\frac{d\theta}{dx} + \frac{1}{U} \frac{dU}{dx} (2\theta + \delta^*) = \frac{\tau_0}{\rho U^2}, \qquad (12\text{–}36)$$

where τ_0 is the shearing stress at the wall. The symbols $\theta(x)$ and $\delta^*(x)$ are called the *momentum* and *displacement thicknesses*, respectively, and are defined by

$$\theta = \int_0^h \frac{u}{U} \left(1 - \frac{u}{U}\right) dy, \qquad (12\text{–}37)$$

$$\delta^* = \int_0^h \left(1 - \frac{u}{U}\right) dy. \qquad (12\text{–}38)$$

Note that both θ and δ^* have the dimension of length. Their physical meanings are as follows. Consider a certain streamline at a distance y_1 from the wall. The total mass flow rate $\dot{m}$ between this streamline and the wall is

$$\dot{m} = \int_0^{y_1} \rho u \, dy.$$

If the flow were frictionless so that the boundary layer did not exist, the same amount of mass would flow between the wall and a streamline at y_2, given by

$$\dot{m} = \int_0^{y_2} \rho U \, dy.$$

Equating the above two equations for $\dot{m}$, we obtain

$$\int_0^{y_1} \rho u \, dy = \int_0^{y_2} \rho U \, dy \quad \text{or} \quad \int_0^{y_2} \rho(U - u) \, dy = \int_{y_2}^{y_1} \rho u \, dy. \quad (12\text{--}39)$$

If y_1 and y_2 are taken sufficiently large such that $u = U$ near y_1 and y_2, Eq. 12–39 can be rewritten as $\rho U \delta^* = \rho U(y_1 - y_2)$ or $y_1 = y_2 + \delta^*$. Hence we see that any streamline outside the boundary layer is displaced by δ^* due to the presence of the boundary layer. The name *displacement thickness* is thus highly appropriate.

By an analysis similar to that presented above, the quantity θ can be shown to represent the difference between the momentum fluxes in the boundary layer for these two cases discussed above. Figure 12–6 shows approximately the relative order of magnitudes of θ and δ.

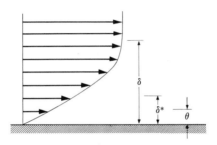

FIGURE 12–6

Example 12–2. Calculate again the skin friction coefficient for a flat plate at zero angle of attack, using the Kármán-Pohlhausen technique.

Since U is constant, Eq. 12–36 becomes

$$\frac{d\theta}{dx} = \frac{\tau_0}{\rho U^2}. \quad (12\text{--}40)$$

Let us assume that the velocity profile can be written in the form

$$\frac{u}{U} = F\left(\frac{y}{\theta(x)}\right).$$

Differentiating, we have

$$\frac{\partial u}{\partial y} = \frac{U}{\theta} \frac{dF}{d(y/\theta)},$$

and consequently $\tau_0 = \mu(\partial u/\partial y)_{y=0} = \mu U F'(O)/\theta$. The constant $F'(O)$ is to be determined from the assumed velocity profile F later.

Using this relation in Eq. 12–40, we have

$$\theta \frac{d\theta}{dx} = \frac{\nu}{U} F'(O) \quad \text{or} \quad \theta = \sqrt{2(\nu x/U)F'(O)}.$$

Thus,

$$\tau_0 = \rho U^2 \sqrt{\frac{\nu}{2Ux}} \, F'(O).$$

The coefficient of skin friction c_f is defined as

$$c_f = \frac{\tau_0}{\frac{1}{2}\rho U^2} \, .$$

In this case,

$$c_f = \sqrt{2F'(O)} \, \sqrt{\nu/Ux}.$$

Hence all we need now is the numerical value of $F'(O)$.

Let us assume that the velocity profile $F(y/\theta)$ is given by

$$F = \sin\left(k\,\frac{y}{\theta}\right) \quad \text{for} \quad 0 < \frac{ky}{\theta} < \frac{\pi}{2} \, ,$$

$$F = 1 \quad \text{for} \quad \frac{ky}{\theta} \geq \frac{\pi}{2} \, .$$

Since $F' = \partial F/\partial(y/\theta)$, we have $F'(O) = k$. Using the above profile in Eq. 12–37, we obtain an equation for $F'(O)$ from which $F'(O)$ can be determined:

$$F'(O) = \int_0^{\pi/2} \sin\alpha(1 - \sin\alpha)\, d\alpha = 1 - \frac{\pi}{4} \, .$$

Using this value for $F'(O)$, we have for the skin friction coefficient

$$c_f = \sqrt{2(1 - \pi/4)} \, \sqrt{\nu/vx} = 0.656\sqrt{\nu/Ux},$$

in agreement with Eq. 12–34.

PROBLEMS

12–5. Assume that the velocity profile is approximately $u = U \sin[(\pi/2)(y/\delta)]$ for $0 \leq y \leq \delta$. Find δ^*/δ and θ/δ.

12–6. Assume that the velocity profile for the flat plate problem is $u/U = 1 - e^{-y/\delta}$. Find the corresponding skin friction coefficient formula.

12–7. The Thwaites method

The example and problems in the last section illustrate the strong and weak points of the momentum integral technique. While the original partial differential equations, Eqs. 12–7, have been reduced to a single ordinary differential equation for $\theta(x)$, no further progress can be made unless some reasonable assumptions are made about the velocity profiles. For the flat-plate problem, we were fortunate that an exact solution was available, so that results from the momentum integral method may be appraised. For a general problem, such appraisal would not be available. This is the greatest drawback of the method. Although in the examples the solution is rather insensitive to the type of velocity profile assumed, some guide to the proper choice of velocity profile in the general case is desirable, where precision in the solution is important.

The method to be presented in this section is called the Thwaites method, named after its author. The presentation below follows closely that given by Cohen and Reshotko,* who extended Thwaites' original idea. The Kármán-Pohlhausen momentum integral equation, Eq. 12–28, is first made dimensionless by multiplying through by $2\theta U/\nu$. After rearrangement, the result is

$$U \frac{d}{dx}\left(\frac{\theta^2}{\nu}\right) = -2\left[\frac{\theta^2}{\nu}\frac{dU}{dx}\left(\frac{\delta^*}{\theta}+2\right) - \frac{\theta}{U}\left(\frac{\partial u}{\partial y}\right)_{y=0}\right]. \qquad (12\text{–}41)$$

Now, we observed that if we introduce the dimensionless variables l, H, and n:

$$l = \frac{\theta}{U}\left(\frac{\partial u}{\partial y}\right)_{y=0}, \qquad (12\text{–}42\text{a})$$

$$H = \frac{\delta^*}{\theta}, \qquad (12\text{–}42\text{b})$$

$$n = -\frac{\theta^2}{\nu}\frac{dU}{dx}, \qquad (12\text{–}42\text{c})$$

Eq. 12–41 may be rewritten as

$$-U \frac{d}{dx}\left[\frac{n}{dU/dx}\right] = 2[n(H+2)+l]. \qquad (12\text{–}43)$$

Since $U(x)$ is assumed known, this is a single equation for the three unknown dimensionless variables l, H, and n. Thwaites then proceeded to collect from various sources all the available solutions of the original laminar boundary-layer equation, and compute from them their corresponding values of l, H, and n. For example, the flat-plate solution gives $l = 0.22$, $H = 2.591$, and $n = 0$. Table 12–1 shows the collection of data (from Cohen and Reshotko's paper).

The right-hand side of Eq. 12–43 was then plotted as a function of n and it was found that it could be accurately approximated by a linear function of n:

$$2[n(H+2)+l] \cong 0.44 + 5.5n.$$

Using this empirical formula, we can write Eq. 12–43 as

$$-U \frac{d}{dx}\left[\frac{n}{dU/dx}\right] = 0.44 + 5.5n. \qquad (12\text{–}44)$$

---

* Cohen, C. B. and Reshotko, E., *NACA Report 1294*, 1956.

TABLE 12–1

n	l	H
0.0681	0.	4.032
0.0487	0.1051	3.094
0.0000	0.220	2.591
−0.0602	0.3220	2.298
−0.8029	0.3556	2.218
−0.1002	0.3808	2.180
−0.1064	0.3892	2.152

Integrating this simple equation, we obtain

$$n = -0.44U^{-5.5} \frac{dU}{dx} \int_0^x U^{4.5}\, dx. \qquad (12\text{–}45)$$

Therefore, once $U(x)$ is known, the distribution of n can be computed easily. Knowing n, Table 12–1 then furnishes the value of l and H. Hence we can solve for the values of θ and $(\partial u/\partial y)$ at $y = 0$. This technique has been applied to a variety of problems with success.

PROBLEMS

12–7. Express $[\nu(\partial u/\partial y)]_{y=0}$, θ, and δ^* in terms of l, H, and n.

12–8. Consider a two-dimensional source flow with $u_r = Ar_0/r$. A thin flat plate is placed radially with its leading edge at $r = r_0$. See Fig. 12–7. Find the distribution of the skin friction coefficient on the plate.

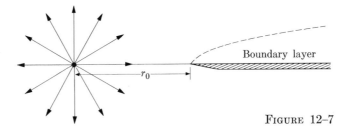

FIGURE 12–7

12–9. Find the skin friction distribution on a thin flat plate at zero angle of attack, using the Thwaites method. [*Hint:* Use Eq. 12–41 to find θ as a function of x.]

12–8. Separation and other remarks

It is clear from the previous section that once the pressure distribution about a solid surface is known, the behavior of the thin boundary layer adjacent to the surface can be calculated. When the pressure falls in the streamwise direc-

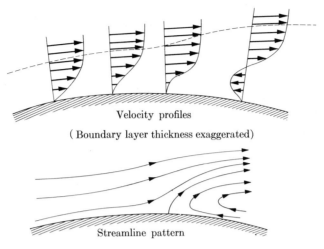

Velocity profiles

(Boundary layer thickness exaggerated)

Streamline pattern

FIGURE 12–8

tion, we say the pressure gradient is favorable. When the pressure rises in the streamwise direction, we say it is unfavorable. Pressure gradients are so classified because their respective consequences on the development of the boundary layers are quite different.

In the case of a favorable pressure gradient, since the pressure gradient tends to accelerate the fluid in the boundary layer, the pressure gradient counteracts partly the retarding action of the viscous stresses. In the case of an unfavorable pressure gradient, however, the problem is more complicated. The fluid near the wall now must not only overcome the wall shear stress, but it must also flow against the unfavorable pressure gradient. In contrast to this, the fluid near the outer edge of the boundary layer needs only to flow against the unfavorable pressure gradient, since the shear stress there is much smaller. As a consequence, the fluid near the wall is decelerated more rapidly. It may eventually be forced to flow backward. At the point where this flow reversal first occurs, the slope of the velocity profile, $\partial u/\partial y$, at the wall must be zero. This point is called the *separation point*, and is illustrated in Fig. 12–8. Obviously, this is what happens to the large Reynolds number flow over such bodies as a circular cylinder. Behind the mid-point of the circular cylinder, the pressure gradient is unfavorable and soon after passing the mid-point the boundary layer separates. In the frictionless solution presented in Chapter 6, it is tacitly assumed that the flow attaches to the surface at all times. The concepts of boundary layer and its separation explain the discrepancy between theory and experiment.

An important and interesting conclusion about the boundary-layer theory can be made here. It is noted that when the boundary-layer equations are written in the form given by Eqs. 12–13 and 12–14, the value of the Reynolds number does not appear. It is also noted that the boundary conditions are also independent of the Reynolds number. This point is also well illustrated in the Thwaites method, where we see that Eq. 12–44 does not involve the Reynolds number. Whenever the solution for n reaches the value 0.0681 at some value of x so that the value of l becomes zero, the flow is said to separate there. The position of this separation point is therefore independent of the Reynolds number. Consequently, if we predict boundary-layer separation to occur on a given body according to the boundary-layer theory, we cannot eliminate or change the location of the separation point by changing the Reynolds number, provided that the flow remains laminar.

13 Turbulent Flow

13–1. Necessity of study of turbulence

In previous chapters, laminar flows of viscous and frictionless fluids have been studied. Although some of the conclusions thus arrived at can describe turbulent flows approximately under restricted conditions (e.g., see Section 5–6), many phenomena in fluid flows cannot be explained without taking turbulence into consideration.

In a turbulent flow, the velocity at a point fluctuates at random with high frequency, as illustrated in Fig. 13–5. As a result, mixing of the fluid is much more intense than in a laminar flow, where mixing is affected by molecular action. Thus physical characteristics at a point (e.g., high temperature, high concentration of dissolved materials, etc.) are spread much more readily to other points in a turbulent flow (compare Figs. 2–5a and b).

Smooth surface   Roughened surface

(a)   (b)

FIGURE 13–1

Similarly, high concentration of momentum is readily dispersed in a turbulent flow, resulting in a velocity distribution much more uniform than that of a laminar flow. This difference in velocity distribution gives rise to a phenomenon of importance in technology. In Fig. 5–11 is shown the separation of a boundary layer in a decelerating flow when the velocity is sufficiently reduced. If the flow in a laminar boundary layer is made turbulent (e.g., by roughening the solid surface), the velocity will become more uniform and higher, and as a result, fluid particles in the boundary layer can move farther downstream before separation takes place. The roughened body will have a smaller wake downstream where the pressure is low, and therefore a considerably smaller drag force (see Fig. 13–1). It is this effect of turbulence on the location of the

253

point of separation that facilitates the higher increase of pressure in diffusers and turbine blades with a turbulent boundary layer.

The energy dissipation in a turbulent flow cannot be computed without consideration of the turbulence. For the flow in a pipe, it has been found that the dissipation is proportional to the mean speed V if the flow is laminar (see Eq. 11–5), while the dissipation is approximately proportional to V^2 if the flow is turbulent. The dissipation in a turbulent flow may be a hundred or a thousand times greater than the value computed by ignoring the turbulence.

In this chapter, several elementary topics of turbulent flow are presented. More advanced topics are beyond the scope of this elementary text.

13–2. Origin of turbulence

There are two main problems in the study of turbulence, namely, the explanation of how a laminar flow becomes turbulent, and the analysis of the flow characteristics after turbulence has developed. The first problem will be commented on briefly in this section. The second problem will be dealt with in subsequent sections of this chapter.

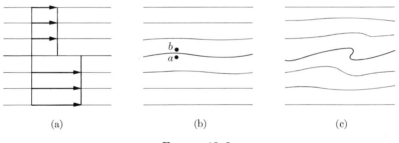

$$(a) \qquad\qquad (b) \qquad\qquad (c)$$

FIGURE 13–2

Several theories have been suggested for the origin of turbulence. Of these, the theory of stability of laminar flows seems to be the most successful. (The reader will recall that two problems of stability have been studied in Sections 3–5 and 3–6.) A motion may satisfy all the equations of motion and is therefore a possible motion, but it may not be stable; i.e., its flow pattern will change if it is disturbed even slightly. An unstable laminar flow may thus turn into a turbulent flow. Take for example Fig. 13–2(a), where two parallel uniform streams of a frictionless fluid are flowing with different speeds. This flow satisfies all equations of motion and is therefore a possible flow. However, if the interface is slightly disturbed, as shown in Fig. 13–2(b), the pressure p_a at point a will increase because of the decrease of speed at this point, while p_b at point b will decrease. As a result, the fluid particle at the interface between these points will be accelerated toward point b by this pressure difference, thus accentuating the disturbance of the interface. The flow shown in Fig. 13–2(a)

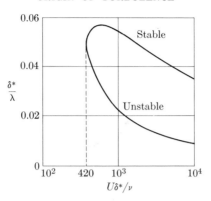

FIG. 13–3. Adapted from W. Tollmien, *NACA Tech. Memo.* *No. 609*, 1931.

is therefore unstable, and eddies will form at the interface, as shown in Fig. 13–2(c). The *theory of stability of laminar flows* assumes that the origin of turbulence is related to the stability of laminar flows. Given a possible laminar flow, the problem is to determine the conditions under which the flow becomes unstable with respect to small disturbances.

The stability of a laminar flow must depend on the linear dimensions of the system, its pressure and velocity distribution, the mechanical properties of the fluid, and the wavelength of the small disturbances. The most successful analysis of stability has been that of the laminar boundary layer of a flat plate under uniform pressure. The velocity profile in this boundary layer being similar at all cross sections (see Fig. 12–4), the condition of the flow at any section can therefore be described by the stream speed U and a representative thickness of the boundary layer there, e.g., the displacement thickness δ^* as defined in Eq. 12–38. The stability of the flow at a section depends therefore on U, δ^*, ρ, μ, and the wavelength λ of the small disturbances, or rather upon two independent dimensionless quantities formed by them, say, $U\delta^*/\nu$ and δ^*/λ (see Section 2–2). The mathematical analysis (by W. Tollmien) is too involved to be presented here, but the result is given in Fig. 13–3. It can be seen that when the Reynolds number $U\delta^*/\nu$ exceeds the critical value of 420, the flow becomes unstable if disturbances of certain wavelengths are present. The curve of neutral stability in Fig. 13–3 has been verified experimentally with disturbances introduced with a vibrating wire near the plate. It should be mentioned that the section where the flow becomes unstable is not the section where the flow becomes turbulent, since the small disturbances must be amplified many times before transition takes place.

However, the relation between the instability of laminar flow and the transition to turbulent flow has not been rigorously established. While it has been successful in several simple cases to predict the initiation of turbulence with the theory of instability of laminar flows, instability is not a sufficient condition for

the transition to turbulent flow; i.e., an unstable laminar flow may turn into another
mode of laminar flow. For example, the flow
between two concentric cylinders (see Fig.
13–4) has been shown (by G. I. Taylor) to
be unstable if the angular speed of the inner
cylinder is sufficiently high. The resultant
motion has been observed to be laminar,
with spiral streamlines around the inner
cylinder (see Fig. 13–4). Only when the
speed of the inner cylinder is further increased will the flow become turbulent.
Furthermore, stability has not been shown
to be a necessary condition for the transition to turbulent flow. It is still an open
question whether it is the only mechanism
for the creation of turbulence.

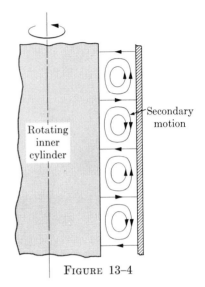

FIGURE 13–4

13–3. The Reynolds equations of motion

The equations of motion derived in previous chapters are valid for the analysis of fluid flow, be it turbulent or laminar. However, since the velocity in a
turbulent flow fluctuates rapidly at random, the solution, even if one could
be obtained, would be so complicated as to be useless. It seems therefore desirable to consider an instantaneous velocity component at a point to be com-
posed of a temporal mean value and fluctuations, e.g., $u(t)$ in Fig. 13–5:

$$u(t) = \bar{u} + u'(t), \tag{13–1}$$

where

$$\bar{u} = \frac{1}{T} \int_{t-T/2}^{t+T/2} u(t)\, dt, \tag{13–2}$$

where the period T of averaging should include a large number of fluctuations.
In a quasi-steady flow, $\bar{u}$ does not change with time, and T can be taken to
infinity. By definition, the temporal mean value of $u'(t)$ is zero:

$$\bar{u'} = \frac{1}{T} \int_{t-T/2}^{t+T/2} u'\, dt$$

$$= \frac{1}{T} \int_{t-T/2}^{t+T/2} (u - \bar{u})\, dt = \bar{u} - \bar{u} = 0. \tag{13–3}$$

The other velocity components and the pressure $p(t)$ at a point can be treated

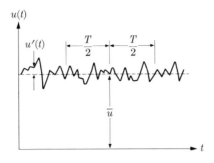

FIGURE 13–5

in a similar manner:

$$v(t) = \bar{v} + v'(t),$$
$$w(t) = \bar{w} + w'(t),$$
$$p(t) = \bar{p} + p'(t),$$

(13–4)

where $\bar{v}$, $\bar{w}$, and $\bar{p}$ are defined by equations similar to Eq. 13–2. Obviously, $\overline{v'} = \overline{w'} = \overline{p'} = 0$. It is then hoped that the equations of motion may perhaps be useful in predicting the distribution of $\bar{u}$, $\bar{v}$, $\bar{w}$, and $\bar{p}$ in a flow.

In expressing the equations of motion in terms of these mean values, the following equalities will be found useful. For functions $a(t)$ and $b(t)$ of time t,

$$\bar{a} = \bar{b}, \quad \text{if } a = b \text{ at any } t,$$

(13–5)

$$\overline{(a + b)} = \bar{a} + \bar{b},$$

(13–6)

$$\overline{(ka)} = k\bar{a} \quad (k \text{ being independent of } t),$$

(13–7)

$$\overline{\left(\frac{da}{dt}\right)} = \frac{d\bar{a}}{dt}.$$

If function a also varies spatially, then

$$\overline{\left(\frac{\partial a}{\partial t}\right)} = \frac{\partial \bar{a}}{\partial t}$$

(13–8)

and

$$\overline{\left(\frac{\partial a}{\partial s}\right)} = \frac{\partial \bar{a}}{\partial s},$$

(13–9)

where s represents any one of the coordinates, for example, x. These equalities can be derived from the definition of the temporal mean in Eq. 13–2. For example, we can show that $\overline{(\partial a/\partial x)}$, the mean of the derivative $\partial a/\partial x$ at a point, is equal to $\partial \bar{a}/\partial x$, the derivative of the mean of a, as follows. By definition of a derivative,

$$\frac{\partial \bar{a}}{\partial x} = \lim_{\Delta x \to 0} \frac{(\bar{a} \text{ at } x + \Delta x) - (\bar{a} \text{ at } x)}{\Delta x}.$$

According to the definition in Eq. 13–2,

$$\bar{a} \text{ at } (x + \Delta x) = \frac{1}{T} \int_{t-T/2}^{t+T/2} (a \text{ at } x + \Delta x) \, dt,$$

$$\bar{a} \text{ at } x = \frac{1}{T} \int_{t-T/2}^{t+T/2} (a \text{ at } x) \, dt.$$

Thus

$$\frac{\partial \bar{a}}{\partial x} = \lim_{\Delta x \to 0} \frac{1}{T} \int_{t-T/2}^{t+T/2} \frac{(a \text{ at } x + \Delta x) - (a \text{ at } x)}{\Delta x} \, dt = \frac{1}{T} \int_{t-T/2}^{t+T/2} \frac{\partial a}{\partial x} \, dt = \overline{\left(\frac{\partial a}{\partial x}\right)}.$$

We shall consider for simplicity quasi-steady incompressible flows where the mean values $\bar{u}$, $\bar{v}$, $\bar{w}$, and $\bar{p}$ do not vary with time. We aim to express the equations of motion in terms of these mean values. First take the equation of continuity,

$$\frac{\partial u}{\partial x} + \frac{\partial v}{\partial y} + \frac{\partial w}{\partial z} = 0,$$

which expresses a condition at a point at any t. Using Eqs. 13–5, 13–6, and 13–9, we have successively for the point under consideration,

$$\overline{\left(\frac{\partial u}{\partial x} + \frac{\partial v}{\partial y} + \frac{\partial w}{\partial z}\right)} = 0,$$

$$\overline{\left(\frac{\partial u}{\partial x}\right)} + \overline{\left(\frac{\partial v}{\partial y}\right)} + \overline{\left(\frac{\partial w}{\partial z}\right)} = 0,$$

$$\frac{\partial \bar{u}}{\partial x} + \frac{\partial \bar{v}}{\partial y} + \frac{\partial \bar{w}}{\partial z} = 0, \qquad (13\text{–}10)$$

which is the desired equation of continuity of the mean flow. Since the equation for the instantaneous flow can also be written as

$$\frac{\partial (\bar{u} + u')}{\partial x} + \frac{\partial (\bar{v} + v')}{\partial y} + \frac{\partial (\bar{w} + w')}{\partial z} = 0,$$

the difference between the last two equations gives

$$\frac{\partial u'}{\partial x} + \frac{\partial v'}{\partial y} + \frac{\partial w'}{\partial z} = 0, \qquad (13\text{–}11)$$

which is the equation of continuity of the instantaneous fluctuations.

Next consider a Navier-Stokes equation from Eq. 11–30:

$$a_x = \frac{\partial u}{\partial t} + u \frac{\partial u}{\partial x} + v \frac{\partial u}{\partial y} + w \frac{\partial u}{\partial z}$$

$$= -\frac{\partial}{\partial x}\left(\frac{p}{\rho} + gh\right) + \nu \left(\frac{\partial^2 u}{\partial x^2} + \frac{\partial^2 u}{\partial y^2} + \frac{\partial^2 u}{\partial z^2}\right).$$

Using Eqs. 13–5 to 13–9, we obtain

$$\overline{a_x} = \frac{\partial \overline{u}}{\partial t} + u\frac{\partial u}{\partial x} + v\frac{\partial u}{\partial y} + w\frac{\partial u}{\partial z} = -\frac{\partial}{\partial x}\left(\frac{\overline{p}}{\rho} + gh\right) + \nu\left(\frac{\partial^2 \overline{u}}{\partial x^2} + \frac{\partial^2 \overline{u}}{\partial y^2} + \frac{\partial^2 \overline{u}}{\partial z^2}\right),$$

where $\partial \overline{u}/\partial t = 0$ for quasi-steady flows, and the elevation h is not a function of time. The mean of the nonlinear inertial terms can be expressed in terms of $\overline{u}$, $\overline{v}$, etc., as follows. With $u = \overline{u} + u'$, and $v = \overline{v} + v'$, where $\overline{u}$ and $\overline{v}$ are independent of time for quasi-steady flows, we have, for example, by using Eqs. 13–6 and 13–7,

$$\overline{v\frac{\partial u}{\partial y}} = \overline{(\overline{v} + v')\frac{\partial(\overline{u} + u')}{\partial y}} = \overline{\overline{v}\frac{\partial \overline{u}}{\partial y}} + \overline{v'\frac{\partial \overline{u}}{\partial y}} + \overline{\overline{v}\frac{\partial u'}{\partial y}} + \overline{v'\frac{\partial u'}{\partial y}}$$

$$= \overline{v}\frac{\partial \overline{u}}{\partial y} + \overline{v'}\frac{\partial \overline{u}}{\partial y} + \overline{v}\frac{\partial \overline{u'}}{\partial y} + \overline{v'\frac{\partial u'}{\partial y}} = \overline{v}\frac{\partial \overline{u}}{\partial y} + \overline{v'\frac{\partial u'}{\partial y}}.$$

We have used the relations that $\overline{u'} = \overline{v'} = 0$ from Eq. 13–3, and that $\overline{u}$ and $\overline{v}$ are independent of time. Thus

$$\overline{a_x} = \overline{u\frac{\partial u}{\partial x}} + \overline{v\frac{\partial u}{\partial y}} + \overline{w\frac{\partial u}{\partial z}}$$

$$= \overline{u}\frac{\partial \overline{u}}{\partial x} + \overline{v}\frac{\partial \overline{u}}{\partial y} + \overline{w}\frac{\partial \overline{u}}{\partial z} + \overline{u'\frac{\partial u'}{\partial x}} + \overline{v'\frac{\partial u'}{\partial y}} + \overline{w'\frac{\partial u'}{\partial z}}.$$

But

$$\overline{u'\frac{\partial u'}{\partial x}} + \overline{v'\frac{\partial u'}{\partial y}} + \overline{w'\frac{\partial u'}{\partial z}}$$

$$= \left[\frac{\partial(u'u')}{\partial x} - u'\frac{\partial u'}{\partial x}\right] + \left[\frac{\partial(u'v')}{\partial y} - u'\frac{\partial v'}{\partial y}\right] + \left[\frac{\partial(u'w')}{\partial z} - u'\frac{\partial w'}{\partial z}\right]$$

$$= \frac{\partial(u'u')}{\partial x} + \frac{\partial(u'v')}{\partial y} + \frac{\partial(u'w')}{\partial z} - u'\left(\frac{\partial u'}{\partial x} + \frac{\partial v'}{\partial y} + \frac{\partial w'}{\partial z}\right).$$

The last parenthesis vanishes by virtue of Eq. 13–11. Using Eqs. 13–5, 13–6, and 13–9, we obtain

$$\overline{u'\frac{\partial u'}{\partial x}} + \overline{v'\frac{\partial u'}{\partial y}} + \overline{w'\frac{\partial u'}{\partial z}} = \frac{\partial \overline{u'u'}}{\partial x} + \frac{\partial \overline{u'v'}}{\partial y} + \frac{\partial \overline{u'w'}}{\partial z}.$$

Thus, the Navier-Stokes equation for quasi-steady incompressible flows can be written as

$$\overline{a_x} = \overline{u}\frac{\partial \overline{u}}{\partial x} + \overline{v}\frac{\partial \overline{u}}{\partial y} + \overline{w}\frac{\partial \overline{u}}{\partial z} + \frac{\partial \overline{u'u'}}{\partial x} + \frac{\partial \overline{v'u'}}{\partial y} + \frac{\partial \overline{w'u'}}{\partial z}$$

$$= -\frac{\partial}{\partial x}\left(\frac{\overline{p}}{\rho} + gh\right) + \nu\left(\frac{\partial^2 \overline{u}}{\partial x^2} + \frac{\partial^2 \overline{u}}{\partial y^2} + \frac{\partial^2 \overline{u}}{\partial z^2}\right). \qquad (13\text{–}12)$$

This is the desired momentum equation of the mean flow for the x-direction. Similarly, equations can be written for the other directions:

$$\overline{a_y} = \bar{u}\frac{\partial\bar{v}}{\partial x} + \bar{v}\frac{\partial\bar{v}}{\partial y} + \bar{w}\frac{\partial\bar{v}}{\partial z} + \frac{\partial\overline{u'v'}}{\partial x} + \frac{\partial\overline{v'v'}}{\partial y} + \frac{\partial\overline{w'v'}}{\partial z}$$

$$= -\frac{\partial}{\partial y}\left(\frac{\bar{p}}{\rho} + gh\right) + \nu\left(\frac{\partial^2\bar{v}}{\partial x^2} + \frac{\partial^2\bar{v}}{\partial y^2} + \frac{\partial^2\bar{v}}{\partial z^2}\right), \tag{13–13}$$

$$\overline{a_z} = \bar{u}\frac{\partial\bar{w}}{\partial x} + \bar{v}\frac{\partial\bar{w}}{\partial y} + \bar{w}\frac{\partial\bar{w}}{\partial z} + \frac{\partial\overline{u'w'}}{\partial x} + \frac{\partial\overline{v'w'}}{\partial y} + \frac{\partial\overline{w'w'}}{\partial z}$$

$$= -\frac{\partial}{\partial z}\left(\frac{\bar{p}}{\rho} + gh\right) + \nu\left(\frac{\partial^2\bar{w}}{\partial x^2} + \frac{\partial^2\bar{w}}{\partial y^2} + \frac{\partial^2\bar{w}}{\partial z^2}\right). \tag{13–14}$$

Equations 13–12 to 13–14 are usually referred to as the *Reynolds equations*. These equations for the mean flow differ from the Navier-Stokes equations for the instantaneous flow due to the presence of the inertial terms involving $\overline{u'u'}$, $\overline{u'v'}$, etc. It has been explained in Section 5–6 that these quantities are not equal to zero, although $\overline{u'} = \overline{v'} = \overline{w'} = 0$. In fact, $\overline{u'u'}$, $\overline{v'v'}$, and $\overline{w'w'}$ are always positive. In other words, the mean value of the particle acceleration at a point cannot be evaluated without considering the velocity fluctuations.

The Navier-Stokes equations have been derived according to momentum equations such as (see Eq. 11–22)

$$a_x = -g\frac{\partial h}{\partial x} + \frac{1}{\rho}\left(\frac{\partial S_{xx}}{\partial x} + \frac{\partial S_{yx}}{\partial y} + \frac{\partial S_{zx}}{\partial z}\right),$$

where S_{xx}, etc., are the viscous stress components. In view of this relation, it is the convention to call the quantities $-\rho\overline{u'u'}$, etc., the *Reynolds stress* components:

$$\begin{bmatrix} T_{xx} & T_{xy} & T_{xz} \\ T_{yx} & T_{yy} & T_{yz} \\ T_{zx} & T_{zy} & T_{zz} \end{bmatrix} = \begin{bmatrix} -\rho\overline{u'u'} & -\rho\overline{u'v'} & -\rho\overline{u'w'} \\ -\rho\overline{v'u'} & -\rho\overline{v'v'} & -\rho\overline{v'w'} \\ -\rho\overline{w'u'} & -\rho\overline{w'v'} & -\rho\overline{w'w'} \end{bmatrix}, \tag{13–15}$$

such that Eq. 13–13 can be written as

$$\bar{u}\frac{\partial\bar{u}}{\partial x} + \bar{v}\frac{\partial\bar{u}}{\partial y} + \bar{w}\frac{\partial\bar{u}}{\partial z}$$

$$= -g\frac{\partial h}{\partial x} + \frac{1}{\rho}\left[\frac{\partial(\bar{S}_{xx} + T_{xx})}{\partial x} + \frac{\partial(\bar{S}_{yx} + T_{yx})}{\partial y} + \frac{\partial(\bar{S}_{zx} + T_{zx})}{\partial z}\right]. \tag{13–16}$$

Thus the mean flow behaves like an instantaneous flow with the addition of the Reynolds stress components. As indicated above, these Reynolds stress components are part of the inertial terms. They do not cause directly the dissipation of mechanical energy as do the viscous stresses.

The three Reynolds equations, Eqs. 13–12 to 13–14, and the equation of continuity, Eq. 13–10, constitute a set of four equations for the mean flow variables $\bar{u}$, $\bar{v}$, $\bar{w}$, and $\bar{p}$. However, due to the presence of the unknown Reynolds stresses, these equations cannot yield the solutions for these variables, unless additional information about these stresses is available, as will be discussed in the next section.

Example 13–1. Study the Reynolds equations for a quasi-steady parallel shear flow with the mean velocity in one direction only. Such a flow is shown in Fig. 13–6 with $\bar{v} = \bar{w} = 0$. The Reynolds stress components and $\bar{u}$ are functions of y only.

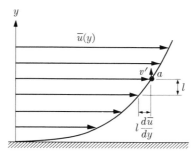

FIGURE 13–6

Since the Reynolds equations alone cannot yield the solutions of $\bar{u}$ and $\bar{p}$, we are only trying to get as much information as possible from these equations. For the given flow, the rate of pressure drop in the direction of flow is constant. Let

$$-\frac{\partial}{\partial x}(p + \rho g h) = \text{constant } c.$$

Equation 13–12 becomes

$$\frac{d}{dy}\left(\mu \frac{d\bar{u}}{dy} - \rho \overline{v'u'}\right) = -c.$$

Thus

$$\mu \frac{d\bar{u}}{dy} - \rho \overline{v'u'} = \tau_0 - cy. \qquad (13\text{–}17)$$

Here $\mu \cdot d\bar{u}/dy$ is the mean viscous shearing stress, and $-\rho \overline{v'u'}$ is the Reynolds shearing stress. Their sum has been found to vary linearly with y in this flow. (This conclusion has been verified experimentally with hot-wire anemometer measurements. See Fig. 13–8 for the flow between two large plates.) In Eq. 13–17, τ_0 is the shearing stress at the wall where $y = 0$ and $v' = u' = 0$.

For the given flow, Eq. 13–13 becomes

$$\frac{\partial}{\partial y}(\bar{p} + \rho g h + \rho \overline{v'v'}) = 0. \qquad (13\text{–}18)$$

The distribution of the mean pressure $\bar{p}$ is therefore not hydrostatic across a cross section, but

$$\bar{p} + \rho g h + \rho\overline{v'v'} = \text{constant at a section.}$$

Thus $\bar{p} + \rho g h$ is higher at the wall (where $\overline{v'v'} = 0$) than in the stream (where $\overline{v'v'} > 0$).

PROBLEMS

13-1. In the accompanying table are shown the instantaneous values of u at point A. Point B is unit distance away in the x-direction. Fill in some arbitrary values for u at point B, and estimate the following. (a) The instantaneous values of $\partial u/\partial x$ in this vicinity. (b) $\overline{(\partial u/\partial x)}$. (c) $\bar{u}$ at A. (d) $\bar{u}$ at point B. (e) $\partial\bar{u}/\partial x$. Does your result substantiate Eq. 13-9?

t	u at A	u at B
1	2	
2	4	
3	3	
4	1	
5	3	

13-2. By using Eq. 13-2, derive Eqs. 13-5 to 13-7.

13-3. (a) In Fig. 13-7 are shown some observed simultaneous values of u' and v' at three points in a flow. Estimate the values of $\overline{u'u'}$, $\overline{v'v'}$, and $\overline{u'v'}$ at each point. (b) The *correlation coefficient R_{xy}*, defined as

$$R_{xy} = \frac{\overline{u'v'}}{\sqrt{\overline{u'u'}} \cdot \sqrt{\overline{v'v'}}},$$

measures a statistical relation between u' and v'. Estimate the value of R_{xy} for each of these three points, and note the effect of the distribution of u' and v' in Fig. 13-7 on the value of R_{xy}. (c) For the case with $u' = a \sin(\omega_1 t)$

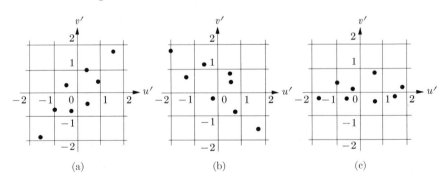

(a)   (b)   (c)

FIGURE 13-7

and $v' = b \sin(\omega_2 t)$ (not shown in Fig. 13–7) show that the correlation co-efficient $R_{xy} = 0$ if $\omega_1 \neq \omega_2$, and $R_{xy} = ab/2$ if $\omega_1 = \omega_2$. [*Hint:*

$$\sin(\omega_1 t) \cdot \sin(\omega_2 t) = \tfrac{1}{2}[\cos(\omega_1 - \omega_2)t - \cos(\omega_1 + \omega_2)t].$$

Since the given u' and v' are not random functions, they do not represent actual turbulent fluctuations.]

13–4. Show that for compressible flows, the equation of continuity can be written as

$$\frac{\partial \bar{\rho}}{\partial t} + \frac{\partial}{\partial x}(\bar{\rho}\bar{u} + \overline{\rho' u'}) + \frac{\partial}{\partial y}(\bar{\rho}\bar{v} + \overline{\rho' v'}) + \frac{\partial}{\partial z}(\bar{\rho}\bar{w} + \overline{\rho' w'}) = 0.$$

13–5. A large horizontal plate is moved relative to another plate as shown in Fig. 13–12. Assuming that this flow is turbulent and $\partial \bar{p}/\partial x = 0$, (a) show that the mean pressure $\bar{p}$ is not hydrostatically distributed, but $p + \rho g h$ is higher at the plates; and (b) show that the mean velocity profile $\bar{u}(y)$ must have a point of inflection (i.e., $d^2\bar{u}/dy^2 = 0$) between the plates.

13–4. The mixing-length hypothesis

In the previous section, four equations of motion, Eqs. 13–10 and 13–12 to 13–14, have been derived for quasi-steady incompressible flows in terms of the mean flow variables $\bar{u}$, $\bar{v}$, $\bar{w}$, and $\bar{p}$. Unfortunately, these equations involve addi-tional unknown quantities, namely, the Reynolds stress components. If these equations are to be used for the solution of the mean flow, additional equations must be made available to relate these stress components to the mean flow velocity components. So far, no such equations are available for general use. However, since the introduction of the hot-wire anemometer as an instrument for measuring electrically the velocity fluctuations, the nature of turbulent flows has been studied extensively. On the one hand, turbulence has been studied statistically. The discussion of this approach, however, is beyond the scope of this text. On the other hand, several empirical equations have been suggested which have been found to yield usable results in many cases. We shall discuss one of these empirical equations presently.

In the following presentation, we shall consider only quasi-steady parallel shear flow as shown in Fig. 13–6. For this flow, the Reynolds equation of mo-tion has been reduced to Eq. 13–17:

$$\mu \frac{d\bar{u}}{dy} - \overline{\rho v' u'} = \tau_0 - cy,$$

where $\bar{u}(y)$ is the temporal mean velocity, $\overline{v'u'}$ is the mean of the product of the velocity fluctuations at a point, τ_0 is the shearing stress at the wall, and $c = -\partial(\bar{p} + \rho g h)/\partial x$. Here $\mu \cdot d\bar{u}/dy$ is the mean viscous stress and $-\overline{\rho v'u'}$ is the Reynolds shearing stress due to turbulence. This equation states that their sum is balanced by the wall shear and the pressure gradient. To use this equation, it is necessary to have information about the unknown quantity $-\rho\overline{v'u'}$. One of the early suggestions for $-\rho\overline{v'u'}$ was due to J. Boussinesq. In view of the following relation for the mean viscous shearing stress,

$$\overline{S}_{yx} = \rho\nu\,\frac{d\overline{u}}{dy},$$

he proposed to express the Reynolds shearing stress as

$$-\rho\overline{v'u'} = \rho\epsilon\,\frac{d\overline{u}}{dy}, \tag{13–19}$$

where ϵ is called the *eddy kinematic viscosity*. However, since ϵ varies with the flow and there is no way of predicting it, the situation has not been improved by the introduction of this equation. Since then, several other equations have been suggested, each being more useful in dealing with a particular group of flow systems. Of these empirical equations, the one based on the *mixing-length hypothesis* of L. Prandtl is among the most useful.

Consider point a in the flow shown in Fig. 13–6. When there is a positive v' at this point, a fluid particle of a layer at a certain distance l below would be brought to point a, causing a fluctuation u' of the order of magnitude of $-l \cdot d\overline{u}/dy$. Similarly, a negative v' would cause a u' of the order of $l \cdot d\overline{u}/dy$. Thus the product $v'u'$ at this point is of the same sign most of the time. Assuming that v' is of the same order of magnitude as u' at this point, we have

$$-\rho\overline{v'u'} \doteq \rho l^2\left(\frac{d\overline{u}}{dy}\right)^2.$$

Since the Reynolds stress should change sign with $d\overline{u}/dy$, Prandtl put

$$-\rho\overline{v'u'} = \rho l^2\left|\frac{d\overline{u}}{dy}\right|\frac{d\overline{u}}{dy}. \tag{13–20}$$

The length l in this equation is called the *mixing length*. This being essentially an empirical equation, its derivation is not intended to be rigorous.

That Eq. 13–20 is superior to Eq. 13–19 in application is due to the fact that it is easier to guess the mixing length l to obtain reasonable results. For example, as u', v', and therefore $\overline{u'v'}$ at a solid surface must be zero, Prandtl assumed that near a solid wall, l is proportional to the distance y from the wall:

$$l = \kappa y, \tag{13–21}$$

where κ is an empirical constant. Another formula for l is suggested by Th. von Kármán for regions away from solid boundaries where l probably depends mostly on the local velocity variations; i.e., $d\overline{u}/dy$, $d^2\overline{u}/dy^2$, etc. Choosing the simplest quantity with the dimension of a length formed by these derivatives, he proposed

$$l = \kappa \left| \left(\frac{d\bar{u}}{dy} \right) \Big/ \left(\frac{d^2\bar{u}}{dy^2} \right) \right|, \qquad (13\text{--}22)$$

where κ is an empirical constant. It can be shown that the constants κ in Eqs. 13–21 and 13–22 are the same (see Problem 13–11). These formulas for l are not intended to be physically correct. For example, one expects that the mixing length in a pipe flow must be influenced by the size of the pipe, but the size is not involved in these formulas. For the center of the pipe where mixing is most intense with $d\bar{u}/dy = 0$, Eq. 13–22 suggests $l = 0$! In spite of these inadequacies, however, these formulas have been used successfully to predict the distribution of the mean velocity. The value of κ in these formulas must be determined experimentally, but is supposed to be essentially the same for turbulent parallel shear flows.

With the mixing-length hypothesis, the equation of motion for parallel shear flows along a plane wall, Eq. 13–17, becomes

$$\mu \frac{d\bar{u}}{dy} + \rho l^2 \left| \frac{d\bar{u}}{dy} \right| \frac{d\bar{u}}{dy} = \tau_0 - cy, \qquad (13\text{--}23)$$

where $c = -\partial(\bar{p} + \rho g h)/\partial x$, τ_0 is the shearing stress at the wall, and y is the distance from the wall. It should be mentioned that Eq. 13–23 can be shown to be valid also for the shear flow in a circular pipe. This may be done by first deriving the Reynolds equations of motion in cylindrical coordinates, but the derivation will not be presented here. Since the total shear at the centerline of a pipe is zero where $y = a$, the radius of the pipe, we have $c = \tau_0/a$ and

$$\mu \frac{d\bar{u}}{dy} + \rho l^2 \left| \frac{d\bar{u}}{dy} \right| \frac{d\bar{u}}{dy} = \tau_0 \left(1 - \frac{y}{a} \right). \qquad (13\text{--}24)$$

Since the shearing force at the wall is balanced by the pressure drop, we have for circular pipes

$$-\frac{\partial}{\partial x} (\bar{p} + \rho g h)\, dx \cdot \pi a^2 = \tau_0 2\pi a\, dx, \qquad \tau_0 = -\frac{a}{2} \frac{\partial}{\partial x} (\bar{p} + \rho g h).$$

In Fig. 13–8 are shown the shearing stresses and the mean flow $\bar{u}$ in a turbulent flow along a smooth wall. It can be seen that at certain distance from the wall, the viscous stresses are relatively small. For this turbulent region, it is probably sufficient to use

$$\rho l^2 \left| \frac{d\bar{u}}{dy} \right| \frac{d\bar{u}}{dy} = \tau_0 - cy. \qquad (13\text{--}25)$$

On the other hand, viscous stresses are predominant near the smooth wall where mixing is inhibited. There is a thin layer at the wall where the flow is

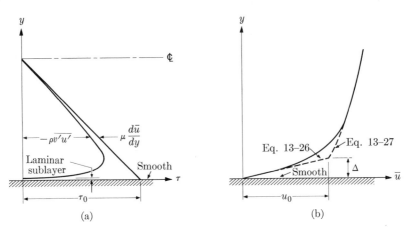

FIGURE 13–8

laminar, which is called the *laminar sublayer*. For this layer where $y \doteq 0$, Eq. 13–23 is reduced to

$$\mu \frac{d\bar{u}}{dy} = \tau_0,$$

$$u = \bar{u} = \frac{\tau_0}{\mu} y. \tag{13–26}$$

(Along a rough wall, there is no continuous laminar sublayer. Equation 13–26 is therefore not applicable.) The application of these equations is demonstrated in the following examples.

Example 13–2. Using Prandtl's formula $l = \kappa y$, find the velocity distribution in the turbulent portion of a flow near a smooth wall where y is so small that cy in Eq. 13–25 is negligible compared with τ_0; i.e., where the shearing stress is practically uniform.

With $l = \kappa y$ and $\tau_0 - cy \doteq \tau_0$, Eq. 13–25 becomes

$$\rho \kappa^2 y^2 \left| \frac{d\bar{u}}{dy} \right| \frac{d\bar{u}}{dy} = \tau_0.$$

For positive $d\bar{u}/dy$, $|d\bar{u}/dy| = d\bar{u}/dy$. Thus

$$\frac{1}{u_*} \frac{d\bar{u}}{dy} = \frac{1}{\kappa y},$$

where $u_* = \sqrt{\tau_0/\rho}$ has been used for convenience. This quantity u_* has the dimensions of a velocity and is called the *friction velocity*. Integrating, we obtain

$$\frac{\bar{u}}{u_*} = \frac{1}{\kappa} \ln y + C. \tag{13–27}$$

The constant of integration C must be determined from the conditions at the wall. For a smooth wall, C can be determined as follows. Let a nominal thickness Δ of the laminar sublayer and the wall velocity u_0 be defined as indicated in Fig. 13–8(b). Then C must be such that

$$\frac{u_0}{u_*} = \frac{1}{\kappa} \ln \Delta + C.$$

To determine u_0 and Δ, we have from Eq. 13–26 for the laminar sublayer,

$$\tau_0 = \mu \frac{u_0}{\Delta}.$$

Also, there must be a critical value of the Reynolds number $\rho u_0 \Delta / \mu$ for the sublayer to remain laminar; i.e.,

$$\frac{\rho u_0 \Delta}{\mu} = \text{constant } N^2,$$

where the constant N is to be determined experimentally. Solving the last two equations for u_0 and Δ, we obtain

$$\frac{u_0}{u_*} = \frac{\rho u_* \Delta}{\mu} = N. \tag{13–28}$$

Thus

$$C = \frac{u_0}{u_*} - \frac{1}{\kappa} \ln \Delta = N - \frac{1}{\kappa} \ln \left(\frac{N\mu}{\rho u_*} \right),$$

and Eq. 13–27 yields the velocity distribution in the turbulent zone:

$$\frac{\bar{u}}{u_*} = G + \frac{1}{\kappa} \ln \left(\frac{\rho u_* y}{\mu} \right), \tag{13–29}$$

where $G = N - (\ln N)/\kappa$.

In Fig. 13–9 is shown the velocity distribution in smooth pipes observed by J. Nikuradse. It can be seen that the flow is laminar as described by Eq. 13–26, where $\rho u_* y / \mu < 5$. It can also be seen that Eq. 13–29 is valid and the flow is completely turbulent at $\rho u_* y / \mu > 70$. From these experimental data of the turbulent zone, one can obtain $\kappa = 0.40$, $N = 11.6$, and $G = 5.5$. Observed velocity distributions over smooth flat plates indicate that $\kappa = 0.42$, $N = 11.7$, and $G = 5.8$. Thus the velocity distribution near a plate is practically the same as that in a pipe. Although Eq. 13–29 has been derived with the assumption of constant shear, it can describe, to one's surprise, the velocity distribution in a pipe where the shear is not uniform.

It has been found empirically that for $\rho u_* y / \mu$ between 70 and about 700, Eq. 13–29 can be approximated by

$$\frac{\bar{u}}{u_*} = 8.74 \left(\frac{\rho u_* y}{\mu} \right)^{1/7}. \tag{13–30}$$

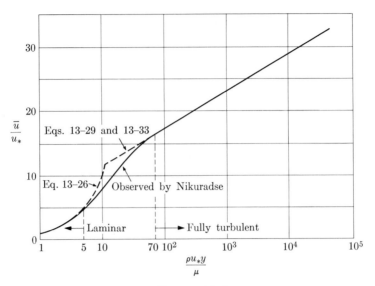

FIGURE 13–9

This is called the 1/7-*th power law* for smooth surfaces, and is often used because of its simplicity.

Example 13–3. Find the velocity distribution in a pipe, using the mixing length in Eq. 13–22 as suggested by von Kármán.

With l from Eq. 13–22, we have from Eq. 13–24 for the turbulent region of the pipe flow,

$$\frac{\rho\kappa^2(d\bar{u}/dy)^3|d\bar{u}/dy|}{(d^2\bar{u}/dy^2)^2} = \tau_0\left(1 - \frac{y}{a}\right) \qquad (0 < y < a),$$

where a is the radius of the pipe. Let $r = a - y$ be the distance measured from the centerline of the pipe. Then $d\bar{u}/dy = -d\bar{u}/dr$. Thus

$$-\frac{d^2\bar{u}/dr^2}{(d\bar{u}/dr)^2} = \frac{\kappa}{u_*}\sqrt{\frac{a}{r}}.$$

Integrating, we obtain

$$\frac{1}{du/dr} = \frac{2\kappa\sqrt{ar}}{u_*} + c_1.$$

The constant c_1 would be infinite if we specify $d\bar{u}/dr = 0$ at the centerline, where $r = 0$. This difficulty arises because Eq. 13–22 is not valid there. We can specify instead that $d\bar{u}/dr$ becomes very large as r approaches a near the wall. Then $c_1 = -2\kappa a/u_*$, and

$$\frac{d\bar{u}}{dr} = -\frac{u_*}{2\kappa\sqrt{a}\,(\sqrt{a} - \sqrt{r})}.$$

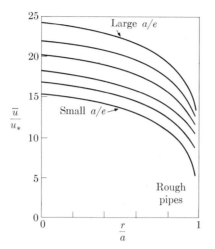

FIGURE 13–10

Integrating again, we have

$$\frac{\bar{u}}{u_*} = \frac{1}{\kappa}\left[\sqrt{\frac{r}{a}} + \ln\left(1 - \sqrt{\frac{r}{a}}\right)\right] + c_2.$$

The constant c_2 can be determined according to the conditions at the wall, as in the previous example.

It is interesting to express c_2 in terms of the maximum $\bar{u}_m$ at the centerline, where $r = 0$. Thus $c_2 = \bar{u}_m/u_*$, and

$$\frac{\bar{u}_m - \bar{u}}{u_*} = -\frac{1}{\kappa}\left[\sqrt{\frac{r}{a}} + \ln\left(1 - \sqrt{\frac{r}{a}}\right)\right]. \tag{13–31}$$

Note that the velocity deficiency $\bar{u}_m - \bar{u}$ in the turbulent region is proportional to the friction velocity u_*, and is independent of the wall conditions. In Fig. 13–10 are shown some observed velocity distributions in pipes with various degrees of roughness. Note the similar shape of the curves. That $(\bar{u}_m - \bar{u})/u_*$ is the same for all turbulent pipe flows can also be shown with Eq. 13–27, which is applicable for all wall conditions. With $\bar{u}_m$ at $r = a - y = 0$, we obtain

$$\frac{\bar{u}_m - \bar{u}}{u_*} = \frac{1}{\kappa}\ln\left(\frac{a}{a - r}\right). \tag{13–32}$$

Equations 13–32 and 13–31 fit the available experimental data equally well.

PROBLEMS

13–6. (a) The drop of pressure in a horizontal smooth 12-in. water pipe is 1.12 lb/ft^2/ft when the mean velocity V is 10 ft/sec. Compute the friction velocity. (b) Repeat when V is 1 ft/sec and the drop of pressure is 0.017 lb/ft^2/ft.

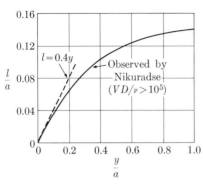

FIGURE 13–11   FIGURE 13–12

13–7. It is possible to compute from Eq. 13–24 the mixing length in pipes from observed velocity distributions. In the turbulent flow described in Problem 13–6(a), it is observed that $d\bar{u}/dy$ is 4.7 ft/sec/ft at a point 3 in. from the centerline. Compute the mixing length there and compare with the value from Fig. 13–11 obtained by Nikuradse from pipes with $VD/\nu > 10^5$.

13–8. From the experimental data for smooth pipes shown in Fig. 13–9, find the values of the constants G and κ in Eq. 13–29.

13–9. Compute the nominal thickness Δ of the laminar sublayer, the thickness of the laminar flow, the diameter of the core where viscous stresses are negligible, and the wall velocity, in the smooth pipe described in Problem 13–6. Note the effects of V on these quantities.

13–10. With Eq. 13–29, compute the values of $\bar{u}$ at the centerline of the pipe in Problem 13–6(a), and at a point 3 in. away. With the same $\bar{u}_m$, compute with Eq. 13–31 the value of $\bar{u}$ at the latter point, and compare the results.

13–11. With $l = \kappa y$, Eq. 13–27 has been derived for a turbulent shear flow near a wall. With this equation, show that $\kappa (d\bar{u}/dy)/(d^2\bar{u}/y^2) = l$. (This indicates that the two constants in Eqs. 13–21 and 13–22 are equal.)

13–12. A plate is moved relative to another under constant shearing stress as shown in Fig. 13–12. Assuming that the viscous stress is negligible in the turbulent zone and $l = \kappa y(H - y)$, (a) show that there is a point of inflection midway between the plates; (b) the plates being smooth so that Eq. 13–28 is valid, show that in the turbulent zone

$$\frac{\bar{u}}{u_*} = N + \frac{1}{\kappa} \ln\left[\left(\frac{\rho u_* y}{N\mu} - \frac{y}{H}\right) \Big/ \left(1 - \frac{y}{H}\right)\right];$$

and (c) show that the relative speed U is given by

$$\frac{U}{2u_*} = N + \frac{1}{\kappa} \ln\left(\frac{\rho u_* H}{N\mu} - 1\right).$$

13–13. A turbulent flow of a liquid down a large plane is shown in Fig. 4–2. The angle between the horizontal and the plane is θ, and the depth of flow is H. (a) Show that τ_0 at the plane is $\rho g H \sin \theta$. (b) Show that the shearing stress $\tau_0 - cy$ in Eq. 13–25 is $\rho g(H - y)\sin\theta$. (c) Assuming that $l = \kappa y$, show that

$$\frac{\bar{u}_m - \bar{u}}{\sqrt{gH\sin\theta}} = \frac{1}{\kappa}\left[\ln\left(\frac{\sqrt{H} + \sqrt{H - y}}{\sqrt{H} - \sqrt{H - y}}\right) - 2\sqrt{1 - \frac{y}{H}}\right].$$

13–14. To show that $(\bar{u}_m - \bar{u})/u_*$ in a turbulent pipe flow is independent of the wall conditions, use $l = \kappa y \cdot f(y/a)$, where a is the radius of the pipe and f is any arbitrary function of y/a. Show that $(\bar{u}_m - \bar{u})/u_*$ is a function of y/a only.

13–5. Turbulent flow in circular pipes

Turbulent flow in circular pipes has been studied extensively because of its practical importance. Near the entrance, the velocity distribution in the pipe varies from section to section as the boundary layer develops as shown in Fig. 13–13. In this section, we discuss the flow with a fully developed velocity profile only. In particular, we are interested in the velocity distribution and the frictional resistance of the flow.

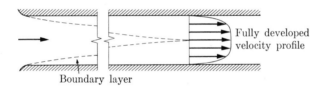

Fully developed
velocity profile

Boundary layer

FIGURE 13–13

Equations 13–27 and 13–31 have been derived for the velocity distribution in the fully turbulent region under all wall conditions. Because of its relative simplicity, we shall use Eq. 13–27 in the following development. It is

$$\frac{\bar{u}}{u_*} = 2.5\ln y + C,$$

where y is the distance from the wall, and the constant C depends on the wall conditions.

For a smooth pipe, C has been determined in Example 13–2, such that

$$\frac{\bar{u}}{u_*} = 5.5 + 2.5\ln\left(\frac{\rho u_* y}{\mu}\right) \qquad \text{for smooth pipes.} \qquad (13\text{–}33)$$

It has also been found experimentally that in the layer where $\rho u_* y/\mu$ is less than 5, the flow is laminar. In the region where $\rho u_* y/\mu$ is larger than 70, the viscous stress is negligible and Eq. 13–33 is applicable (see Fig. 13–9).

Turbulent flows have been observed to be affected by the roughness of the wall surface, i.e., by the size and shape of the protrusions and their spacing over the wall surface. To avoid the difficulty of dealing with many combinations of these parameters, only surfaces with tightly spaced protrusions of irregular shapes are considered here (e.g., concrete and cast-iron surfaces). To describe such wall roughness, only a statistical size e of the protrusions is usually sufficient. Obviously, whether a surface is to be considered smooth or rough, so far as fluid flow is concerned, depends on the value of e relative to the thickness of the laminar sublayer in the pipe. This thickness has been observed to be $5\mu/\rho u_*$ (see Fig. 13–9) which decreases with an increase of the flow in the pipe. It has been experimentally verified that surface roughness can be classified as follows.

(a) The surface is *smooth* if $\rho u_* e/\mu < 5$. The protrusions being contained in the laminar sublayer in this case, the size e has no influence on the flow.

(b) The surface is called *rough* if $\rho u_* e/\mu > 70$. In this case, all protrusions reach the completely turbulent region. The resistance to the flow is mainly due to the form drag of the protrusions. The effect of viscosity becomes negligible.

(c) With $\rho u_* e/\mu$ between 5 and 70, resistance is partly due to viscous shear at the wall and partly due to the form drag of the protrusions.

Thus a pipe may behave as a smooth pipe at low discharge (small τ_0 and u_*) and as a rough pipe at a higher discharge.

To determine the constant C in Eq. 13–27 for a rough pipe ($\rho u_* e/\mu > 70$), let u_0 be the wall velocity at $y = e$. It can be reasoned that u_0 depends on τ_0 at the wall, ρ, and e (μ being unimportant in this case). These quantities can form only one independent dimensionless product, say $u_0/\sqrt{\tau_0/\rho}$ or u_0/u_*. Thus $u_0/u_* = $ constant M (see Eq. 2–5), and

$$C = \frac{u_0}{u_*} - 2.5 \ln e = M - 2.5 \ln e.$$

M has been found experimentally to be 8.5 (see Fig. 13–14). Thus Eq. 13–27 becomes

$$\frac{\bar{u}}{u_*} = 8.5 + 2.5 \ln \left(\frac{y}{e}\right) \qquad \text{for rough pipes.} \qquad (13\text{–}34)$$

In the general case, the wall velocity u_0 must depend not only on τ_0, ρ, and e, but also on the viscosity μ. These quantities can form two independent dimensionless products, say u_0/u_* and $\rho u_* e/\mu$. Thus $u_0/u_* = F(\rho u_* e/\mu)$,

$$C = \frac{u_0}{u_*} - 2.5 \ln e = F\left(\frac{\rho u_* e}{\mu}\right) - 2.5 \ln e,$$

and

$$\frac{\bar{u}}{u_*} = F\left(\frac{\rho u_* e}{\mu}\right) + 2.5 \ln \left(\frac{y}{e}\right). \qquad (13\text{–}35)$$

Computed values of function F from observed $\bar{u}$ are shown in Fig. 13–14. For

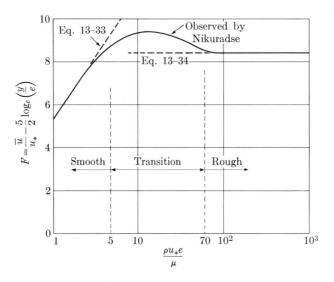

FIGURE 13–14

rough pipes, $F = 8.5$ and Eq. 13–35 becomes Eq. 13–34. For smooth pipes, Eqs. 13–33 and 13–35 become identical with

$$F = 5.5 + 2.5 \ln \left(\frac{\rho u_* e}{\mu} \right).$$

It is of practical interest to relate the mean speed V to the friction velocity u_*. This relation can be obtained from

$$V = \frac{1}{\pi a^2} \int_0^a \bar{u} \cdot 2\pi (a - y) \, dy.$$

With $\bar{u}$ from Eq. 13–33, we have

$$\frac{V}{u_*} = 1.75 + 2.5 \ln \left(\frac{\rho u_* a}{\mu} \right) \qquad \text{for smooth pipes.} \qquad (13\text{–}36)$$

With $\bar{u}$ from Eq. 13–34, we obtain

$$\frac{V}{u_*} = 4.75 + 2.5 \ln \left(\frac{a}{e} \right) \qquad \text{for rough pipes.} \qquad (13\text{–}37)$$

The frictional resistance to pipe flow is usually expressed in terms of the dimensionless *friction coefficient f* defined by

$$- \frac{\partial}{\partial x} (\bar{p} + \rho g h) = \frac{f}{2D} \rho V^2, \qquad (13\text{–}38)$$

where $D = 2a$ is the diameter of the pipe, and x is the direction of flow. For

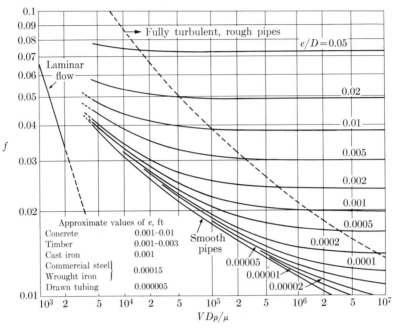

FIGURE 13–15

laminar flow, it has been shown in Example 11–2 that $f = 64/(VD/\nu)$. The experimental values of f for turbulent flow are shown in Fig. 13–15. It is now possible to derive the values of f for smooth pipes and for rough pipes. It has been shown in connection with Eq. 13–24 that

$$\tau_0 = - \frac{D}{4} \frac{\partial}{\partial x} (\bar{p} + \rho g h). \qquad (13\text{–}38a)$$

Thus

$$f = \frac{8\tau_0}{\rho V^2} = \frac{8u_*^2}{V^2}. \qquad (13\text{–}38b)$$

With this equation and another relating V and u_*, f can be found in terms of V by eliminating u_* in the two equations. In this manner, we obtain with Eq. 13–36 and Eq. 13–27, respectively,

$$\frac{1}{\sqrt{f}} = 2.0 \log \left(\frac{\rho V D}{\mu} \sqrt{f} \right) - 0.8 \qquad \text{for smooth pipes,} \qquad (13\text{–}39)$$

$$\frac{1}{\sqrt{f}} = 2.0 \log \left(\frac{a}{e} \right) + 1.74 \qquad \text{for rough pipes.} \qquad (13\text{–}40)$$

(The numbers 0.8 and 1.74 have been adjusted slightly to get a closer agreement with experimental results.)

For the transition between the smooth and the rough pipes, an empirical formula f has been suggested by Colebrook and White. This formula has been found to describe satisfactorily the transition of commercial pipes:

$$\frac{1}{\sqrt{f}} = 1.74 - 2.0 \log \left(\frac{e}{a} + \frac{18.7}{\sqrt{f}} \frac{\mu}{\rho V D} \right). \tag{13-41}$$

With $e = 0$, this equation transforms into Eq. 13-39 for smooth pipes. As $\rho V D/\mu$ becomes large, the pipe behaves as a rough pipe, and Eqs. 13-40 and 13-41 become identical. For smooth pipes, f is a function of the Reynolds number $\rho V D/\mu$ only, while f for rough pipes depends only on the relative roughness e/D. In the transition, f varies with both of these parameters.

PROBLEMS

13-15. What are the maximum allowable sizes of the protrusions at the wall surface so that the pipe described in Problem 13-6(a) and (b), respectively, can be considered to be smooth?

13-16. A liquid flows down a large plane as shown in Fig. 4-2. Examine the derivation of Eqs. 13-33 and 13-34 and decide whether they are also valid for a turbulent flow down a smooth plane and a rough plane, respectively.

13-17. When a liquid flows down a large plane as shown in Fig. 4-2, the shearing stress at the plane is $\rho g H \sin \theta$, where H is the depth of flow and θ is the angle between the plane and the horizontal. Assuming that Eq. 13-34 is applicable, show that the mean speed is

$$V = \left[6.0 + 2.5 \ln \left(\frac{H}{e} \right) \right] \sqrt{gH \sin \theta}.$$

13-18. Derive Eqs. 13-39 and 13-40.

13-19. For turbulent flow in smooth pipes with $\rho V D/\mu < 10^5$, the 1/7-power velocity distribution in Eq. 13-30 covers the whole pipe adequately. (a) Find V/u_*; and (b) show that f is inversely proportional to $(\rho V D/\mu)^{1/4}$, and compare with Blasius' empirical formula for smooth pipes:

$$f = 0.3164 \left(\frac{\mu}{\rho V D} \right)^{1/4}. \tag{13-41a}$$

13-6. Empirical formulas for pipes and open channels

Although the velocity distribution and the frictional resistance in circular pipes are understood in a somewhat satisfactory manner, many other problems of pipe flow in technology remain to be approached empirically.

In pipes of noncircular cross section, secondary currents have been observed, as shown in Fig. 13-16. While the velocity distribution in such pipes remains

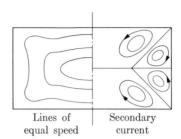

$$\underset{\text{FIGURE 13–16}}{} \quad \begin{matrix} \text{Lines of} \\ \text{equal speed} \end{matrix} \quad \begin{matrix} \text{Secondary} \\ \text{current} \end{matrix}$$

to be studied, it has been found that for pipes with reasonably uniform velocity distribution (e.g., pipes with square or oval cross section), the frictional resistance can be estimated as follows. Let τ_0 be the average shearing stress at the wall necessary to balance the pressure gradient:

$$\tau_0 = - \frac{A}{P} \frac{\partial}{\partial x} (\overline{p} + \rho gh) = - m \frac{\partial}{\partial x} (\overline{p} + \rho gh),$$

where P and A are the wetted perimeter and the area of the cross section, respectively, and $m = A/P$ is called the *hydraulic radius* of the cross section. For a circular cross section, $P = \pi D$, $A = \pi D^2/4$, and $m = D/4$. It is assumed that the average τ_0 for all these pipes can be used in the same formulas of frictional resistance. Then a cross section with a hydraulic radius m is equivalent in frictional resistance to a circular section with a diameter $D = 4m$. The pressure drop can then be computed from Eq. 13–38, which can be written as

$$- \frac{\partial}{\partial x} (\overline{p} + \rho gh) = \frac{f}{8m} \rho V^2, \tag{13–42}$$

where f is to be found from Fig. 13–15 with the Reynolds number $4\rho Vm/\mu$ and the relative roughness $e/4m$.

At every fitting of a pipe line (e.g., elbow, expansion, contraction, valve, etc.), additional mechanical energy is dissipated. For example, take the sudden expansion shown in Fig. 9–16. Due to separation of the flow at the expansion, the intensity of turbulence and therefore the kinetic energy of eddies are increased. This increase of energy of turbulence is then dissipated through viscous action. The additional drop of $(\overline{p}/\rho g) + h + (\alpha V^2/2g)$ caused by a fitting is called its *head loss* h_L or loss of mechanical energy per unit weight. (Here α is the kinetic energy correction factor defined in Eq. 5–19). It is the convention to express h_L with a dimensionless coefficient K:

$$h_L = K \frac{V^2}{2g}, \tag{13–43}$$

where V is the mean speed in the smaller pipe connected to the fitting. The value of K depends on the shape of the fitting, the relative roughness of the surface, and the Reynolds number of the pipe flow. In practical problems, the

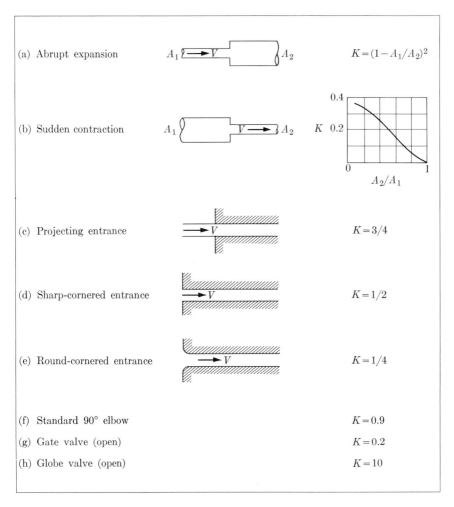

FIGURE 13–17

Reynolds number is usually so high that the effect of viscosity on K is negligible. Since a fitting is usually relatively short, the effect of surface roughness cannot be fully developed. The value of K for fittings of the same shape is then identical. The empirical values of K for several common fittings are shown in Fig. 13–17.

In the analysis of a steady flow in a pipe line with fittings, we compare two points in the flow, and write (see Eq. 9–14)

$$\frac{\overline{p}_1}{\rho g} + h_1 + \frac{\alpha_1 V_1^2}{2g} = \frac{\overline{p}_2}{\rho g} + h_2 + \frac{\alpha_2 V_2^2}{2g} + \sum h_f + \sum h_L, \qquad (13\text{–}44)$$

where point 1 is upstream of point 2, $\sum h_f$ is the sum of the frictional losses, and $\sum h_L$ is the sum of the additional losses at fittings between the two points. For a straight pipe of length L, we have from Eq. 13–38

$$h_f = f \frac{L}{D} \frac{V^2}{2g} \quad \text{or} \quad f \frac{L}{4m} \frac{V^2}{2g}. \tag{13–45}$$

For an example of analysis of a pipe line, see Example 13–4 below.

Another turbulent flow system of importance in technology is the flow in *open channels* (e.g., flumes, canals, rivers, etc.). We shall discuss only those cases where the change of the cross section of flow is so gradual that only frictional resistance need be considered. Equation 13–44 is also valid for such gradually varied flows. The sum $\overline{p}/\rho g + h$ at a cross section can be written as $Y + h_0$, where Y is the depth of flow and h_0 is the elevation of the bottom of the section (see Example 5–6). Thus Eq. 13–44 can be rewritten for steady open-channel flow as

$$Y_1 + h_{01} + \frac{\alpha_1 V_1^2}{2g} = Y_2 + h_{02} + \frac{\alpha_2 V_2^2}{2g} + \int \frac{fV^2}{8gm} \, dx. \tag{13–46}$$

In a long, straight prismatic channel, a steady flow will have constant Y and V through its course except near the two ends. A flow with constant Y and V is called a *uniform flow*. For a uniform flow, Eq. 13–46 becomes

$$\frac{fV^2}{8gm} = -\frac{dh_0}{dx} = s,$$

or

$$V = \sqrt{8gms/f}, \tag{13–47}$$

where s is the slope of the channel bottom in the direction of the flow. Equation 13–47 is known as *Chézy's formula*, usually written as $V = C\sqrt{ms}$ with $C = \sqrt{8g/f}$. It can be seen that the velocity in uniform flow in a given channel varies with m and therefore with the depth of flow. The values of f for uniform flow must be determined experimentally. In most practical cases, the Reynolds number is so high that the channel may be considered as rough. The value of f and Chézy's C then depends only on the relative roughness e/m. The most commonly used empirical formula for C is *Manning's formula*:

$$\sqrt{8g/f} = C = 0.262\sqrt{g} \, m^{1/6}/n \quad \text{for rough channels}, \tag{13–48}$$

where the *roughness coefficient* n has the dimension of (length)$^{1/6}$, and is presumably proportional to $e^{1/6}$. The values of n in ft$^{1/6}$ for some common channel surfaces are listed in Table 13–1. Combining Eqs. 13–47 and 13–48, we have for uniform flow

$$V = \frac{1.486}{n} m^{2/3} s^{1/2} \quad \text{(ft-sec units).} \tag{13–49}$$

For an example of application of this equation, see Example 13–5 below.

TABLE 13-1

MANNING'S n, $ft^{1/6}$

Surface material	Approximate n	Surface material	Approximate n
Timber	0.012 to 0.013	Riveted steel	0.018
Concrete	0.012 to 0.014	Corrugated metal	0.022
Cast iron	0.015	Earth	0.025
Brick	0.016	Excavated rock	0.040

For a steady but nonuniform flow with a constant discharge Q, Eq. 13–46 can be rewritten as

$$\frac{d}{dx}\left(Y + h_0 + \frac{Q^2}{2gA^2}\right) = -\frac{fQ^2}{8gmA^2}.\qquad(13\text{--}50)$$

Since A of a cross section is a known function of the location x of the cross section and the depth Y, this equation involves variables x and Y only. Upon integration, it yields the surface profile $Y(x)$. In most cases, the integration must be performed step by step. For an example of integration of a similar equation, see Section 10–6. Several typical surface profiles have been discussed in Sections 10–5 and 10–6.

Example 13-4. In Fig. 13–18 is shown a cast-iron water pipe with a nozzle. Assuming $K = 0.1$ for the nozzle, find the distance H required to deliver 0.2 cfs.

The governing equation is Eq. 13–44. First compute the speeds and the losses in the flow. From the equation of continuity (see Eq. 5–17)

$$VA = Q,$$

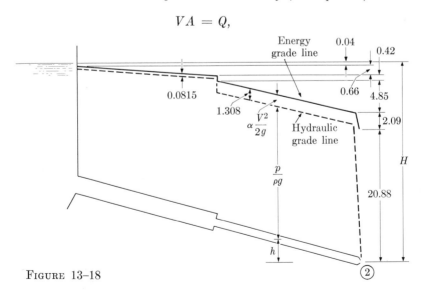

FIGURE 13–18

where Q is the discharge of 0.2 cfs, we have

for the 4-in. pipe: $A = 0.0873$ ft^2, $V = 0.2/0.0873 = 2.29$ ft/sec,
$$V^2/2g = 0.0815 \text{ ft};$$
for the 2-in. pipe: $V^2/2g = 1.308$ ft;
for the 1-in. jet: $V^2/2g = 20.88$ ft.

The frictional loss h_f in each pipe is computed from Eq. 13–45 by first finding the coefficient f from Fig. 13–15. Using $e = 0.001$ ft for cast iron and $\nu = 10^{-5}$ ft^2/sec for water, we have for the 4-in. pipe:

$$\frac{e}{D} = 0.003, \qquad \frac{VD}{\nu} = 7.63 \times 10^4, \qquad f = 0.027,$$

$$h_f = 0.027 \times \frac{100}{\frac{1}{3}} \times 0.0815 = 0.66 \text{ ft}.$$

Similarly, for the 2-in. pipe, $h_f = 4.85$ ft.

The losses at fittings are computed with values of K from Fig. 13–17:

Sharp-cornered entrance: $K = 0.50$, $h_L = 0.50 \times 0.0815 = 0.04$ ft.

Sudden contraction: for area ratio $= 0.25$, $K = 0.323$,

$$h_L = 0.323 \times 1.308 = 0.42 \text{ ft}.$$

Nozzle: given $K = 0.1$, $h_L = 0.1 \times 20.88 = 2.09$ ft.

With point 1 at the reservoir surface ($p_1 = 0$, $h_1 = H$, $V_1 = 0$) and point 2 at the jet ($p_2 = 0$, $h_2 = 0$, $V_2^2/2g = 20.88$ ft), Eq. 13–44 becomes

$$H = \frac{V_2^2}{2g} + \Sigma h_f + \Sigma h_L$$

$$= 20.88 + (0.66 + 4.85) + (0.04 + 0.42 + 2.09)$$

$$= 28.94 \text{ ft}.$$

It is instructive to plot the sum $(\bar{p}/\rho g) + h + (\alpha V^2/2g)$ along the flow. This line is called the *energy grade line*, as shown in Fig. 13–18. It can be constructed by starting from the reservoir surface and deducting h_L or h_f at each fitting and pipe. This line shows the distribution of the potential energy at the reservoir between the kinetic energy of the jet and the dissipation in the pipe line. If the sum $\bar{p}/\rho g + h$ is plotted along the flow, the line is called the *hydraulic grade line*. This line can be constructed by deducting the appropriate value of $\alpha V^2/2g$ from the energy grade line, as shown in Fig. 13–18. The hydraulic grade line is useful because it shows graphically how the pressure $\bar{p}$ varies along the flow.

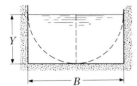
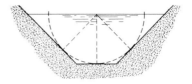

FIGURE 13–19

Example 13–5. Find the size of a rectangular concrete flume required to deliver in uniform flow 50 cfs of water at a slope of 0.00007. In order to promote deposition of suspended materials in the water, it is desirable to have V not larger than 1.4 ft/sec.

The governing equations of a uniform flow in an open channel are

$$Q = VA, \tag{13–51}$$

$$V = \frac{1.486}{n} m^{2/3} s^{1/2} \quad \text{(ft-sec units)}, \tag{13–52}$$

where Q is the discharge. Since area A and hydraulic radius m depend on the width B and the depth Y of the cross section, there are six parameters involved in these two equations, namely, Q, V, B, Y, n, and s. In this problem, only Q, s, and n (0.013 for concrete) are specified. With two equations and three unknowns (V, B, and Y), there is an infinite number of possible solutions. It can be shown that of all the possible solutions, the channel with the sides tangent to a semicircle entails the minimum cross-sectional area and wall surfaces (see Problem 13–28 and Fig. 13–19). Such a cross section is called an *economical section*. Therefore, try to use $B = 2Y$. Then Eqs. 13–51 and 13–52 become

$$50 = V(2Y^2)$$

and

$$V = \frac{1.486}{0.013} \left(\frac{Y}{2}\right)^{2/3} \sqrt{0.00007} = 0.603 Y^{2/3}.$$

Solving simultaneously, we obtain $Y = 4.04$ ft and $V = 1.53$ ft/sec. Therefore $B = 2Y = 8.08$ ft. However, this economical section cannot be used because V is specified to be not larger than 1.4 ft/sec.

The solution is to be obtained with $V = 1.4$ ft/sec. With Q, s, n, and V specified, Eqs. 13–51 and 13–52 give B and Y. With the specified values, these equations become

$$50 = 1.4\, BY$$

and

$$1.4 = \frac{1.486}{0.013} \left(\frac{BY}{B + 2Y}\right)^{2/3} \sqrt{0.00007}.$$

Eliminating B and simplifying, we obtain

$$Y^2 - 10.1Y + 17.85 = 0, \qquad Y = 2.28 \text{ or } 7.81 \text{ ft.}$$

With $Y = 2.28$ ft, we obtain $B = 15.7$ ft. For $Y = 7.81$ ft, $B = 4.57$ ft. Both cross sections satisfy the specified conditions.

PROBLEMS

13–20. An oil of $\nu = 0.0001$ ft^2/sec flows with $V = 10$ ft/sec in a cast-iron 8-in. pipe. Find the frictional loss per 1000 ft of pipe.

13–21. In a horizontal 12-in. cast-iron water pipe, the allowable pressure drop is 10 psi per 1000 ft. Find the maximum discharge in this pipe. [*Hint:* Solve with trial values of f, and then check with Fig. 13–15.]

13–22. It is required to deliver 10 cfs of an oil with $\nu = 10^{-4}$ ft^2/sec in a steel pipe. The allowable frictional loss is 8 ft-lb/lb per 1000 ft. Determine the required diameter of the pipe. [*Hint:* Solve with trial values of f, and check with Fig. 13–15.]

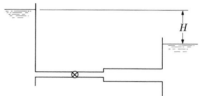

FIGURE 13–20

13–23. A 2-in. pipe and a 4-in. pipe, each 10 ft long and made of cast iron, are connected in series between two large tanks of water. There is a gate valve in the 2-in. pipe, as shown in Fig. 13–20. (a) When the valve is fully open, the speed V in the 4-in. pipe is 2 ft/sec. Find the distance H. Construct the energy grade line and the hydraulic grade line, showing dimensions. (b) The valve is then partially closed such that $V = 1$ ft/sec in the 4-in. pipe under the same H. Find the head loss across the valve under this condition, and construct the grade lines. (c) Sketch the grade lines when the valve is completely closed.

13–24. A semicircular corrugated metal flume of 3-ft radius is used to convey water at a bottom slope of 0.0009. Find the discharge when the flume is flowing full.

13–25. A rectangular concrete flume, 5 ft in width, is to convey 25 cfs of water at a speed $V = 2$ ft/sec. Find the necessary bottom slope.

13–26. A timber rectangular flume, 6 ft in width, delivers 150 cfs of water at a slope of 0.002. How deep should this flume be? [*Hint:* You may have to solve an equation by trials.]

13–27. Show (a) that V in a circular channel of diameter D is higher when $Y = 0.8D$ than when $Y = D$; and (b) that Q is larger when $Y = 0.9D$ than when $Y = D$. [*Hint:* $A = 0.674D^2$ and $P = 2.22D$ when $Y = 0.8D$; $A = 0.745D^2$ and $P = 2.50D$ when $Y = 0.9D$.]

13-28. With given values of Q, s, and n of a rectangular channel, it is desired to determine the width B and the depth Y. (a) Show that all possible cross sections have the same values of $P^{0.4}/A$ and $P^{0.4}/Y(P - 2Y)$. (b) Show that of all these possible cross sections, the one with $Y = P/4$ or $Y = B/2$ has the minimum P and A.

13-29. Design a rectangular corrugated metal flume to carry 50 cfs at a slope of 0.004.

13-30. A canal with $n = 0.02$ is trapezoidal in cross section with the side walls at 45° with the horizontal. It is to convey 1000 cfs at a slope of 0.0001. To prevent scouring, the speed V should not exceed 2.5 ft/sec. Design the cross section. [*Hint:* Use the economical section shown in Fig. 13-19 if possible.]

13-7. Turbulent boundary layer

When a viscous fluid passes a solid body at a high Reynolds number, a thin boundary layer develops at the solid surface. Near the upstream end of the body, the flow in the layer is usually laminar, as dicussed in Chapter 12. As the thickness δ of the layer increases downstream, the flow becomes turbulent where $\rho U \delta / \mu$ becomes sufficiently large. The turbulent boundary layer can be studied by using the momentum theorem as presented in Sections 12-5 and 12-6 with an approximate velocity profile in the boundary layer and a formula for computing the wall shear τ_0.

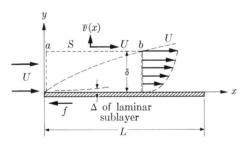

FIGURE 13-21

We shall consider the turbulent boundary layer of a flat plate at zero incidence as shown in Fig. 13-21. This problem is of technical importance, since the result can be used to estimate the skin friction of bodies without separation. First consider a *smooth* flat plate. For simplicity, we assume that the boundary layer becomes turbulent at the leading edge. We shall use the momentum theorem with an approximate velocity profile in the boundary layer. (See Example 12-1 for a similar analysis of a laminar boundary layer.) For this purpose, we choose the 1/7-th power velocity profile in Eq. 13-30 for its simplicity:

$$\frac{\bar{u}}{u_*} = 8.74 \left(\frac{\rho u_* y}{\mu} \right)^{1/7}. \tag{13-53}$$

Since $\bar{u} = U$ at $y = \delta$, we have

$$\frac{U}{u_*} = 8.74 \left(\frac{\rho u_* \delta}{\mu}\right)^{1/7},$$ (13–54)

from which we can obtain u_* $(= \sqrt{\tau_0/\rho})$ and then $\tau_0(x)$ in terms of $\delta(x)$:

$$u_* = 0.150 U \left(\frac{\mu}{\rho U \delta}\right)^{1/8},$$

$$\tau_0 = \rho u_*^2 = 0.0225 \rho U^2 \left(\frac{\mu}{\rho U \delta}\right)^{1/4}.$$ (13–55)

Another equation relating τ_0 and δ is obtained with the momentum theorem. Consider the control volume S shown in Fig. 13–21. With constant ρ, Eq. 9–5 gives

$$-f = \rho \int_0^\delta \bar{u}^2 \, dy + \rho U \int_{ab} \bar{v} \, dx - \rho U^2 \delta,$$

where f is the wall shear per unit width over the length of the control volume S. For continuity,

$$\int_{ab} \bar{v} \, dx = U\delta - \int_0^\delta \bar{u} \, dy.$$

Thus

$$f = \rho U \int_0^\delta \bar{u} \, dy - \rho \int_0^\delta \bar{u}^2 \, dy.$$

With the 1/7-th power velocity profile (Eqs. 13–53 and 13–54),

$$\bar{u} = U \left(\frac{y}{\delta}\right)^{1/7},$$

we obtain

$$f = \frac{7}{72} \rho U^2 \delta.$$

Since $\tau_0(x) = df/dx$, we have

$$\tau_0 = \frac{7}{72} \rho U^2 \frac{d\delta}{dx}.$$ (13–56)

From Eqs. 13–55 and 13–56, $\delta(x)$ and $\tau_0(x)$ can be obtained. Eliminating τ_0, we have

$$\frac{7}{72} \frac{d\delta}{dx} = 0.0225 \left(\frac{\mu}{\rho U \delta}\right)^{1/4}.$$

With the assumed condition that $\delta = 0$ at $x = 0$, the solution of this equation is

$$\frac{\delta}{x} = 0.37 \left(\frac{\mu}{\rho U x}\right)^{1/5}.$$ (13–57)

Thus the thickness increases with $x^{4/5}$, whereas it increases with $x^{1/2}$ in the laminar boundary layer. With this δ in Eq. 13–55, we obtain the distribution of the shearing stress $\tau_0(x)$:

$$\frac{\tau_0}{\frac{1}{2}\rho U^2} = 0.0592 \left(\frac{\mu}{\rho U x}\right)^{1/5} \tag{13–58}$$

Thus τ_0 is inversely proportional to $x^{1/5}$. (Here the coefficient has been adjusted slightly to fit experimental data.) For a plate of length L and width b, the total skin friction is $D = \int_0^L \tau_0 b\, dx$. It is simple to show that the *coefficient of skin friction* is

$$c_f = \frac{D}{\frac{1}{2}\rho U^2 A} = 0.074 \left(\frac{\mu}{\rho U L}\right)^{1/5}. \tag{13–59}$$

This result has been found experimentally to be valid for $\rho U L/\mu$ between 5×10^5 and 10^7 (see Fig. 13–22).

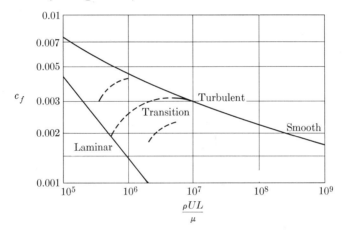

FIG. 13–22. Adapted from Th. von Kármán, *J. Aero. Sci.*, Jan., 1934.

For a plate with a surface roughness e, the situation is more complicated. Near the upstream end, where u_* is large and the thickness Δ of the laminar sublayer is small compared with e, the plate may behave as a rough surface. As u_* decreases and Δ increases along the plate, the downstream portion of the plate may behave as a smooth surface. The velocity profile in Eq. 13–35 can be used with the momentum theorem for this case. However, the computation is involved, since u_* and therefore the function $F(\rho u_* e/\mu)$ in Eq. 13–35 vary along the plate.

The formulas above have been derived with the assumption that the boundary layer is turbulent throughout its length. When there is a laminar portion near the leading edge, the skin friction can be estimated by first finding the skin friction for a turbulent boundary layer over the whole plate, and then subtracting from it the skin friction of a turbulent layer for a length up to the point of transition, and adding to it the skin friction of a laminar boundary layer of the same length. The position of the point of transition varies with the condition of the leading edge and the intensity of turbulence in the free stream, and is usually located at $\rho U x/\mu$ between 3×10^5 and 3×10^6. The skin friction of a plate with a transition lies between those of a wholly laminar layer and a wholly turbulent layer, as shown in Fig. 13–22.

The analysis above has been performed for a flat plate at zero incidence. The result has been found useful for estimating the skin friction of bodies such as ship hulls and airfoils before the point of separation, if any. As mentioned in Section 13–1, the turbulence in a boundary layer has a decided effect on the location of the point of separation and therefore the size of the wake and the magnitude of the form drag. For example, in the case of a sphere in a stream, there is a sudden decrease of the value of the drag coefficient at $U d\rho/\mu$ between 10^5 and 10^6 (see Fig. 2–10) when the boundary layer becomes turbulent before the point of separation. The value of $U d\rho/\mu$ when this occurs depends on the roughness of the surface and the intensity of turbulence of the stream.

14 Thermodynamics and Fluid Flows

14-1. Introduction

In our studies so far, we have confined our attention completely to incompressible fluid flows. One may recall at this point that we have really never explicitly applied the concept of conservation of energy. The Bernoulli equation,

$$p + \tfrac{1}{2}\rho(u^2 + v^2 + w^2) = \text{constant},$$

which is valid for steady, frictionless, incompressible fluid flows in the absence of gravitational force, is sometimes interpreted as an energy equation (e.g., Section 5–5) although it is derived solely from momentum considerations. From this chapter onward, we shall deal with compressible fluid flow problems. We shall see that the concept of conservation of energy plays an important role in the discussions. We shall derive an energy equation based on the concept of conservation of energy, and this independent equation along with momentum and continuity equations completes the governing set of equations for general compressible flows. As it turns out, when the flow is steady, incompressible, and frictionless, the energy equation reduces to Bernoulli's equation under similar assumptions. We shall have more to say on this point later.

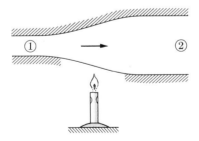

FIGURE 14–1

To illustrate the point that for compressible fluid flows the concept of energy is relevant, let us consider the flow in a straight but variable-area pipe, as shown in Fig. 14–1.

If the fluid involved is a liquid whose density is known to vary within only narrow limits, then the velocity at station 2 can be directly related to that at station 1 by the continuity equation, regardless of whether heat energy has been added between the two stations. We have

$$\rho_1 A_1 V_1 = \rho_2 A_2 V_2. \qquad (14\text{--}1)$$

Or, since ρ is practically a constant, we have

$$V_2 = \frac{A_1}{A_2} V_1. \qquad (14\text{--}2)$$

However, if the fluid under consideration is a gas, we can no longer draw the same conclusion. Equation 14–2 now becomes

$$V_2 = \frac{\rho_1}{\rho_2} \frac{A_1}{A_2} V_1. \qquad (14\text{--}3)$$

Hence, V_2 cannot be determined unless ρ_1/ρ_2 is known. It is obvious that ρ_1/ρ_2 will depend on the amount of heat energy being added between the two stations. As a matter of fact, we shall see later (Chapter 15) that even in the absence of heat addition, ρ_1/ρ_2 is in general different from unity for a compressible flow. In any case, it is clear that since addition of heat energy affects the flow, the concept of energy must be brought into the discussion.

We shall ignore the effects of gravitational and other body forces in the following sections. This simplification is generally used in ordinary gas dynamics.

14–2. Equation of state

First let us study how the density of a gas varies with its other physical properties. For a given sample of gas, its pressure, density, and temperature can readily be measured. Obviously, given two samples of a certain gas in equilibrium which are at the same pressure, density, and temperature, we say that the two gas samples are in the same *state*. Pressure, density, and temperature are therefore called *state variables*. There are many other state variables besides the three mentioned above, and they will be introduced in the following sections.

It is an experimental fact that for a given gas sample in equilibrium, only two state variables are independent. That is, if two samples of the same gas have the same pressure p and density ρ, their temperatures T are *necessarily* equal. Hence, it is possible to construct experimentally, for each gas of interest, a relation of the form

$$T = \text{function of } (p, \rho) = T(p, \rho) \qquad (14\text{--}4)$$

by performing a series of careful experiments. Such a relation as Eq. 14–4 is called an *equation of state*. Different materials have different equations of state and data must be individually obtained. However, for most gases under ordinary conditions, it has been found that the *empirical* relation

$$T(°F) = \frac{p}{\rho} \frac{\mathfrak{M}}{\mathfrak{R}} - 459.7°F \tag{14-5}$$

is an adequate approximation. In Eq. 14–5, temperature is measured in degrees fahrenheit, pressure in pounds per square foot, density in slugs per cubic foot; $\mathfrak{M}$ is the molecular weight of the gas in question, and $\mathfrak{R}$ is called the *universal gas constant;* its value is experimentally found to be

$$\mathfrak{R} = 1545.4 \times 32.174 \text{ ft-lb}^2/\text{slug-mole-}°F. \tag{14-6}$$

Equation 14–5 is usually rewritten as

$$p = \rho R T, \tag{14-7}$$

where R is defined by

$$R = \frac{\mathfrak{R}}{\mathfrak{M}} \tag{14-8}$$

and is called the *gas constant*, and T is understood to be the absolute temperature, measured in degrees rankine, $°R$

$$T = T(°R) = 459.7 + T(°F). \tag{14-9}$$

Throughout the rest of the book, T shall always denote absolute temperature. For values in cgs units, see Problem 14–1. For the properties of air, see Problem 14–2 and the Appendix.

When we accept Eq. 14–7 as an adequate approximation to the actual experimentally constructed equation of state, it is said that we have adopted the *perfect gas assumption*. Equation 14–7 is sometimes referred to as the *perfect gas law*. Because of its simplicity and generally acceptable accuracy, the perfect gas law is extensively used in ordinary gas dynamics. We shall adopt it throughout our discussions in this and the following chapters.

PROBLEMS

14–1. In cgs units, the absolute temperature is measured in degrees kelvin (°K), defined by

$$T(°K) = T(°C) + 273.2.$$

If p is measured in dyne-cm^{-2}, ρ is measured in gm-cm^{-3}, find the magnitude of the universal gas constant $\mathfrak{R}$ in cgs units (ergs/mole-°K).

14–2. The molecular weight of air is 28.97. Find the density of air in slugs per cubic foot under standard sea level conditions: $p = 14.7$ lb/in^2, $T(°F) = 60°F$.

14–3. The first law of thermodynamics

As indicated in Section 14–1, the concept of energy must be introduced in the study of compressible flow. For this purpose, let us consider a thermodynamic system such as a certain parcel of fluid which is in thermodynamic equilibrium at state 1. If an amount of heat energy ΔQ is added to the system by its environment and an amount of work ΔW is performed by the system on its environment, and as a consequence of such energy exchanges, the state of the system changes from 1 to 2, the first law of thermodynamics states that *an entity denoted by ΔE exists such that*

$$\Delta E = \Delta Q - \Delta W \qquad (14\text{–}10)$$

and the magnitude of ΔE is independent of the energy-exchange processes involved, and is dependent only on the initial and final states of the system.

The above statement of the first law appears rather abstract. Actually, the first law merely summarizes an accumulation of certain experiences dealing with the effects of energy exchanges on systems. Instead of compiling a list of such experiences, we choose to state the first law in this abstract way so that these experiences can be deduced as the consequences of the statement. We shall deduce here the following consequences.

(1) All forms of energy are equivalent. This is so because for a given change of state of a system from 1 to 2, the magnitude of ΔE is a constant. Thus we can accomplish this change of state either by exchanging heat energy alone or by exchanging any other form of energy or work. Thus all forms of energy are equivalent in the sense that they produce the same effects. For example, it is found that when 778.2 ft-lb of mechanical work is performed on one pound of water (e.g., by stirring) at 62°F and atmospheric pressure, the water temperature will rise by 1°F. The same change of state can of course be accomplished by heating. Note that Eq. 14–10 also allows us to establish experimentally the conversion factors between the various units of energy. The engineering unit for heat energy is the B.T.U., or the British Thermal Unit, which is the amount of heat energy required to raise the temperature of one pound of water at 62°F at atmospheric pressure by 1°F. Hence we have

$$778.2 \text{ ft-lb} = 1 \text{ B.T.U.} \qquad (14\text{–}11)$$

If the water is to be heated by an electrical heater, it is found that 0.2930 watt-hour of electrical energy is consumed. Hence

$$0.2930 \text{ watt-hour} = 1 \text{ B.T.U.}$$

Similarly, other conversion factors have been determined experimentally. In Eq. 14–10, it is understood that all energies are expressed in the same units by the use of such conversion factors. For our purposes, we shall express all energies in mechanical units, e.g., ft-lb, unless otherwise stated.

(2) Energy can neither be created nor destroyed. Since $\Delta Q - \Delta W$ is the net amount of energy added to the system, we can introduce the concept of conservation of energy by saying that this energy is considered stored in the system. For example, let us bring a system from state 1 to state 2 so that an amount of energy $\Delta E_{12} (= \Delta Q_{12} - \Delta W_{12})$ is added to the system. This amount of energy is now considered stored in the system. Now, let us bring the system from state 2 back to state 1. In doing so, an amount of energy $\Delta E_{21} (= \Delta Q_{21} - \Delta W_{21})$ is again added to the system. The total amount of energy added to the system is therefore $\Delta E_{12} + \Delta E_{21}$. Now, since ΔE is independent of the energy exchange processes involved, we must have

$$\Delta E_{11} = \Delta E_{12} + \Delta E_{21},$$

where ΔE_{11} is the value of ΔE for any process which has identical initial and final states (at state 1). The simplest process to accomplish no change in state is of course not to exchange energy at all, so that

$$\Delta E_{11} = 0.$$

Hence we have

$$\Delta E_{12} = -\Delta E_{21}.$$

In other words, the amount of energy ΔE_{12} is indeed stored in the system by the first process, for it can be recovered by the second process.

(3) An entity E, called the *internal energy*, exists and is a state variable. To show this, all we need is to demonstrate that for a given system at a specific state of interest, this internal energy E can be uniquely measured. We can do this by assigning an arbitrary value E_1 to E at a fixed reference state 1. Now by various conceivable energy-exchange processes, the system can be brought to any specified state of interest. From the first law, we know that the value of ΔE is the same regardless of the processes involved. We now define

$$E = E_1 + \Delta E.$$

Hence we see that the value of E for any state of interest of the system can be uniquely determined. Hence E is a state variable and thus has equal status with any other state variable such as density and temperature. As an example, let us assign $E_1 = 0$ for water at 62°F and atmospheric pressure. We can now describe the state of one pound of water at 63°F at atmospheric pressure by saying $E = 778.2$ ft-lb and $p = 14.7$ lb/in². This description is completely adequate and unambiguous. In general, it is advantageous to deal with internal energy per unit mass, or *specific internal energy*, defined by

$$e = \frac{E}{m},$$

where m is the mass of the system. For a homogeneous material such as a gas, once the reference state is specified, the relation

$$e = \text{function of } (\rho, T) = e(\rho, T) \tag{14-12}$$

can be experimentally constructed once and for all, and this relation is also called an *equation of state*, with equal status and importance as Eq. 14–4.

Generally, the first law of thermodynamics is referred to as the law of conservation of energy. We see from the above discussion that the concept of conservation is only one of the assertions of the first law.

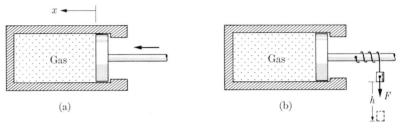

FIGURE 14–2

Example 14–1. Consider a cylinder filled with a gas, as shown in Fig. 14–2(a). One end of the cylinder is fitted with a piston. The mass of the gas is m, the initial volume occupied by the gas is V_1, and the internal energy of the gas is E_1.

(a) If the piston is pushed inward with a constant force of 100 lb for a distance of 0.1 ft, what is the final value of E? Assume no heat energy is exchanged; i.e., the system is insulated.

(b) If the piston is pushed very slowly inward for an infinitesimal distance dx such that the pressure of the gas is uniform in space at all time, and that no heat energy is exchanged between the gas and its environment, what is the change of the internal energy?

(c) If the piston is allowed to rotate briefly as shown by the arrangement in Fig. 14–2(b) such that the string is pulled with a constant force F for a distance h, what is the final value of E for the gas? We shall assume that there is no friction between the piston and the cylinder walls. The gas inside the cylinder will, however, be set into motion by viscosity but will eventually come to rest again.

For part (a), the answer is very simple. Since $\Delta Q = 0$, we have $\Delta E = -\Delta W$, and thus $E = E_1 - \Delta W$. Note that the work performed by the gas on its environment is simply the negative of the work done by the environment to the gas:

$$\Delta W = -(100 \times 0.1) = -10 \text{ ft-lb.}$$

Thus the answer is $E = E_1 + 10$ ft-lb.

For part (b), we proceed as follows. Since the motion of the piston is very slow, then the force acting on the piston must be very close to pA, where p is the pressure of the gas and A is the area of the piston. Thus for ΔW we have

$\Delta W = -PA\,dx$. But $A\,dx$ is merely $-dV$, where V denotes the volume of the gas. Since $\Delta Q = 0$, then from the first law we have

$$\Delta E = -\Delta W = -p\,dV.$$

Here, we see that for a slow process involving volume change, the value of ΔW is $p\,dV$. We shall have occasions to use this result later. The quantity $p\,dV$ is commonly referred to as *pressure work*.

For part (c), the external work done is Fh; hence $\Delta W = -Fh$. The answer is therefore $E = E_1 + Fh$.

PROBLEM

14-3. A rigid container is insulated from its environments, and is divided into two equal volumes. Suppose that one part of the container is filled with a gas with internal energy $E = E_1$ and the other part is a vacuum, and suddenly the partition between the two parts is removed. After the gas settles down and fills the whole volume, what is the final internal energy?

14-4. Specific internal energy

Strictly speaking, whenever we deal with a problem involving a gas, we must first search for available experimental data on the equations of state $T = T(p, \rho)$ and $e = e(\rho, T)$ for that particular gas of interest. We have shown that under ordinary conditions, the perfect gas law in Eq. 14-7 suffices as an adequate approximation relating T, p, and ρ. All we need to know is the molecular weight of the gas for the purpose of computing the gas constant R. We shall show in this section that analogous approximations are available under ordinary conditions for the relation $e = e(\rho, T)$.

First of all, it is observed experimentally that under ordinary conditions, the specific internal energy e is quite insensitive to ρ. In other words, once $e = e(T)$ has been experimentally determined for a given gas at a fixed value of ρ, the same formula serves with good accuracy at other values of ρ also. As a matter of fact, it can be shown mathematically that if the gas obeys the perfect gas law, then its specific internal energy is necessarily independent of ρ. (See F. W. Sears, *Thermodynamics* 2nd ed., Addison-Wesley Publishing Company, Reading, Mass., 1953, p. 62.) Since we shall adopt the perfect gas assumption throughout our discussions, we shall for the sake of consistency consider $e = e(T)$ only.

We can rewrite the relation $e = e(T)$ in the following form:

$$e = e_0 + \int_{T_0}^{T} C_v(T)\,dT, \qquad (14\text{-}13)$$

where

$$C_v = \frac{de}{dT}, \qquad (14\text{-}14)$$

and e_0 is a constant assigned arbitrarily to be the value of e at some reference temperature $T = T_0$. The quantity C_v is called the *specific heat at constant volume* of the gas. Experimentally it is found that within reasonable ranges of temperature, it is a good approximation to consider C_v a constant. Thus we have

$$e = (e_0 - C_v T_0) + C_v T = \text{constant} + C_v T. \tag{14-15}$$

At this point, we see that once C_v is known, the relation between e and T is determined except for an additive constant. Now, by collecting and inspecting data on C_v for various gases, the following astonishing rules can be arrived at. For monatomic gases, (e.g., helium), C_v is very nearly $3R/2$. For diatomic gases, (e.g., oxygen, nitrogen), C_v is very nearly $5R/2$, where R is the gas constant. The same rule also holds for nonreacting mixtures of monatomic *or* diatomic gases (e.g., air). For more complicated molecules, analogous rules also exist, but they do not hold with sufficient accuracy to be meaningful and will not be presented here. It suffices to say that for such gases, $C_v > 5R/2$. It may be mentioned in passing that such rules relating C_v and R can actually be derived from the kinetic theory of gases and therefore they should not be considered as completely fortuitous.

Generally, in a practical problem, one should use the experimentally determined values for C_v whenever possible. However, the simple rule given above is extremely useful, since it gives the order of magnitude and the qualitative dependence of C_v on the type of gas molecules involved.

PROBLEM

14–4. Find the value of C_v for the following gases and check your numerical values with experimental data available in most engineering handbooks.

Gas	Molecular weight	Gas	Molecular weight
H_2	2.016	A	39.95
O_2	32.000	Air	28.97
CO	28.000	He	4.002
N_2	28.016		

14–5. The energy equation

We are now ready to derive the energy equation for general fluid flows. We shall first apply the first law of thermodynamics to an element of a general fluid, and then specialize the result to a perfect gas later.

Consider a certain element of fluid in the flow field at a certain time t. We shall denote the volume of the fluid by τ and suppose that the volume is so small that the density ρ of the fluid in it can be considered uniform. Then the mass m of the fluid element is given by

$$m = \rho\tau, \qquad (14\text{–}16)$$

and its total internal energy is

$$E = me = \rho\tau e. \qquad (14\text{–}17)$$

We shall consider this fluid element as our thermodynamic system of interest and shall apply the first law to it as it undergoes changes of state and moves about the flow field. At time $t + \Delta t$, the total internal energy of the fluid element is, for small Δt,

$$E + \frac{DE}{Dt}\,\Delta t.$$

Here we use the particle derivative since we are considering a fluid element. Thus, between time t and time $t + \Delta t$, the first law, Eq. 14–10, asserts that

$$\frac{DE}{Dt}\,\Delta t = \Delta Q - \Delta W, \qquad (14\text{–}18)$$

where ΔQ is the net amount of heat energy added, and ΔW is the net amount of work performed by the fluid element during the interval Δt. Let us first consider the term ΔW. If the fluid is compressible, the volume occupied by the fluid element may change during the time interval Δt. Let p be the pressure of the fluid element. Since the volume of the element changes by the amount $(D\tau/Dt)\,\Delta t$ during the interval Δt, the work done by the element will be given by

$$+p\,\frac{D\tau}{Dt}\,\Delta t$$

(see Example 14–1b). We can thus write

$$\Delta W = p\,\frac{D\tau}{Dt}\,\Delta t + \Delta W',$$

where $\Delta W'$ now represents all other work (such as work done against frictional forces) performed by the fluid elements *except* the pressure work discussed above. Equation 14–18 can now be written as

$$\frac{DE}{Dt} + p\,\frac{D\tau}{Dt} = \frac{1}{\Delta t}\,(\Delta Q - \Delta W').$$

Using Eqs. 14–16 and 14–17 to eliminate E and τ in favor of e, m, and ρ, we obtain

$$\frac{De}{Dt} + p\,\frac{D}{Dt}\left(\frac{1}{\rho}\right) = \dot{Q} - \dot{W}', \qquad (14\text{–}19)$$

where for brevity we denoted by $\dot{Q}$ and $\dot{W}'$ the net heat added and net *non-*

pressure work performed per unit time by a fluid element of unit mass. Equation 14–19 is the general energy equation.

For a heat-conducting and viscous fluid, $\dot{Q}$ and $\dot{W}'$ are proportional to the heat conductivity and viscosity coefficients, respectively. If we consider a *frictionless, non-heat-conducting fluid*, the energy equation becomes simply

$$\frac{De}{Dt} + p \frac{D}{Dt}\left(\frac{1}{\rho}\right) = 0. \tag{14–20a}$$

This equation can be rewritten in a more usable form. Let us first add and subtract the term $(1/\rho)(Dp/Dt)$ on its left-hand side. We have

$$\frac{De}{Dt} + p \frac{D}{Dt}\left(\frac{1}{\rho}\right) + \frac{1}{\rho}\frac{Dp}{Dt} - \frac{1}{\rho}\frac{Dp}{Dt} = \frac{D}{Dt}\left(e + \frac{p}{\rho}\right) - \frac{1}{\rho}\frac{Dp}{Dt} = 0.$$

Now by definition, we have

$$\frac{Dp}{Dt} = \frac{\partial p}{\partial t} + u\frac{\partial p}{\partial x} + v\frac{\partial p}{\partial y} + w\frac{\partial p}{\partial z}.$$

We can eliminate $\partial p/\partial x$, $\partial p/\partial y$, $\partial p/\partial z$ by means of the *frictionless* momentum equations, Eqs. 5–1, 5–2, and 5–3. Neglecting the body force terms, we obtain, for example, from Eq. 5–1,

$$u\frac{\partial p}{\partial x} = -u\rho\frac{Du}{Dt} = -\rho\frac{D}{Dt}\left(\frac{u^2}{2}\right).$$

Thus we can show

$$\frac{Dp}{Dt} = \frac{\partial p}{\partial t} - \rho\frac{D}{Dt}\left(\frac{u^2}{2} + \frac{v^2}{2} + \frac{w^2}{2}\right).$$

Using this expression for Dp/Dt, we obtain,

$$\frac{D}{Dt}\left(e + \frac{p}{\rho} + \frac{q^2}{2}\right) = \frac{1}{\rho}\frac{\partial p}{\partial t}, \tag{14–20b}$$

where $q^2 = u^2 + v^2 + w^2$. Equation 14–20b is most useful in steady flow problems when $\partial p/\partial t = 0$. Then we have

$$e + \frac{p}{\rho} + \tfrac{1}{2}q^2 = \text{constant} = h^0 \tag{14–21a}$$

along any streamline. The quantity $e + p/\rho$ is of course a state variable, and is called the *specific enthalpy* of the gas. It is denoted by the symbol h:

$$h = e + \frac{p}{\rho}. \tag{14–22}$$

Thus, Eq. 14–21a becomes

$$h + \tfrac{1}{2}q^2 = \text{constant} = h^0. \qquad (14\text{–}21\text{b})$$

The quantity h^0 is called the *stagnation enthalpy* of the gas, since h^0 is the value of h at the stagnation point of a steady, frictionless, and non-heat-conducting flow.

If the fluid is incompressible, frictionless and non-heat conducting, Eq. 14–20a gives $De/Dt = 0$, or $e = $ constant following a given particle. Using this result in Eq. 14–21 under the steady flow assumption, we have

$$\frac{p}{\rho} + \tfrac{1}{2}q^2 = h^0 - e = \text{constant}$$

along a streamline. This is of course simply the Bernoulli equation for an incompressible flow (without body force).

PROBLEM

14–5. What is the temperature at the stagnation point of a supersonic jet airliner flying at velocity of 1000 miles per hour at a high altitude where the ambient temperature is $-67°F$?

14–6. Specific enthalpy

We shall give in this section a physical interpretation for the new state variable h defined in Eq. 14–22. Consider a vessel containing a unit mass of gas. If the vessel is rigid, we can raise the temperature of the gas by ΔT by simply adding an amount of heat $C_v \Delta T$ to the gas. During this process, the pressure of the gas will change while the volume does not. Since no work is done, Eq. 14–10 gives $\Delta Q = \Delta e$. Now, if the vessel is a flexible one, and it is desired to raise the gas temperature by ΔT so that during the process the pressure of the gas does not change, what is the amount of heat energy now required? If τ is the volume of the gas and $\Delta \tau$ is the volume change of the gas during the process, then, since pressure is constant, the amount of work done by the gas is $p \Delta \tau$. Applying the first law, we have

$$\Delta e = \Delta Q - p \Delta \tau,$$

or

$$\Delta Q = \Delta(e + p\tau) = \Delta \left(e + \frac{p}{\rho} \right) = \Delta h.$$

We see that in the constant-pressure process, heat energy added is used to increase the internal energy of the gas *and* to allow the gas to perform work in changing its volume. Comparing the corresponding expressions for ΔQ, it is clear that h serves the same role in a constant-pressure process as e does in a constant-volume process.

The *specific heat at constant pressure*, C_p, is defined analogously to C_v:

$$C_p = \left(\frac{\partial h}{\partial T}\right)_p . \tag{14-23}$$

In general, C_p can be a function of ρ and T, although again for a perfect gas it is a function of T only. Note that in general any state variable can be considered as a function of any two other state variables. For example, $h = h(p, T)$ or $h = (\rho, T)$. Hence, in differentiating one state variable with respect to another, it is important always to specify by a subscript the third state variable which is being held constant. Thus for a general gas the expression $\partial h/\partial T$ would be meaningless, for we would not know whether it is $(\partial h/\partial T)_p$ or $(\partial h/\partial T)_\rho$ or something else. In the same spirit, Eq. 14–14 for a general gas should be written as

$$C_v = \left(\frac{\partial e}{\partial T}\right)_\rho ,$$

so that there is no ambiguity.

Now, for a perfect gas, using the definition of h and C_p, we have

$$C_p = \left(\frac{\partial h}{\partial T}\right)_p = \left(\frac{\partial e}{\partial T}\right)_p + \left(\frac{\partial (p/\rho)}{\partial T}\right)_p = C_v + R, \tag{14-24}$$

since $e = e_0 + \int^T C_v\, dT$ and $p/\rho = RT$. Thus, C_v, C_p, and R are related. The ratio of C_p to C_v is called the *ratio of specific heats*, and is denoted by γ:

$$\gamma = \frac{C_p}{C_v} . \tag{14-25}$$

Using our simple rules on C_v, we see that $\gamma = \frac{5}{3}$ for monatomic gases and $\gamma = \frac{7}{5}$ for diatomic gases including air.

There are several other useful and interesting relations among C_p, C_v, γ, and the gas constant R for a perfect gas. Since

$$C_p = C_v + R = \gamma C_v,$$

we have

$$C_v = \frac{R}{\gamma - 1}, \tag{14-26}$$

and thus

$$C_p = \frac{\gamma R}{\gamma - 1} . \tag{14-27}$$

It should be emphasized that Eqs. 14–24 to 14–27 are valid even if C_v and C_p are functions of temperature. For perfect gases with constant C_v and γ, any two of the parameters C_p, C_v, γ, and R completely describe the gas. Parallel to Eq. 14–15, one can write

$$h = \text{constant} + C_p T$$

as a useful approximation when C_p is approximately constant.

14–7. Isentropic relations

For a perfect gas with constant specific heats, Eq. 14–20, which is valid under the frictionless and non-heat-conducting assumptions, can be integrated immediately to yield a simple relation between p and ρ. We proceed as follows. Using Eqs. 14–15 and 14–7 in Eq. 14–20a to eliminate e in favor of p and ρ, we have

$$\frac{De}{Dt} + p \frac{D}{Dt}\left(\frac{1}{\rho}\right) = \frac{C_v}{R} \frac{D}{Dt}\left(\frac{p}{\rho}\right) + p \frac{D}{Dt}\left(\frac{1}{\rho}\right) = 0.$$

Simplifying, we obtain

$$\frac{1}{p}\frac{Dp}{Dt} = \frac{\gamma}{\rho}\frac{D\rho}{Dt}, \tag{14–28}$$

where $\gamma = (R/C_v) + 1$ is the ratio of specific heats. Equation 14–28 can be integrated to give

$$\frac{p}{\rho^\gamma} = \text{constant} \tag{14–29}$$

for a given fluid element. This relation is called the *isentropic relation* for a perfect gas.

Equation 14–29 is extremely interesting. Suppose we know that at a certain moment an element of fluid has $p = p_0$ and $\rho = \rho_0$. Then for this particular fluid element, its pressure and density will be related by

$$\frac{p}{\rho^\gamma} = \frac{p_0}{\rho_0^\gamma} \tag{14–30}$$

for all later time. In other words, at any time, if we wish to identify the state of a certain fluid element in a frictionless, non-heat-conducting flow, we need only to specify one state variable such as ρ (or p) and the value of p_0/ρ_0^γ at some earlier time. Thus, the quantity p/ρ^γ must be a state variable which has the interesting property that it remains a constant in such a flow.

We shall see that p/ρ^γ is related to a new state variable called specific entropy. It is denoted by s and is sometimes referred to simply as entropy. We shall also see that if the flow is frictionless and non-heat-conducting, but the gas itself is not a perfect gas with constant specific heats, Eq. 14–29 will not hold. The concept of entropy, which is introduced by the second law of thermodynamics, will allow us to find the corresponding isentropic relations for such cases.

In Section 5–5, the Bernoulli equation was derived for an incompressible flow. When the flow is isentropic, a corresponding Bernoulli equation can be obtained. Neglecting gravity forces, Eq. 5–8 is

$$q \frac{\partial q}{\partial l} = -\frac{1}{\rho}\frac{dp}{dl},$$

where the flow has been assumed to be steady and frictionless. Integrating with respect to l, we have

$$\frac{q^2}{2} = -\int \frac{1}{\rho} \frac{\partial p}{\partial l} \, dl = -\int \frac{1}{\rho} \, dp.$$

Using Eq. 14–30, the integral can be evaluated to give

$$\frac{\gamma p}{(\gamma - 1)\rho} + \frac{q^2}{2} = \text{constant}.$$

Using Eqs. 14–7 and 14–27, we have

$$C_p T + \frac{q^2}{2} = \text{constant}.$$

For constant C_p, this equation is equivalent to

$$h + \frac{q^2}{2} = \text{constant},$$

in agreement with Eq. 14–21b, which was derived from energy considerations. Hence we see that in steady-flow problems, the frictionless momentum equation supplemented by the isentropic relation together imply the energy equation. It is important to note that the converse is not true; Eq. 14–21b can remain valid even when the flow is nonisentropic.

PROBLEM

14–6. Using Eq. 14–30, find the isentropic relations between p and T and between T and ρ.

14–8. The second law of thermodynamics

For compressible flows, density and temperature are additional unknowns. The equation of state of the fluid and the energy equation are now available to supplement the continuity and momentum equations to form a complete set of governing equations. It is possible, at the present level of sophistication, to proceed with the study of compressible fluid mechanics without introducing the second law of thermodynamics. However, the second law asserts the existence of a new state variable S called entropy which turns out to be the easiest and most convenient to monitor in a large class of important problems. These are the isentropic flow problems in which the specific entropy s (entropy per unit mass) is a constant. In this section, we shall discuss the second law with the primary objective of defining entropy and setting forth conditions under which the entropy of a fluid element will remain constant. The reader, however, must not be misled to the conclusion that such applications constitute the major significance of the second law. The second law has far-reaching consequences in many fields of science, including fluid mechanics. We shall deal with only some of its elementary applications.

Before we proceed further, we must first define what is meant by reversible and irreversible processes. A reversible process is any process which, during its course, produces no lasting changes of any kind in the environment if the process is allowed to go forward and then back to its original state. Actually, a reversible process can never be achieved in reality, and all physical processes are irreversible to some extent. For example, all processes involving friction are irreversible. A reversible process is merely an idealized process with which we can perform conceptual, idealized experiments. Processes (a) and (c) in Example 14–1 are examples of irreversible processes, while process (b) is a reversible process. To show that process (b) is a reversible process, we proceed as follows. Since the volume change is small and the process is performed slowly, the pressure p will remain essentially constant. Hence the work ΔW done by the gas in volume change from V_1 to $V_1 + \Delta V$ is

$$\Delta W = p \int_{V_1}^{V_1 + \Delta V} dV.$$

If the piston is moved slowly forward and then slowly backward so that the volume goes back to V_1, the net work done ΔW by the gas is evidently zero. Since no heat energy is exchanged, we see that no lasting changes of any kind have been produced. Hence this is a reversible process. Processes (a) and (c) in Example 14–1 are clearly irreversible processes for it is impossible to return the gas to its initial state by applying the reverse process.

The second law asserts that all thermodynamic systems possess a state variable called *entropy* S which is defined by

$$\Delta S = \int_{\text{rev}} \frac{dQ}{T}. \tag{14–31}$$

The formula is interpreted to mean that the change of entropy S between two states is the sum of all dQ/T incurred during any reversible process which joins the two states. The second law also further asserts that for an insulated system, the entropy of the system never decreases. The above assertions can all be deduced from the general statement of the second law of thermodynamics: *No system can pass through a complete cycle of states and deliver positive work to the environment while exchanging heat with only a single source of heat at uniform temperature.* It is possible to deduce from this abstract statement that S as defined by Eq. 14–31 is indeed a state variable. The deduction is, however, rather subtle and involved and we shall not attempt to present it here. (See, for example, Chapter 8 of *Thermodynamics* by F. W. Sears.)

The assertion that the entropy of an insulated system never decreases is rather easily proved. Let us consider a system which undergoes some process *while insulated* and its state undergoes a change. We shall specify the initial state of the system by its internal energy E_1 and its entropy S_1. If the process involved is a reversible one, then obviously S remains constant at S_1 by the definition of entropy in Eq. 14–31. Now suppose that the process is an irreversible one and we further suppose that the entropy of the system changes by the process from S_1 to $S_1 + \Delta S$. Now let us add heat to the system reversibly and isothermally by the amount $-T \Delta S$. By the definition of S, we see that the entropy of the system recovers the value S_1. Now let us insulate the system and somehow extract work reversibly from the system so that the value of E returns to its initial value E_1. The value of S of course remains unchanged at S_1 by definition. Now the system has returned to its initial state, since we have $E = E_1$, $S = S_1$, and is ready to repeat the whole sequence of processes. Applying the first law to the system between the initial and the final states, we have

$$\Delta E = \Delta Q - \Delta W = 0.$$

Now the total amount of heat energy added is $\Delta Q = -T \Delta S$. Hence the total amount of work performed by the system is

$$\Delta W = -T \Delta S.$$

Now if ΔS is negative, it would mean that $\Delta Q > 0$ and $\Delta W > 0$. In other words, this sequence of processes enables the system to deliver positive work while exchanging heat with a single source of heat at uniform temperatures, and it can operate cycle after cycle. This is in clear contradiction to the statement of the second law, and ΔS must be positive. Hence the proof is complete.

Example 14–2. Find the entropy change of the gas undergoing the process described in Problem 14–3. Assume that the gas is a perfect gas.

First of all, we should recognize that the process involved in Problem 14–3 is an irreversible one. To compute the entropy change, we must imagine some reversible process that would bring the gas from its initial state to its final state so that we can apply the definition of entropy change given in Eq. 14–31. The following is one such possible reversible process. Instead of removing the partition suddenly, let us add heat slowly to the gas and at the same time allow the partition to move outward slowly so that the temperature of the gas remains constant in both space and time (constant-temperature processes are called *isothermal processes*). We continue with this reversible process until the gas fills the whole volume V_2 of the container. Since we are dealing with a perfect gas and consequently its specific internal energy is a function of T only, we see that the specific energy of the gas remains a constant in this process. We have thus successfully brought the gas to the desired final state, since the final values of E and ρ are now achieved. The amount of work done by the gas on its environment during this reversible process is

$$\Delta W = \int p \, dV.$$

For a perfect gas, we have $p = \rho R T$. Since $\rho = m/V$, where m is the mass of the gas, the integral for the work done can be written as

$$\Delta W = \int_1^2 mRT \frac{dV}{V} = mRT \, [\ln V]_1^2 = mRT \ln 2.$$

Since $\Delta E = 0$ between the initial and final states, the first law states that $\Delta Q = \Delta W$. Hence we have

$$\Delta Q = mRT \ln 2 \qquad \text{or} \qquad \frac{\Delta Q}{T} = mR \ln 2.$$

Since the process considered is a reversible one, then by Eq. 14–31 we have

$$\Delta S = \int_{\text{rev}} \frac{dQ}{T} = mR \ln 2,$$

which is the desired result.

14–9. Entropy and elementary compressible flows

If a thermodynamic system, such as an element of fluid in a flow field, is assumed to have no heat exchange with its environment and undergoes only reversible processes, then by the definition of S, the entropy of the system remains constant. From this viewpoint, when we discuss and manipulate with the energy equation in Section 14–5 for a frictionless, nonconducting fluid, it becomes immediately clear without further discussion that the entropy of each fluid element must remain constant—since in the absence of friction and heat conduction the flow process is reversible. Since entropy is a state variable, it follows again immediately that for each element of fluid in such isentropic flows, only one other state variable can be independent, and hence it should come as no surprise that a definite relation between p and ρ is found by the analysis. Now, when we deal with a gas which is not adequately described by the perfect gas law and whose specific heats are not constants, we can conclude from the discussions in this section that the equation corresponding to $p/\rho = p_0/\rho_0^\gamma$ would be a relation of the form $p = p(\rho, s)$ with the understanding that s is the specific entropy of the gas and is a constant for each fluid element. The relation $p = p(\rho, s)$ itself is of course simply another equation of state for this particular gas and can be experimentally constructed once and for all.

The equation of state relating p, ρ, and s for a perfect gas with constant specific heat is derived as follows. Let us first write down the first law for an infinitesimal change of state. We have, for a unit mass of the gas,

$$de = dQ - dW.$$

We now consider a reversible process joining the states. Hence $dQ = dQ_{\mathrm{rev}}$ and $dW = pd(1/\rho)$. Using Eqs. 14–7 and 14–15 for the perfect gas, we have

$$C_v \, dT = dQ_{\mathrm{rev}} - RT \, d(1/\rho).$$

Dividing by T and rearranging, we have

$$ds = \frac{dQ_{\mathrm{rev}}}{T} = C_v \frac{dT}{T} + R \, d\left(\frac{1}{\rho}\right).$$

Since C_v is assumed constant, we can integrate immediately to yield

$$s = C_v \ln T - R \ln \rho + \text{constant}. \tag{14–32a}$$

Eliminating T by the perfect gas law again in favor of p and ρ, we have

$$s = C_v \ln \frac{p}{\rho^\gamma} + \text{constant}, \tag{14–32b}$$

which is the desired result. Note that from the analysis in Section 14–7 we had anticipated that s would be a function of p/ρ^γ.

PROBLEM

14–7. Use Eq. 14–32 and repeat the calculation of Example 14–2.

14–10. The differential equation of state

Consider a unit mass of gas at state $e = e_1$, $\rho = \rho_1$. Suppose that by some arbitrary process, which may be reversible or irreversible, the state is brought to an adjacent state $e = e_1 + de$, $\rho = \rho_1 + d\rho$. Obviously, if its original specific entropy is s_1, its final specific entropy would also be changed to $s = s_1 + ds$. We shall now derive a relation relating de, $d\rho$, and ds for gases with a general equation of state.

To do this we shall consider a reversible process joining the original state and the final state. Writing the first law from Eq. 14–10, we have

$$de = dQ - dW. \tag{14–33}$$

Now since the process being considered is reversible, we have by definition of entropy

$$dQ = dQ_{\mathrm{rev}} = T \, ds. \tag{14–34}$$

The work done by the gas on its environment would be that caused by the pressure work and is given by

$$dW = p \, d(1/\rho).$$

Substituting dQ and dW from the above into Eq. 14–33, we have

$$de = T \, ds + p \, d(1/\rho), \tag{14–35}$$

which is the relation desired. Equation 14–35 relates de, ds, and $d\rho$ between *any* two adjacent states of a given gas. Hence it remains correct even if the gas actually undergoes an irreversible process in changing from one state to the other.

Equation 14–35 is an extremely useful relation. For example, for a given fluid element in a flow problem, regardless of whether the fluid is a perfect gas or the flow is reversible, we can write

$$\frac{De}{Dt} = T \frac{Ds}{Dt} - p \frac{D}{Dt}\left(\frac{1}{\rho}\right). \tag{14–36}$$

Using the definition of h, Eq. 14–22, we can rewrite Eq. 14–36 as

$$\frac{Dh}{Dt} = T \frac{Ds}{Dt} + \frac{1}{\rho}\frac{Dp}{Dt}. \tag{14–37}$$

The above relation is valid in general. Hence the general energy equation, Eq. 14–19, derived in Section 14–5 can be rewritten as

$$T \frac{Ds}{Dt} = \dot{Q} - \dot{W}'. \tag{14–38}$$

Therefore, it becomes apparent that when Q and W' are zero, we have $Ds/Dt = 0$. In other words, specific entropy s remains constant for a given particle. Such flows are appropriately called *isentropic flows*. When a flow field has uniform entropy in space and time, it is sometimes called a *homentropic flow*. In the technical literature, however, the term isentropic flow generally means homentropic flow unless it is specifically noted. When entropy remains constant for any given particle but is not uniform in space and time, it is sometimes called *particle isentropic flow*.

15 One-Dimensional Steady Compressible Flow

15-1. Introduction

In this chapter, we study compressible fluid mechanics under a very general simplification—the one-dimensional assumption. Under this simplifying assumption we can study a wide range of problems which are of practical interest, and in so doing the main effects of compressibility will be brought clearly into focus. We shall deal exclusively with "internal aerodynamics," i.e., flows inside ducts or tubes. In this class of problems we are normally not interested in what happens at a point, but rather at an axial station. In applying the one-dimensional assumption, it is assumed that all quantities of interest are uniform across the cross section at the station of interest. In essence, it considers the whole flow as a giant stream tube. We may consider the study of one-dimensional flow as essentially a study of the behavior of a stream tube under various conditions. In practical situations, however, flow conditions in a tube are seldom uniform. For example, the velocity profiles near the walls of the tube are necessarily nonuniform because of the presence of viscous boundary layers. In applying the one-dimensional assumption to practical problems, one simply has to have other assurances that the flow conditions are reasonably uniform at the stations of interest. Within this limitation it is a useful assumption, as demonstrated in Sections 5–7 and 9–2.

15-2. The speed of sound

Before we proceed with the study of steady one-dimensional compressible flows, we shall first study the speed of propagation of a sound wave. This is important for we shall see very shortly that meaningful interpretations of most compressible effects depend on the thorough understanding of the role played by the speed of sound.

Let us consider the following idealized situation. A long straight tube containing a fluid at rest is fitted at one end with a flat piston which is free to slide longitudinally. At time $t = 0$ the piston suddenly starts from rest and moves or vibrates with some velocity $U_p(t)$. It is desired to know what kind of disturbances will be produced in the fluid. Before we proceed with the analysis, let us first consider what happens physically. The fluid immediately adjacent to the piston must move with the piston. If the fluid is incompressible, then the whole body of fluid in the tube will have to move with the piston. Any motion of the piston is therefore immediately felt at all points. If we define the speed of sound as the speed with which a weak disturbance travels, then under the incompressible assumption, the speed of sound will be infinite (we shall make clear what is considered a "weak" disturbance later). When the fluid is compressible, the fluid at some station away from the piston need not move with the piston, since the fluid density between the piston and the station of interest would change. Thus if the piston makes a weak disturbance, the station of interest will not be disturbed immediately, but at some later time. From this simple consideration, we see clearly that compressibility has a great deal to do with our problem. In other words, the phenomena of sound is a compressibility effect.

We shall proceed with the following simplifying assumptions. Firstly, we assume that internal friction is not of importance and can be ignored. Secondly, we assume that the one-dimensional assumption applies so that all quantities of interest are functions of x and t only, where x is axial distance measured from the original position of the piston and t is time. The continuity and momentum equations are therefore,

$$\frac{\partial \rho}{\partial t} + \frac{\partial}{\partial x}(\rho u) = 0, \tag{15–1}$$

$$\frac{\partial u}{\partial t} + u \frac{\partial u}{\partial x} = -\frac{1}{\rho}\frac{\partial p}{\partial x}, \tag{15–2}$$

where we have ignored body forces. Thirdly, we assume that heat-conduction effects are not important. From the discussion of Section 14–9, we conclude that in the absence of friction and heat conduction, the entropy of each fluid element will remain a constant. Since the entropy of the fluid is originally uniform, it will remain uniform even after the piston moves. For fluids in general, pressure, density, and entropy are related by an equation of state of the form

$$p = p(\rho, s), \tag{15–3a}$$

which is understood to be an experimentally obtained equation. Since s remains constant at the value of s_0, we have a unique relation between p and ρ:

$$p = p(\rho, s_0). \tag{15–3b}$$

Equations 15–1, 15–2, and 15–3b constitute the governing equations, and the remaining task is to find the solutions p, ρ, and u as functions of x and t as a result of the piston motion.

Let us first take advantage of Eq. 15–3b to eliminate pressure in favor of density from Eq. 15–2. Using the chain rule of calculus, we have

$$\frac{\partial p}{\partial x} = a^2 \frac{\partial \rho}{\partial x}, \tag{15–4}$$

where a^2 is defined as

$$a^2 = \left(\frac{\partial p}{\partial \rho}\right)_s \tag{15–5}$$

and is itself a state variable expressible as a function of ρ and s. The subscript s indicates a derivative at constant entropy.

For a perfect gas, the equation of state is

$$p = \rho R T, \tag{15–6}$$

and the isentropic relation, corresponding to Eq. 15–3b, is

$$\frac{p}{p_0} = \left(\frac{\rho}{\rho_0}\right)^{\gamma}, \tag{15–7}$$

where p_0 and ρ_0 are the undisturbed values of p and ρ, respectively. Using Eqs. 15–6 and 15–7 in Eq. 15–5, we have for a perfect gas

$$a^2 = \gamma \frac{p_0}{\rho_0^{\gamma}} \rho^{\gamma-1} = \gamma \frac{p}{\rho} = \gamma R T. \tag{15–8}$$

Equation 15–8 clearly substantiates the previous statement that a is a state variable, being uniquely determined by the state variable T.

With Eq. 15–4, Eq. 15–2 becomes, for fluids in general,

$$\frac{\partial u}{\partial t} + u \frac{\partial u}{\partial x} + \frac{a^2}{\rho} \frac{\partial \rho}{\partial x} = 0. \tag{15–9}$$

Let us now take advantage of the fact that we are dealing with weak disturbances. To this end, we first nondimensionalize all our dependent variables as follows:

$$u = a_0 u', \qquad \rho = \rho_0(1 + \rho'), \qquad a = a_0(1 + a'), \tag{15–10}$$

where a_0 is the undisturbed value of a, which has the dimension of a velocity. For weak disturbances, the dimensionless variables, u', a', and ρ' are assumed to be small compared with unity. Substituting Eq. 15–10 into Eqs. 15–1 and 15–9, we have

$$\frac{1}{a_0} \frac{\partial \rho'}{\partial t} + \frac{\partial}{\partial x} [(1 + \rho')u'] = 0, \tag{15–11a}$$

$$\frac{1}{a_0} \frac{\partial u'}{\partial t} + \frac{(1 + a')^2}{1 + \rho'} \frac{\partial \rho'}{\partial x} + \frac{\partial}{\partial x}\left(\frac{u'^2}{2}\right) = 0. \tag{15–11b}$$

These equations look more complicated than the original ones, but the fact that the primed quantities are considered small allows us certain simplifications.

By neglecting the higher-order terms involving products of the primed quantities, we have, as an approximation to Eqs. 15–11, the following equations:

$$\frac{1}{a_0}\frac{\partial \rho'}{\partial t} + \frac{\partial u'}{\partial x} = 0, \tag{15-12a}$$

$$\frac{1}{a_0}\frac{\partial u'}{\partial t} + \frac{\partial \rho'}{\partial x} = 0. \tag{15-12b}$$

Differentiating the first equation with respect to x and the second equation with respect to $a_0 t$ and subtracting, we have

$$\frac{1}{a_0^2}\frac{\partial^2 u'}{\partial t^2} - \frac{\partial^2 u'}{\partial x^2} = 0. \tag{15-13}$$

This equation is commonly known as the wave equation (compare with Eq. 10–15). A variety of standard methods are available for solving this equation, e.g., the method of separation of variables. The boundary condition for our problem is that at the piston, the gas velocity must be equal to the piston velocity. Now if the amplitude of motion of the piston is small, then the piston location is given approximately at $x = 0$. Let us suppose that the piston vibrates sinusoidally so that our boundary condition is

$$u = \epsilon a_0 \sin \omega t \qquad \text{at } x = 0, \quad t > 0,$$

or

$$u' = \epsilon \sin \omega t \qquad \text{at } x = 0, \quad t > 0, \tag{15-14}$$

where ϵa_0 is the maximum piston speed, and ω is the circular frequency of vibration of the piston. Using the method of separation of variables, the solution of Eq. 15–13 is found to be

$$u' = \epsilon \sin\left[\omega\left(t - \frac{x}{a_0}\right)\right]. \tag{15-15}$$

Or, in terms of the actual dimensional velocity, it is

$$u(x, t) = \epsilon a_0 \sin\left[\omega\left(t - \frac{x}{a_0}\right)\right]. \tag{15-16}$$

Equation 15–15 satisfies Eq. 15–13 and the boundary condition given by Eq. 15–14, as can be verified immediately upon substitution. The density perturbation ρ' can now be found by substituting Eq. 15–15 into either 15–12a or 15–12b and integrating the resulting equation. This gives

$$\rho' = \epsilon \sin\left[\omega\left(t - \frac{x}{a_0}\right)\right], \tag{15-17}$$

indicating that ρ' and u' are identical.

We can now see that the speed of sound is a_0. For if we observe the disturbances from a coordinate system which moves in the x-direction with constant velocity a_0, the disturbances would appear stationary. It can easily be verified that for any arbitrary piston motion $U_p(t)$ the solution to Eq. 15–13 is

$$u' = \frac{1}{a_0} U_p \left(t - \frac{x}{a_0} \right). \tag{15-18}$$

(See also the discussion of Eq. 10–16 on shallow-water waves.)

We can now state more explicitly what is meant by a weak disturbance. Since u' was assumed to be small compared with unity, then from Eq. 15–14, ϵ must be a small number. Hence the disturbance created by the vibrating piston can be considered weak only if the maximum speed of the piston is small compared with the speed of sound a_0 of the undisturbed gas. In air at 68°F, the speed of sound is 1117 ft/sec.

We can summarize the important and interesting consequences of our analysis as follows. First of all, the speed of sound of the gas is found to depend only on the state of the gas. For a general gas, it is given by Eq. 15–5, while for a perfect gas it is given by Eq. 15–8. In particular, it is important to note that the speed of sound of a perfect gas can be determined from the temperature of the gas alone. Secondly, the composition of the gas (whether it be air or pure hydrogen, for example) enters only through the values of γ and R (which depend on the molecular weight). Most important of all, it is independent of the frequency of vibration (so that for a concert-goer in the back of a concert hall, the sounds from the violin and the cello will not be out of step). Thirdly, the compressibility of a fluid introduces the speed of sound as a characteristic speed. We therefore have a reference value to gage whether a gas is moving fast or slow. The dimensionless ratio formed by dividing the speed q by a at a point is called the Mach number at this *point:*

$$M = \frac{q}{a}. \tag{15-19}$$

This is one of the most important parameters in compressible fluid mechanics.

We have obtained the speed of sound in gases. For liquids, this speed can be found by using the appropriate equation of state. In terms of the bulk modulus K, the speed of sound can be shown to be $\sqrt{K/\rho}$ (see Problem 15–2) and the Mach number can also be written as $q/\sqrt{K/\rho}$ (see Eq. 2–9).

PROBLEMS

15–1. If we assume blindly that instead of isentropic transformations, the perfect gas undergoes isothermal changes in this problem, show that the speed of sound is $a_0^2 = RT$.

15–2. Assume that for a liquid the bulk modulus $K = \rho(dp/d\rho)$ is constant (see Problem 1–5). Show that the speed of sound is $\sqrt{K/\rho_0}$, where ρ_0 is the undisturbed density. [*Hint:* Since $K = \rho \cdot dp/d\rho$, then $\partial p/\partial x = (\partial \rho/\partial x)K/\rho$ which may be used to eliminate $\partial p/\partial x$ from the momentum equation.]

15–3. The stagnation quantities

We shall consider now only steady flow problems. If we wish to describe the flow condition at a given point of a flow, we may give the values of its pressure, temperature, and velocity. However, it turns out that for most engineering applications, it is more convenient to give instead the values of the stagnation pressure, stagnation temperature, and Mach number. Stagnation temperature of a particle is the temperature that one would measure if the particle is isentropically brought to rest. Stagnation pressure is the corresponding pressure. The Mach number is defined as the ratio of local gas speed to the local speed of sound.

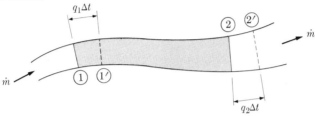

FIGURE 15–1

To clarify the above we shall need the energy equation for one-dimensional flow. Instead of using results from Chapter 14, we shall give an alternative derivation of the energy equation here from the steady, one-dimensional flow viewpoint. Let us consider a stream tube in steady flow between two stations 1 and 2, as shown in Fig. 15–1. At any given instant t_0, the amount of energy contained in the fluid in the stream tube between stations 1 and 2 is

$$E(t_0) = \int_1^2 (e + \tfrac{1}{2}q^2)\rho A \, dl, \tag{15–20}$$

where e is internal energy per unit mass, $\tfrac{1}{2}q^2$ is kinetic energy per unit mass, and l is distance measured along the stream tube. At time $t_0 + \Delta t$, the fluid of interest is now between stations 1' and 2', and the amount of energy it contains is

$$E(t_0 + \Delta t) = \int_{1'}^{2'} (e + \tfrac{1}{2}q^2)\rho A \, dl. \tag{15–21}$$

Now the first law of thermodynamics states that the increase of this energy must equal the net heat energy added minus the net work done by the fluid on its surroundings. Denoting by $\dot{Q}$ and $\dot{W}$ the net heat added per unit time and net work done by the fluid of interest per unit time, we then have

$$E(t_0 + \Delta t) - E(t_0) = \int_{1'}^{2'} (e + \tfrac{1}{2}q^2)\rho A \, dl - \int_1^2 (e + \tfrac{1}{2}q^2)\rho A \, dl$$

$$= \dot{Q} \, \Delta t - \dot{W} \, \Delta t. \tag{15–22}$$

But the difference of the two integrals can be evaluated in the following manner. Since we are dealing with steady flow problems, conditions between stations $1'$ and 2 are independent of time. Furthermore, the distance dl between stations 1 and $1'$, and 2 and $2'$ is simply $q_1 \, \Delta t$ and $q_2 \, \Delta t$, respectively. Thus

$$\int_2^{2'} \rho A(e + \tfrac{1}{2}q^2) \, dl - \int_1^{1'} (e + \tfrac{1}{2}q^2)\rho A \, dl = \dot{m} \, \Delta t[(e_2 + \tfrac{1}{2}q_2^2) - (e_1 + \tfrac{1}{2}q_1^2)],$$

$$(15\text{-}23)$$

where $\dot{m} = \rho_1 A_1 q_1 = \rho_2 A_2 q_2$ is the mass flow rate through the stream tube. Next we shall compute the amount of work that the fluid of interest has done on its surroundings during the time interval Δt. At station 1, the pressure is p_1, and therefore the force exerted by the fluid under consideration against the surrounding is $p_1 A_1$ directed against the flow. The work done by the fluid during the Δt at station 1 is therefore

$$-p_1 A_1 \, dl = -p_1 A_1 q_1 \, \Delta t = - \frac{p_1}{\rho_1} \dot{m} \, \Delta t. \qquad (15\text{-}24)$$

Similarly, the work done by the fluid at station 2 is

$$+p_2 A_2 \, dl = \frac{p_2}{\rho_2} \dot{m} \, \Delta t. \qquad (15\text{-}25)$$

Substituting Eqs. 15–23, 15–24, and 15–25 into Eq. 15–22, we have

$$\left(\frac{p_1}{\rho_1} + e_1 + \tfrac{1}{2}q_1^2\right) + \left(\frac{p_2}{\rho_2} + e_2 + \tfrac{1}{2}q_2^2\right) - \frac{\dot{Q}}{\dot{m}} + \frac{\dot{W}'}{\dot{m}} = 0, \qquad (15\text{-}26)$$

where we have written for $\dot{W} \, \Delta t$

$$\dot{W} \, \Delta t = \left(\frac{p_2}{\rho_2} - \frac{p_1}{\rho_1}\right) \dot{m} \, \Delta t + \dot{W}' \, \Delta t,$$

and thus $\dot{W}'$ is the net amount of work done by the fluid of interest per unit time on its environment exclusive of the pressure work. For example, if an electric motor is placed inside the stream tube, $\dot{W}'$ would be equal to the negative of the electrical energy input rate. When frictional forces are present, $\dot{W}'$ is nonzero only when such frictional forces do work on the environment. Now, by definition, $e + p/\rho$ is h, the enthalpy per unit mass. Thus, Eq. 15–26 can be written as

$$h_1 + \tfrac{1}{2}q_1^2 = h_2 + \tfrac{1}{2}q_2^2 - \frac{\dot{Q} - \dot{W}'}{\dot{m}}, \qquad (15\text{-}27)$$

and Eq. 15–27 is our energy equation (compare with Eq. 14–21b).

Now, to find the stagnation temperature of a particle, imagine a flow where a particle with the same enthalpy h and speed q is brought to rest isentropically at a stagnation point, and apply Eq. 15–27. Since the process employed is isentropic, $\dot{Q} = 0$, and since no other work except pressure work has been done by the system, $\dot{W}' = 0$. Then, denoting by superscript 0 the conditions at the stagnation point, we have

$$h^0 = h + \tfrac{1}{2}q^2; \tag{15–28}$$

the quantity h^0 is called the stagnation enthalpy per unit mass or simply stagnation enthalpy of the particle of interest. If we are dealing with a perfect gas with constant specific heats, $h = C_p T + \text{constant}$, Eq. 15–28 can be rewritten as

$$C_p T^0 = C_p T + \tfrac{1}{2}q^2, \tag{15–29}$$

where T^0 is then the stagnation temperature. Another convenient expression for T^0 is obtained by dividing Eq. 15–29 through by C_p, and noting that

$$C_p = \frac{\gamma R}{\gamma - 1}, \tag{15–30a}$$

$$M^2 = \frac{q^2}{a^2} = \frac{q^2}{\gamma R T}. \tag{15–30b}$$

Equation 15–29 becomes

$$T^0 = T\left(1 + \frac{\gamma - 1}{2} M^2\right). \tag{15–31}$$

The stagnation pressure p^0 is related simply to the stagnation temperature by the isentropic relation

$$\frac{p}{p_0} = \left(\frac{T}{T_0}\right)^{\gamma/(\gamma-1)}.$$

Therefore we have, using Eq. 15–31,

$$p^0 = p\left(1 + \frac{\gamma - 1}{2} M^2\right)^{\gamma/(\gamma-1)}. \tag{15–32}$$

Thus we see that if p^0, T^0, and M are given at any point in the flow field of a perfect gas with constant specific heats, the values of p, T, and q there can be computed from Eqs. 15–30b, 15–31, and 15–32, and vice versa. This computation can be performed whether the actual flow is isentropic or not. We simply prefer to describe the conditions at a point with its values of p^0, T^0, and M instead of p, T, and q.

The reasons that it is preferable to use p^0, T^0, and M to describe flow conditions are many. First of all, the value of T^0 is easily monitored. For example, if no heat is added and no work is extracted from the flow, Eq. 15–27 indicates that h^0 in Eq. 15–28 and T^0 in Eq. 15–29 are constant through the flow. Secondly, if the flow is known to be isentropic between stations 1 and 2, then entropy will be a constant, which allows us to relate p^0 to T^0 by the isentropic relation

$$\frac{p_1^0}{p_2^0} = \left(\frac{T_1^0}{T_2^0}\right)^{\gamma/(\gamma-1)}. \tag{15-33}$$

This relation is especially useful when external work has been isentropically added to the flow. In general, the variation of T^0 and entropy are readily monitored. Thirdly, the magnitude of the Mach number, whether it is greater or less than unity, determines to a large extent the general behavior of the flow. This we shall demonstrate in the next section.

Note that the stagnation temperature is a function of T and M only, and is defined regardless of whether the flow under consideration is itself isentropic or not. For a flow with heat conduction and nonpressure work, Eq. 15–27 can be written as

$$T_1^0 = T_2^0 - \frac{\dot{Q} - \dot{W}'}{\dot{m}C_p}, \tag{15-34}$$

which serves as the energy equation for a perfect gas. Equations 15–27 and 15–34 are valid regardless of whether frictional effects are present. Furthermore, even if the fluid under consideration is viscous, we would still have $\dot{W}' = 0$ if no (frictional) work has been performed by the fluid on its environment.

Example 15–1. A stream of perfect gas of speed U, temperature T_1, and pressure p_1, flows isentropically over a solid body. Find the temperature and pressure distribution along the small stream tube passing point 2, the stagnation point (see Fig. 5–8).

Along the stream tube between points 1 and 2 in this isentropic flow, we have from Eq. 15–27 constant $h^0 = h_1^0$ or $T^0 = T_1^0$, and from Eq. 15–33 constant $p^0 = p_1^0$. With $M_1^2 = U^2/\gamma R T_1$, we have from Eqs. 15–31 and 15–32

$$T^0 = T_1^0 = T_1 \left(1 + \frac{\gamma - 1}{2} \frac{U^2}{\gamma R T_1}\right),$$

and

$$p^0 = p_1^0 = p_1 \left(1 + \frac{\gamma - 1}{2} \frac{U^2}{\gamma R T_1}\right)^{\gamma/(\gamma-1)}.$$

With these values of T^0 and p^0, the values of M, q, T, and p at any point along the stream tube can be computed from Eqs. 15–30b, 15–31, and 15–32, if any one of them is known.

At the stagnation point, we have $q_2 = 0$. Thus $M_2 = 0$,

$$T_2 = \frac{T^0}{1 + \frac{\gamma - 1}{2} M_2^2} = T_1 + \frac{\gamma - 1}{2\gamma R} U^2,$$

and

$$p_2 = \frac{p^0}{\left(1 + \dfrac{\gamma - 1}{2} M_2^2\right)^{\gamma/(\gamma-1)}} = p_1 \left(1 + \frac{\gamma - 1}{2} \frac{U^2}{\gamma R T_1}\right)^{\gamma/(\gamma-1)}.$$

For air with $U = 100$ mph, $T_2 - T_1$ is about 1°C, and $p_2 - p_1$ is about 0.17 psi under atmospheric pressure at 68°F. If the solid body is an instrument for measuring T_1 and p_1, these differences of temperature and pressure will cause errors in the measurements.

PROBLEMS

15–3. In a one-dimensional steady, isentropic, flow problem, we have $\Delta Q = \Delta W' = 0$. (a) Given p_1, T_1, and q_1, write down the expression for p_1^0, T_1^0, and M_1.

(b) When $M_2 = 1$, find p_2/p_1^0 and T_2/T_1^0 in terms of M_1.

15–4. In a one-dimensional steady flow problem, if work is isentropically done on the fluid by means of an ideal compressor (see Fig. 15–2) so that the stagnation temperature of the gas is doubled, how will the stagnation pressure change? (Assume $\gamma = 1.4$.)

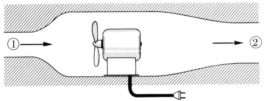

FIGURE 15–2

15–5. (a) Show that when $M = 1$, the speed of sound is (denoted by a^* and is called the critical speed of sound) related to the stagnation temperature by the relation

$$a^{*2} = \frac{2}{\gamma + 1} \gamma R T^0.$$

(b) Show that for a given T^0 the maximum velocity the gas may achieve is given by

$$q_{max} = \sqrt{\frac{2R}{\gamma - 1} \gamma T^0}.$$

15–4. Isentropic nozzle flows

Let us consider now the steady flow of a gas in a long tube whose cross-sectional area $A(x)$ is a known function of the axial distance x. Given the flow conditions at the upstream end of the tube, our problem is to find the flow conditions at any station downstream. For the present section, we confine our attention to cases where the flow can be considered isentropic.

We assume that at any station of interest, the flow is sufficiently uniform to warrant the one-dimensional assumption.   We shall see that under the isentropic assumption, we can analyze the problem by using continuity and energy considerations only.   Momentum consideration is needed only when the process is nonisentropic or when the thrust of the nozzle is involved.   (The reader will recall that a constant-density flow can be analyzed by continuity and momentum considerations only.)   At the upstream station, we can monitor the values of p^0, T^0, and M, and we can also monitor the mass flow rate $\dot{m}$.   Now the law of conservation of mass requires that for steady flow the same mass flow rate exists at any station:

$$\dot{m} = \rho u A = \text{constant}, \tag{15--35}$$

where u is the velocity of the gas. With the help of the equation of state of a perfect gas, Eq. 14–7, the formula for the Mach number, Eq. 15–30b, and the speed of sound, Eq. 15–5, we can rewrite Eq. 15–35 as

$$\dot{m} = \frac{p}{RT} M \sqrt{\gamma RT}\, A. \tag{15--36}$$

But p and T can be written in terms of p^0, T^0, and M through the use of Eqs. 15–31 and 15–32. Thus Eq. 15–36 becomes

$$\dot{m} = \sqrt{\frac{\gamma}{R}} \, \frac{p^0 A D}{\sqrt{T^0}}\,, \tag{15--37}$$

where

$$D(M, \gamma) = \frac{M}{\left(1 + \dfrac{\gamma - 1}{2} M^2\right)^{(\gamma+1)/2(\gamma-1)}}\,, \tag{15--38}$$

and a plot of $D(M, \gamma)$ as a function of M is shown in Fig. 15–3. It is immediately clear that with Fig. 15–3 available, the problem we set out to study becomes very simple. If no energy is added or taken away from the gas, T^0 will remain constant. If in addition to this we assume isentropic flow, p^0 will also be constant (see Eq. 15–33). With $\dot{m}$ also known from the upstream station, the continuity equation becomes

$$A D = \dot{m} \sqrt{\frac{R}{\gamma}} \, \frac{\sqrt{T^0}}{p^0} = \text{constant}. \tag{15--39}$$

Therefore at any station of interest, D can be computed if A is known. From Fig. 15–3 we can get the Mach number M, and consequently all other quantities such as p, ρ, T, and u. Note that M has two possible values for a given value of D. The decision of which value to use is guided by the physical consideration that M should be continuous along the flow.

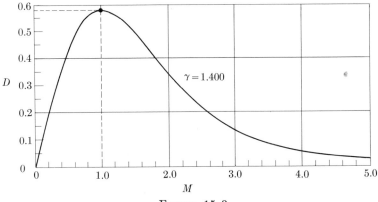

FIGURE 15–3

Equations 15–37, 15–38, and Fig. 15–3 furnish a great deal more interesting information on compressible flows in general. First of all, let us ask: Under what condition can a flow be assumed incompressible? If ρ is constant, Eq. 15–35 gives

$$u = \frac{\dot{m}}{\rho A} \qquad (\rho = \text{constant}). \qquad (15\text{–}40)$$

It can easily be verified that when $M^2 \ll 1$, we have $T^0 \to T$, $p^0 \to p$, and $D \to M$ from Eqs. 15–31, 15–32, and 15–38, and Eq. 15–37 reduces to Eq. 15–40. Therefore, for a gas in isentropic steady flow, the incompressible assumption is justified when $M^2 \ll 1$ everywhere. This is our first important conclusion.

We further observe that $D(M, \gamma)$ has a maximum at $M = 1$, and its value is

$$D = \left(\frac{2}{\gamma + 1}\right)^{(\gamma+1)/2(\gamma-1)} \qquad (\text{when } M = 1). \qquad (15\text{–}41)$$

This fact gives rise to very interesting and important results. Let us suppose that at the upstream station the Mach number is M_1 and the area is A_1. Since we assume p^0, T^0, and $\dot{m}$ to be constant at any downstream station, the value of $D_2 = D(M_2, \gamma)$ at a station where the area is A_2 can be found from Eq. 15–39:

$$D_2 = \frac{A_1}{A_2} D_1. \qquad (15\text{–}42)$$

We see that if the tube contracts, i.e., $A_1/A_2 > 1$, then we have $D_2 > D_1$. From Fig. 15–3 we see that a contraction of cross-sectional area tends to drive the Mach number toward unity. Conversely, if the tube expands in cross-sectional area, the Mach number tends to move away from unity. We thus see that flows at *supersonic speeds* $(M > 1)$ respond to area changes in an opposite manner from those at *subsonic speeds* $(M < 1)$. For example, in a divergent channel, the speed increases for supersonic flow in contrast to subsonic flow

Parameter	Subsonic	Supersonic	Subsonic	Supersonic
Mach number	Increase	Decrease	Decrease	Increase
Velocity	Increase	Decrease	Decrease	Increase
Pressure	Decrease	Increase	Increase	Decrease
Temperature	Decrease	Increase	Increase	Decrease
Density	Decrease	Increase	Increase	Decrease
Entropy	Constant	Constant	Constant	Constant

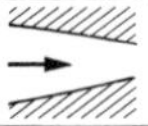

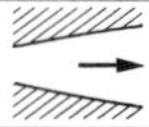

FIGURE 15–4

behavior. This is our second important result. Some of the important qualitative behavior of isentropic flows are summarized in Fig. 15–4.

What happens if for a given upstream condition A_1 and M_1, the tube area continues to contract? The Mach number will, according to our above discussion, move toward unity and will reach unity where the area ratio is, according to Eqs. 15–42 and 15–41,

$$\frac{A_2}{A_1} = \left(\frac{\gamma + 1}{2}\right)^{(\gamma+1)/2(\gamma-1)} D_1. \tag{15-43}$$

If the cross-sectional area is further contracted below this value, Fig. 15–3 shows that no such steady flow solution corresponding to the given upstream conditions is possible, i.e., the given upstream conditions are not realistic. Thus for a given upstream Mach number, whether it be supersonic or subsonic, there is an area contraction ratio at which the flow is said to "choke." This is our third important result.

What happens if the tube contracts to this choking area ratio (Eq. 15–43) and then expands again, as shown in Fig. 15–5? With $M_2 = 1$ at the "throat" of such a convergent-divergent tube, usually called a DeLaval nozzle, the flow at any station 3 downstream can be either supersonic ($M_3 > 1$) or subsonic ($M_3 < 1$); either case permits a continuous variation of M in the tube. Thus a continuous transition from supersonic ($M_1 > 1$) to subsonic flow ($M_3 < 1$), or vice versa ($M_1 < 1$, $M_3 > 1$), is possible if the tube has a throat, and if the Mach number M_2 there is unity. Conversely, a continuous transition is impossible if no throat is present. This is our fourth important result. We shall need one more important analysis in the next section before we can tie all the loose ends together.

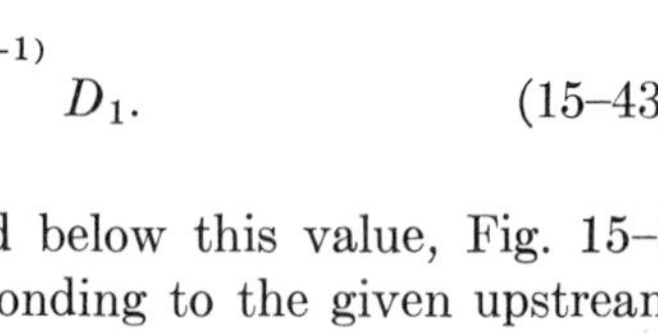

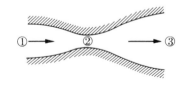

FIGURE 15–5

It may be instructive to remark here on why momentum consideration is not directly involved in the analysis of isentropic flows. Once the assumption of isentropic flow is applied, it is implicitly assumed that the flow is frictionless. Thus in an indirect way, momentum consideration is involved. If the flow is not frictionless, we must immediately abandon the assumption of isentropic flow. For nonisentropic problems, momentum consideration must be invoked. See Sections 15–5 and 15–9.

Example 15–2. A perfect gas at T_1 and p_1 in a large tank flows isentropically through a convergent nozzle of outlet area A_2 into the atmosphere of pressure p_a. Find the mass flow rate $\dot{m}$.

The mass flow rate $\dot{m}$ as defined in Eq. 15–35 can be computed from Eq. 15–37. With $M_1 = 0$ in the tank, we obtain from Eqs. 15–31 and 15–32, $T^0 = T_1$ and $p^0 = p_1$. These stagnation quantities remain constant through the isentropic flow. Thus from Eq. 15–37,

$$\dot{m} = \sqrt{\frac{\gamma}{R}} \frac{p_1 A_2 D_2}{\sqrt{T_1}}, \tag{15–43a}$$

where D_2 as defined in Eq. 15–38 depends on M_2, which can be determined from Eq. 15–32:

$$\left(1 + \frac{\gamma - 1}{2} M_2^2\right)^{\gamma/(\gamma-1)} = \frac{p_1}{p_2}. \tag{15–43b}$$

First, assume p_2 in the jet is equal to the surrounding atmospheric p_a. If M_2 computed from Eq. 15–43b is not greater than unity, then D_2 can be computed from Eq. 15–38 and $\dot{m}$ from Eq. 15–43a.

If p_1 is greater than the critical value

$$\text{critical } p_1 = p_a \left(1 + \frac{\gamma - 1}{2}\right)^{\gamma/(\gamma-1)},$$

the computed M_2 based on the assumption that $p_2 = p_a$ will exceed unity. However, according to Eq. 15–42 and Fig. 15–3, a subsonic flow cannot become supersonic in a convergent nozzle. The value of M_2 can only reach the value of unity, and the pressure p_2 in the jet, as given by Eq. 15–43b with $M_2 = 1$, is higher than the atmospheric p_a. This jet of higher pressure will expand as will be described in Chapter 16. With $M_2 = 1$, D_2 is given by Eq. 15–41, and

$$\dot{m} = \left(\frac{2}{\gamma + 1}\right)^{(\gamma+1)/2(\gamma-1)} \sqrt{\frac{\gamma}{R}} \frac{p_1 A_2}{\sqrt{T_1}}.$$

PROBLEMS

15–6. Show that for any flow which is initially supersonic, when the tube area expands indefinitely, far downstream the gas velocity is finite although the Mach number becomes infinite.

15–7. Verify several cases of Fig. 15–4.

15–8. Assume isentropic flow with p^0 = constant, T^0 = constant. Given $M_2 = 1$, find A_1/A_2 and A_3/A_2 such that (a) $M_1 = \frac{1}{2}$, $M_3 = 2$, (b) $M_1 = 2$, $M_3 = \frac{1}{2}$. (c) With A_3/A_1 given by $M_3 = 2$, what other (subsonic) M_3 can be produced at station 3?

15–9. Given $A_2/A_1 = \frac{1}{2}$, find M_2 for $\gamma = 1.4$. (a) $M_1 = 0.25, 0.333$. (b) $M_1 = 2, 3.0$. If steady solution is not possible, what should M_1 be? [*Hint:* $M_2 = 1$.]

15–5. Shock waves

One of the important results in the previous section is that a continuous transition from supersonic to subsonic flow, or vice versa, is possible only at a throat. The purpose of this section is to investigate the possibility of discontinuous transition. By a discontinuous transition we mean a sudden change in the velocity, pressure, density, etc., in the flow. We shall see that only discontinuous transitions from supersonic flow to subsonic flow are possible. These are called *shock waves*.

Let us consider a constant-area tube with a gas flowing in it (see Fig. 15–6). We denote by subscripts 1 and 2 any two stations in the tube as shown. Now if we assume the flow is isentropic, and that $T_1^0 = T_2^0$, $p_1^0 = p_2^0$, then since area A is constant, Eq. 15–42, the continuity equation, gives $D_1 = D_2$. Figure

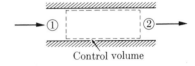

Control volume

FIGURE 15–6

15–3 shows that (assuming $M_1 \neq 1$) there are two possible values for M_2, namely, either $M_2 = M_1$ or M_2 would take on the other value corresponding to the same value of D_1. In the previous section it was stated that M should vary continuously in the tube; hence $M_2 = M_1$ would be the correct solution. *If* this other solution is also permissible, we would have a discontinuous transition, since the values of M_1 and M_2 straddle the value 1.0, and stations 1 and 2 may be as close to each other as we wish.

To check whether the previously suggested discontinuous switch in Mach number is permissible, we apply now directly the law of conservation of momentum. We shall assume there is *no wall friction*. Drawing a control volume as shown in Fig. 15–6 and applying the law of conservation of momentum, we have simply, from Eq. 9–5,

$$p_1 A + \rho_1 u_1^2 A = p_2 A + \rho_2 u_2^2 A. \tag{15–44}$$

By using the equation of state of a perfect gas and by changing variables to p^0 and M through Eqs. 15–30b and 15–32, we have

$$p + \rho u^2 = p \left(1 + \frac{u^2}{RT} \right) = \frac{p^0(1 + \gamma M^2)}{\left(1 + \dfrac{\gamma - 1}{2} M^2 \right)^{\gamma/(\gamma-1)}} = p^0 G, \tag{15–45}$$

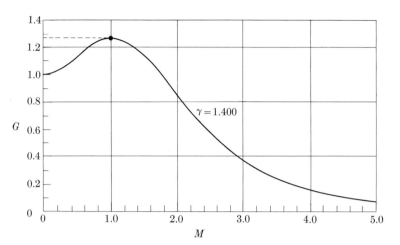

FIGURE 15–7

where $G(M, \gamma)$ is defined as

$$G(M, \gamma) = \frac{1 + \gamma M^2}{\left(1 + \dfrac{\gamma - 1}{2} M^2\right)^{\gamma/(\gamma-1)}} \cdot$$

Then the momentum equation, Eq 15–44, can be written as

$$p_1^0 G_1 = p_2^0 G_2, \qquad (15\text{–}46)$$

which is valid with or without energy transfer. It is, however, valid only for a straight tube with no external force and if wall friction is neglected. The function $G(M, \gamma)$ is plotted in Fig. 15–7. It is thus clear from this diagram that a switch of Mach number dictated by $D_1 = D_2$ will in general result in a change in stagnation pressure in accordance with Eq. 15–46. Since the original suggestion of switching M was made with the assumption that

$$p_1^0 = p_2^0,$$

we have arrived at a contradiction.

It is suggested by the above analysis that a discontinuous transition from supersonic to subsonic flow is accompanied by a change in p^0. Since T^0 is a constant from energy considerations, a change in p^0 would indicate that the flow process is not isentropic. We shall now reconsider the whole problem without making the isentropic assumption. We can eliminate the variable p^0 between Eqs. 15–37 and 15–46, which are continuity and momentum equations, respectively. We obtain

$$\sqrt{T_1^0}\, \frac{G_1}{D_1} = \sqrt{T_2^0}\, \frac{G_2}{D_2}, \qquad (15\text{–}47)$$

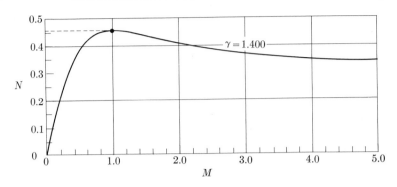

FIGURE 15-8

using the fact that $\dot{m}$ and A are constants. Equation 15–47 can be rewritten as

$$\frac{N_1}{\sqrt{T_1^0}} = \frac{N_2}{\sqrt{T_2^0}}, \tag{15-48}$$

where $N(M, \gamma)$ is defined by

$$N(M, \gamma) = \frac{D(M, \gamma)}{G(M, \gamma)} = \frac{M}{1 + \gamma M^2} \sqrt{1 + \frac{\gamma - 1}{2} M^2} \tag{15-49}$$

and is plotted in Fig. 15–8. Now Eq. 15–48 is valid for a steady flow in a constant-area tube with no external force and no wall friction, and it involves no isentropic assumption. When there is no energy transfer ($T_1^0 = T_2^0$), Eq. 15–48 gives

$$N_1 = N_2. \tag{15-50}$$

Figure 15–8 shows that either $M_1 = M_2$, or a switch is possible, at least in some range of M_1. This kind of switch is dynamically possible, since the three basic conservation laws are satisfied. However, a simple calculation will show that a switch of an originally subsonic flow to supersonic flow causes a *decrease* in specific entropy of the gas, and it is therefore excluded by the second law of thermodynamics. Only a supersonic to subsonic switch is both dynamically and thermodynamically permissible. Such discontinuous transitions with no change in T^0 from supersonic flow to subsonic flow are called *shock waves* or simply *shocks*. One-dimensional shock waves where the velocities before and after the shock are parallel are called normal shocks.

PROBLEMS

15–10. If M_1 is very large, show that the value of M_2^2 after a shock is

$$(\gamma - 1)/2\gamma.$$

15–11. Show that the specific entropy of the gas decreases if a subsonic to supersonic transition occurs.

15–12. If $M_1 = 3.0$, $T_1 = 1000°R$, $p_1 = 0.1$ atmosphere, $\gamma = 1.4 =$ constant, find (a) q_1, the velocity before the shock; (b) M_2, the Mach number after the shock; (c) p_2, the pressure after the shock; and (d) q_2, the velocity after the shock.

15–6. Normal shock relations

We have seen that by means of a normal shock, a sudden transition from a supersonic flow to a subsonic flow is possible. The Mach numbers before and after the shock are related by Eq. 15–50. Because of the great importance of shock waves in supersonic flow, we shall give in this section explicit formulas relating pressure, temperature, velocity, and entropy across the normal shock.

We shall follow the notation used in the previous section and denote by subscripts 1 and 2 conditions before and after the shock, respectively. For simplicity we shall continue to assume that the gas is a perfect gas with constant specific heat ratio γ. From Eq. 15–50, we can solve for M_2 in terms of M_1. After some algebraic manipulations, we obtain

$$M_2^2 = \frac{M_1^2 + \dfrac{2}{\gamma - 1}}{\dfrac{2\gamma}{\gamma - 1} M_1^2 - 1}. \tag{15–51}$$

Utilizing the fact that across a shock wave energy is conserved, $T_1^0 = T_2^0$, we obtain the static temperature ratio across a shock by applying Eqs. 15–51 and 15–31a:

$$\frac{T_2}{T_1} = \frac{T_2^0}{T_1^0} \frac{1 + \dfrac{\gamma - 1}{2} M_1^2}{1 + \dfrac{\gamma - 1}{2} M_2^2} = \frac{2(\gamma - 1)}{(\gamma + 1)^2} \left(1 + \frac{\gamma - 1}{2} M_1^2\right)\left(\frac{2\gamma}{\gamma - 1} - \frac{1}{M_1^2}\right), \tag{15–52}$$

which is also the square of the ratio of the speeds of sound, $(a_2/a_1)^2$. The ratio of velocities across the shock is then

$$\frac{u_2}{u_1} = \frac{M_2\sqrt{\gamma R T_2}}{M_1\sqrt{\gamma R T_1}} = \frac{\gamma - 1}{\gamma + 1}\left(1 + \frac{2}{\gamma - 1}\frac{1}{M_1^2}\right). \tag{15–53}$$

However, from continuity considerations, we have $\rho_1 u_1 = \rho_2 u_2$. Hence

$$\frac{\rho_2}{\rho_1} = \frac{u_1}{u_2} = \frac{\gamma + 1}{\gamma - 1}\left[\frac{1}{1 + \dfrac{2}{\gamma - 1}\dfrac{1}{M_1^2}}\right]. \tag{15–54}$$

Knowing ρ_2/ρ_1 and T_2/T_1, we can obtain the pressure ratio across the shock from the equation of state:

$$\frac{p_2}{p_1} = \frac{\rho_2}{\rho_1}\frac{T_2}{T_1} = \frac{2\gamma}{\gamma+1}M_1^2 - \frac{\gamma-1}{\gamma+1}. \tag{15-55}$$

The entropy rise across the shock can be calculated as follows. Since from Eq. 14-32a, we can write

$$s = C_p \ln T - R \ln p + \text{constant},$$

then

$$s_2 - s_1 = C_p \ln \frac{T_2}{T_1} - R \ln \frac{p_2}{p_1}$$

or

$$\frac{s_2 - s_1}{R} = \frac{\gamma}{\gamma-1} \ln \frac{T_2}{T_1} - \ln \frac{p_2}{p_1} = -\ln \frac{p_2^0}{p_1^0}. \tag{15-56}$$

Substituting p_2/p_1, T_2/T_1 from Eqs. 15-55 and 15-52, we have

$$\frac{s_2 - s_1}{R} = \frac{\gamma}{\gamma-1}\ln\left(\frac{2}{\gamma+1}\frac{1}{M_1^2}+\frac{\gamma-1}{\gamma+1}\right) + \frac{1}{\gamma-1}\ln\left(\frac{2\gamma}{\gamma+1}M_1^2 - \frac{\gamma-1}{\gamma+1}\right). \tag{15-57}$$

Equations 15-51 to 15-55 and 15-57 are functions of M_1 and γ only. Of special interest is Eq. 15-57, which gives the entropy rise across a shock. It is seen that if $M_1 < 1$, the quantity $(s_2 - s_1)/R$ is negative, indicating that it is impossible to have a shock wave in a subsonic flow. Furthermore, if M_1 is close to unity, we see that $(s_2 - s_1)/R$ will be very small, and indeed approaches zero in the limit as $M_1 = 1$. This observation is very useful for it allows one to assume isentropic flow as a simplifying assumption whenever the Mach number M_1 is close to unity.

It is clear that the strength of a shock wave is determined by the Mach number M_1. For M_1 close to unity, the shock wave is called weak, and the flow is nearly isentropic. For very large M_1, the shock wave is called strong, and the entropy rise across the shock is large. Note that across a shock, pressure, density, temperature, and entropy all increase, while the velocity decreases.

Now that we have derived explicit formulas relating conditions across a shock, the relevant questions here are what is a shock wave and what are the physical mechanisms that cause it? In particular, what caused the increase in specific entropy, since we had allowed no external energy transfer and wall friction? The answer is internal friction and heat conduction. Actually, the flow conditions do not jump discontinuously, but they do change extremely rapidly over a very short distance. In a thin transitional region, the gradients of temperature and velocity are extremely large, and these gradients give rise to dissipative processes that increase the specific entropy of the gas. However, dissipation does not cause energy transfer out of the fluid but converts some mechanical energy into heat energy in the system.

Generally, the assumption of negligible heat conduction and frictional effects is made based on the justification that the Reynolds number of the flow is large (see Section 12–1). We have here the interesting situation that independent of how large the Reynolds number is, a shock wave will appear in the flow field when it is called for. The magnitude of the Reynolds number affects only the thickness of the shock but does not affect the shock relations.

Since shock waves are in effect discontinuities, we can relax one assumption underlying our analysis—that the tube has constant area. If the area of the tube varies gradually, the flow conditions *immediately* in front and behind the shock are still related in the same way. The verification of this statement is left as a problem.

PROBLEMS

15–13. Show that for a nonconstant-area tube, Eq. 15–50 still holds across a shock wave of infinitesimal thickness. [*Hint:* First show that the pressure forces acting on the wall of the tube are negligible when the thickness of the shock is infinitesimal.]

15–14. Compute the fractional drop of stagnation pressure across a shock as a function of M_1 for a range of $M_1 > 1$. Plot your results.

15–15. Show that for M_1 close to 1.0, $(s_2 - s_1)/R$ is proportional to $(M_1^2 - 1)^3$.

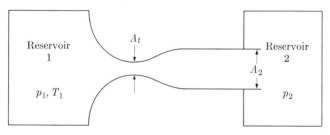

FIGURE 15–9

15–7. Nozzle flow with normal shocks

Let us now consider a practical problem and see how the previous sections contribute toward its solution. Suppose we have two large reservoirs connected by a tube which is smoothly contoured, as shown in Fig. 15–9. Let us also suppose that the reservoirs are equipped with pumps so that reservoir 1 is fixed at a constant pressure p_1, and the pressure in reservoir 2 can be maintained at any pressure p_2 desired. For simplicity we shall consider only $p_1 > p_2$. The temperature of the gas in reservoir 1 is also maintained at a constant value T_1 through additional equipment. Our problem is: What is the flow condition in the tube at any desired value of p_2? In particular what is the mass flow rate? We shall assume that the flow is isentropic except across any shock that may occur.

Since reservoir 1 is assumed to be large, the gas velocity in it must be very low, and consequently the Mach number at any point inside is practically zero. Thus, the stagnation pressure and temperature of the gas in the reservoir will be

$$p_1^0 = p_1, \qquad T_1^0 = T_1. \qquad (15\text{--}58)$$

Under the isentropic assumption, the stagnation pressure p^0 and temperature T^0 will remain constant when there is no shock. When a shock occurs, T^0 remains constant but p^0 will change according to Eq. 15–46.

Let us assume tentatively that no shock is present in the flow so that $p^0 = p_1^0$ everywhere in the tube. Knowing p_2 and $p_2^0 = p^0$, we can compute the Mach number at station 2:

$$M_2 = \sqrt{\frac{2}{\gamma - 1}\left[\left(\frac{p_1^0}{p_2}\right)^{(\gamma-1)/\gamma} - 1\right]}. \qquad (15\text{--}59)$$

If $M_2 < 1$, then D_2 can be computed from Eq. 15–38, the mass flow rate through the tube is given by Eq. 15–37 as

$$\dot{m} = \sqrt{\frac{\gamma}{R}}\,\frac{p_1^0}{\sqrt{T_1^0}}\,A_2 D_2,$$

and the Mach number distribution in the tube is given implicitly by

$$D = \frac{A_2}{A}\,D_2. \qquad (15\text{--}60)$$

The pressure distribution can then be computed from Eq. 15–32, since p^0 and M are now known. From Fig. 15–3 we can see easily that the maximum Mach number in the flow will occur at the minimum area which we denote by A_t in

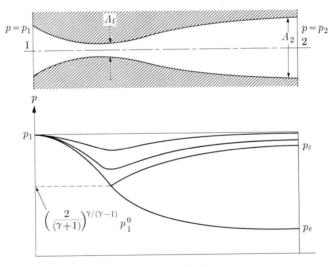

FIGURE 15–10

Fig. 15–9. If this maximum Mach number is less than unity, then we are assured that no shock will appear in the flow. Figure 15–10 shows schematically the pressure distribution in the tube for various p_2. Note that as p_2 is lowered, a critical value $p_2 = p_c$ will be reached when the Mach number at the throat equals unity, and the flow is said to have choked. The pressure at the throat will be exactly (since $M = 1$),

$$\text{choking } p_t = \left(\frac{2}{\gamma + 1}\right)^{\gamma/(\gamma-1)} p_1^0$$

(see Eq. 15–32). The mass flow rate $\dot{m}$ increases with decreasing p_2, as one would expect. It will be seen to be convenient to plot mass flow rate as a function of p_t/p^0, as shown in Fig. 15–11. We see that the mass flow rate reaches a maximum when p_t reaches the choking pressure in the above expression.

For $p_c \leq p_2 \leq p_1$, the flow is subsonic everywhere. The qualitative behavior of the flow is similar to incompressible flow. If p_2 is further lowered below p_c, we see from Fig. 15–10 that the only possible continuous solution is when $p_2 = p_e$, and the flow then is everywhere supersonic after the throat. If $p_c > p_2 > p_e$, an analysis following Eqs. 15–32, 15–59, and 15–60 will show that no continuous solution is possible. For $p_2 = p_e$, the flow conditions in the duct upstream of the throat are seen to be *identical* with the case when $p_2 = p_c$. Furthermore, the mass flow rate remains at the same value as when $p_2 = p_c$, since p_t did not change. All these are new and surprising results which are qualitatively different from what one expects from incompressible flow behavior.

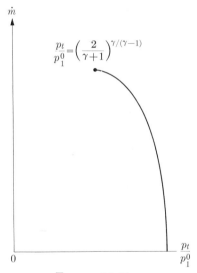

FIGURE 15–11

We next ask what happens if p_2 falls between p_c and p_e. Since no continuous solution is possible, we shall explore the possibility that a shock exists in the tube. Since shocks can occur only in supersonic flow, this shock if it exists must be somewhere between the throat and station 2. As a consequence, we see that for any p_2 below p_c, the value of p_t will not be affected, and hence the mass flow rate becomes independent of p_2. We thus have an important conclusion: for a given p_1^0 and T_1^0, the maximum mass flow rate through a tube under isentropic conditions is given by Eq. 15–37 with $D(1, \gamma)$ given by Eq. 15–41:

$$\dot{m}_{\max} = \sqrt{\frac{\gamma}{R}} \frac{p_1^0}{\sqrt{T_1^0}} A_t \left(\frac{2}{\gamma + 1}\right)^{(\gamma+1)/2(\gamma-1)}, \qquad (15\text{–}61)$$

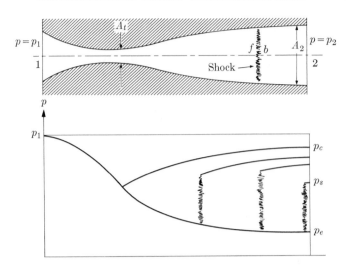

FIGURE 15–12

FIGURE 15–12

where A_t is the minimum cross-sectional area of the tube. This $\dot{m}_{max}$ is achieved as soon as the flow is choked at the throat. Also, conditions at any point before the shock are identical with those when $p_2 = p_c$.

Let the subscripts f and b denote conditions immediately in front and behind the shock, respectively. By assuming that the shock stands at some station between the throat and station 2, we can find out the corresponding value of p_2. As a start, we suppose that the shock stands at the very end of the tube, just in front of station 2. In front of the shock, we then have $p_f = p_e$. The values of p_b and M_b can be found as follows. Since $M_f = M_e$ is known from the shockless solution (See Problem 15–16), M_b can be found from the shock relation Eq. 15–51:

$$M_b^2 = \frac{M_f^2 + \dfrac{2}{\gamma - 1}}{\dfrac{2\gamma}{\gamma - 1} M_f^2 - 1}.$$

Then Eq. 15–46 gives the value of p_b^0:

$$p_b^0 = p_f^0 \frac{G_f}{G_b} = p_1^0 \frac{G_f}{G_b}.$$

where G is a function of M and γ. Then p_b, which in this case equals p_2, can be found from Eq. 15–32:

$$p_2 = p_b = \frac{p_b^0}{\left(1 + \dfrac{\gamma - 1}{2} M_b^2\right)^{\gamma/(\gamma - 1)}}.$$

This value of p_2 will be intermediate between p_c and p_e. We shall denote this value of p_2 by p_S (see Fig. 15–12).

When the shock stands at some station between the throat and station 2, an analysis similar to that above can be made, except that p_2 and p_b must be related by an isentropic flow analysis between the shock and station 2. In general, the value of p_2 will be intermediate between p_c and p_S. Figure 15–12 shows a plot of pressure distributions for various p_2 in this range.

When p_2 falls below p_S, we are again faced with the problem that no solution is possible. We cannot resolve the problem here. We shall show in the next chapter that one-dimensional assumptions are no longer adequate, and for p_2 in this range, the flow at the exit of the tube is definitely not one-dimensional.

The behavior of the compressible flow discussed so far bears a striking resemblance to the shallow-water channel flow presented in Section 10–6. In fact, if we replace γ by 2 in the equations here, it can be shown that the relations become identical, with the depth of flow being analogous to $\sqrt{p}$, and the Froude number being analogous to M. Compare Figs. 15–12 and 10–16.

PROBLEMS

15–16. Given $M_t = 1$ at the throat, find p_c and p_e in the reservoir in Fig. 15–12 for $A_2/A_t = 4.0$. Also find M_c and M_e.

15–17. Given p^0 and T^0 in a nozzle and the external pressure p_2, as shown in Fig. 15–13, find the mass flow rate $\dot{m}$ as a function of p_2.

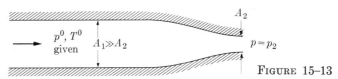

A_2

p^0, T^0 given $A_1 \gg A_2$ $p = p_2$

FIGURE 15–13

15–8. Thrust

It is of practical interest to find the thrust exerted by a one-dimensional flow. Let us consider here a tube, as shown in Fig. 15–14. The flows at stations 1 and 2 are assumed one-dimensional and, for simplicity, we also assume that the flow directions at 1 and 2 are parallel. If conditions at stations 1 and 2 are completely known, we can apply the law of conservation of momentum to compute the force acting on the tube. Consider the control volume indicated by the dashed line in Fig. 15–14. If F is the resultant of the forces exerted by the fluid on the surroundings between stations 1 and 2, then the law of conservation of momentum (Eq. 9–5) gives

$$p_1 A_1 - p_2 A_2 - F = \dot{m}u_2 - \dot{m}u_1$$

or

$$(p_1 A_1 + \dot{m}u_1) - (p_2 A_2 + \dot{m}u_2) = F. \tag{15–62}$$

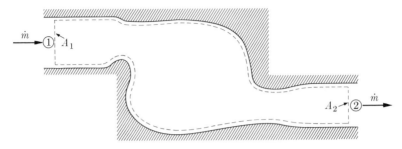

FIGURE 15–14

The quantity $pA + \dot{m}u$ is called *stream thrust*, and can be written as

$$pA + \dot{m}u = p^0 A G, \tag{15–63}$$

using the definition of $G(M, \gamma)$ in Eq. 15–45. Hence Eq. 15–62 becomes

$$p_1^0 A_1 G_1 - p_2^0 A_2 G_2 = F. \tag{15–64}$$

Thus, if the values of the stagnation pressure, cross-sectional area, and Mach number are known at two stations, the net force acting on the tube by the flow between the two stations can be computed by using Eq. 15–64. Note that Eq. 15–64 is valid regardless of whether the flow between the two stations is isentropic or not.

When the process between the two stations is isentropic, then conditions at 1 and 2 can be related by the isentropic analysis given in the previous sections, and Eq. 15–64 serves to give the thrust acting on the tube. When the process between the two stations is not isentropic, Eq. 15–64 serves as an additional relation between stations 1 and 2, as will be seen in Section 15–9.

PROBLEM

15–18. Consider a one-dimensional flow in a tube of constant area A. A compressor is operating in the tube, as shown in Fig. 15–15. If $p_1^0 = 20$ psi, $M_1 = 0.3$, $p_2^0 = 30$ psi, $M_2 = 0.4$, and $A = 100$ in^2, find (a) the force acting on the compressor, (b) the ratio of stagnation temperature T_2^0/T_1^0 (assume $\gamma = 1.4$), and (c) the entropy change $(s_2 - s_1)/R$.

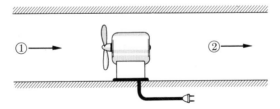

FIGURE 15–15

15–9. Effects of friction and energy exchange in tubes of constant area

We now study the effects of friction and energy exchange in a tube of constant area. No mass is added between stations 1 and 2. The continuity equation, Eq. 15–37, is simply

$$\frac{p_1^0 D_1}{\sqrt{T_1^0}} = \frac{p_2^0 D_2}{\sqrt{T_2^0}}. \tag{15–65}$$

The energy equation, Eq. 15–34, is

$$T_1^0 = T_2^0 - \frac{\dot{Q} - \dot{W}'}{\dot{m} C_p}, \tag{15–66}$$

where $\dot{Q}$ is the net heat added to the flow per unit time, $\dot{W}'$ is the net non-pressure work obtained from the flow per unit time, and $\dot{m}$ is the total mass flow rate through the tube. The momentum equation, Eq. 15–64, is

$$p_1^0 G_1 - p_2^0 G_2 = \frac{F}{A}, \tag{15–67}$$

where F is the force exerted on the tube by the flow.

Since the tube has constant area, F is the resultant of the frictional forces acting on the walls of the tube and any force that may be exerted by the fluid on the tube walls between the two stations, e.g., the force exerted by the fluid on the compressor in Problem 15–18. Now, if conditions at station 1 are completely known, then p_2^0, T_2^0, and M_2 can be computed by means of Eqs. 15–65, 15–66, and 15–67, provided that the values of the external force F and the net energy exchange $(\dot{Q} - \dot{W}')$ between the two stations are known.

To solve for M_2, we eliminate T_2^0 and p_2^0 from Eq. 15–65 by using Eqs. 15–66 and 15–67. We then have

$$N_2 = N_1 \frac{\left(1 + \dfrac{\dot{Q} - \dot{W}'}{\dot{m} C_p T_1^0}\right)^{1/2}}{1 - \dfrac{F}{p_1^0 A G_1}}, \tag{15–68}$$

where $N(M, \gamma)$ was first defined in Eq. 15–49 and is plotted in Fig. 15–8 for $\gamma = 1.4$. Equation 15–68 furnishes a great deal of information, in addition to allowing us to determine M_2. First of all, it shows that the effects of external force F and energy exchange $\dot{Q} - \dot{W}'$ are measured by the dimensionless quantities $F/p_1^0 A G_1$ and $(\dot{Q} - \dot{W}')/\dot{m} C_p T_1^0$, respectively. Thus in a practical problem, if one wishes to decide, for example, whether frictional and heat exchange effects are negligible, one needs only to estimate these two quantities and see whether they are small compared with unity. Secondly, with the help of the $N(M, \gamma)$ curve plotted in Fig. 15–8 we can deduce easily the effects of F and $\dot{Q} - \dot{W}'$. If F represents a frictional force on the tube, it will be positive, i.e., directed along the flow. Since the maximum of $N(M, \gamma)$ occurs at $M = 1$, the effect of friction is to make $N_2 > N_1$, or in other words, M_2 will be closer to unity than M_1. Similarly, if $\dot{Q} - \dot{W}'$ is positive, which means energy has been added to the flow, M_2 will again be closer to unity than M_1. If the computed value of N_2 is larger than the maximum value N at $M = 1$, then the flow is said to have choked, indicating that the conditions specified at station 1 are not compatible with the problem and no one-dimensional steady solution is possible.

PROBLEMS

15–19. A rocket engine is operating under the following condition. The chamber pressure is 300 psi abs, the chamber temperature is 1200°R, the area ratio of chamber to throat is 9.0, the area ratio of exit area to throat is 25.0, the ambient pressure is atmospheric, the gas has molecular weight of 30, and $\gamma = 1.4$. Find the thrust.

15–20. If a straight constant-area tube 10 ft long lets an $M_1 = 0.8$ flow come in and the flow chokes at the exit, what is the average friction per foot? The stagnation pressure is 100 psi at the inlet and the cross-sectional area of the tube is 100 in^2.

16 Shock Waves and Expansion Fans

16–1. General character of supersonic flow

As we have seen in the study of one-dimensional flows, supersonic flows behave in a decidedly different manner from subsonic flows. We shall see in this chapter that this difference in behavior is even more striking in two- or three-dimensional flows.

Consider a small body moving with velocity U along the x-axis in a fluid originally at rest. If the speed of sound of the undisturbed fluid is a, then the disturbances generated by the motion of the body will be propagated with speed a in all directions away from the body. The disturbances issued at time t when the body is at x will be confined to a sphere of radius $a\,\Delta t$ at time Δt later with its center at x. The body, in the meantime, has traveled a distance $U\,\Delta t$. In Fig. 16–1(a) we show the body at several values of t and the associated spheres of disturbances when $U < a$, that is, the motion is subsonic. We see that the fluid in front of the body is disturbed by the motion of the body. When the motion is supersonic, however, the situation is quite different, as is shown in Fig. 16–1(b). The body speed U being faster than the undisturbed speed of sound a, the disturbances will be confined to a circular cone of half apex angle β given by

$$\beta = \sin^{-1}\frac{a}{U} = \sin^{-1}\frac{1}{M}. \tag{16-1}$$

Outside of this cone, the fluid is undisturbed. This undisturbed region is usually called the *zone of silence*. The region inside the cone is usually called the *zone of influence*. The cone itself is called the *Mach cone*. In two-dimensional flows, the zone of influence has the shape of a wedge. A line drawn in the flow field which at every point makes an angle β with the local streamline is called a *Mach line*.

In Chapter 15, we learned that in nozzle flows, shock waves can occur under certain conditions. In general supersonic flow over solid bodies, we shall see that shock waves will generally occur. As an example, let us consider the supersonic flow about a blunt-nosed body (see Fig. 16–2). For a blunt-nosed body,

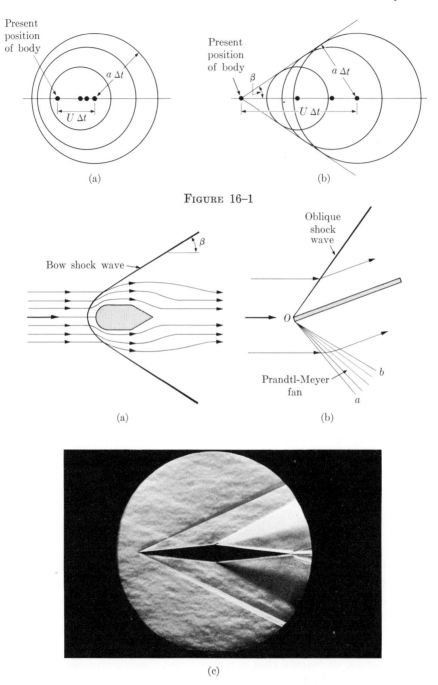

Present
position
of body

$a\ \Delta t$

$U\ \Delta t$

(a)

Present
position
of body

β

$a\ \Delta t$

$U\ \Delta t$

(b)

FIGURE 16–1

Bow shock wave

β

(a)

Oblique
shock
wave

O

Prandtl-Meyer
fan

b

a

(b)

(c)

FIG. 16–2.(c) A schlieren photograph of supersonic flow. $M_\infty = 3.0.$ (Courtesy of
Gas Dynamics Laboratory, Princeton University, Princeton, New Jersey.)

there must be a stagnation point on the body, and the fluid velocity there is zero. Hence there must be a region of subsonic flow near the nose even though the free-stream flow is supersonic. A *bow shock wave* is found to be present in the flow field, as shown in Fig. 16–2(a). The flow field in front of the bow shock wave is completely undisturbed. At large distances from the body, the bow shock wave will eventually degenerate into the Mach cone mentioned earlier, and all disturbances are confined behind this bow shock-Mach cone configuration.

In Fig. 16–2(a), we see that except for the streamline which reaches the stagnation point, all other streamlines bend abruptly at the bow shock. Shock waves which are not oriented normal to the oncoming flow velocity are called *oblique shock waves*. Oblique shock relations are to be studied in Sections 16–2 and 16–3. It suffices to mention here that while the flow behind a normal shock is always subsonic, the flow behind an oblique shock can either be supersonic or subsonic.

Let us now consider the supersonic flow about a thin two-dimensional flat plate at an angle of attack as shown in Fig. 16–2(b). On the windward side of the plate, an oblique shock issues from the leading edge which abruptly deflects the oncoming flow. On the leeward side, a fanlike region issues from the leading edge and turns the flow gradually. In contrast to the oblique shock across which the pressure rises, the pressure decreases across the fanlike region. This is known as the *Prandtl-Meyer fan* and shall be studied in detail in Section 16–4. Oblique shocks and expansion fans occur in supersonic flows over bodies of other shapes. In Fig. 16–2(c) is shown a photograph of the flow over a diamond-shaped body at $M = 3.0$, obtained by the schlieren method in which the variation in the refractive index of the gas is made visible by suitable illumination.

16–2. Oblique shocks

In this section we shall derive relations which hold across oblique shocks. We shall state the problem in the following manner. Given a supersonic flow with Mach number $M_1 > 1$, find p_2/p_1, ρ_2/ρ_1, T_2/T_1, and $(s_2 - s_1)/R$ across the oblique shock which makes an angle σ with respect to the oncoming stream. Subscripts 1 and 2 denote conditions in front of and behind the oblique shock, respectively. The case of normal orientation, $\sigma = \pi/2$, reduces to the problem of normal shocks studied in Sections 15–5 and 15–6. Figure 16–3 shows the geometry of the symbols involved.

We shall denote by δ the angle between the oncoming velocity $\mathbf{q}_1$ and the emerging velocity $\mathbf{q}_2$. Then the angle between the emerging velocity $\mathbf{q}_2$ and the oblique shock is $\sigma - \delta$. We first notice that the oncoming velocity $\mathbf{q}_1$ can be resolved into two components, one perpendicular to the shock and one parallel to the shock. The component parallel to the shock is $q_1 \cos \sigma$. Now, with re-

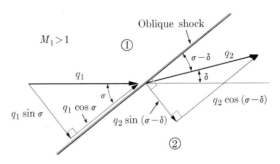

$$\textsc{Figure}\ 16\text{--}3$$

spect to an observer who travels along the oblique shock with speed $q_1 \cos \sigma$ (traveling upward in Fig. 16–3), the oncoming stream appears to flow toward the shock perpendicularly. So far as this observer is concerned, therefore, the shock is a normal shock with oncoming Mach number $(q_1 \sin \sigma)/a$ or $M_1 \sin \sigma$. Thus, all normal shock relations derived in Sections 15–5 and 15–6 can be used. The emerging velocity will also appear normal to the shock. Consequently, the component of the emerging velocity parallel to the shock must be again $q_1 \cos \sigma$. In other words, oblique shocks are simply normal shocks observed in a moving coordinate system. The quantity $M_1 \sin \sigma$ is referred to as the *normal Mach number*. Since the normal Mach number is supersonic, we have $q_1 \sin \sigma > a$. Thus $\sin \sigma > a/q_1 = \sin \beta$; i.e., the shock angle σ is always larger than the Mach angle β.

For a normal shock with oncoming Mach number $M_1 \sin \sigma$, the solutions for p_2/p_1, ρ_2/ρ_1, T_2/T_1, and $(s_2 - s_1)/R$ are obtained from Section 15–6:

$$\frac{p_2}{p_1} = \frac{2\gamma}{\gamma + 1} M_1^2 \sin^2 \sigma - \frac{\gamma - 1}{\gamma + 1}, \tag{16–2a}$$

$$\frac{\rho_2}{\rho_1} = \frac{1}{\dfrac{\gamma - 1}{\gamma + 1} + \dfrac{2}{\gamma + 1} \dfrac{1}{M_1^2 \sin^2 \sigma}}, \tag{16–2b}$$

$$\frac{T_2}{T_1} = \frac{2(\gamma - 1)}{(\gamma + 1)^2} \left(1 + \frac{\gamma - 1}{2} M_1^2 \sin^2 \sigma\right) \left(\frac{2\gamma}{\gamma - 1} - \frac{1}{M_1^2 \sin^2 \sigma}\right), \tag{16–2c}$$

$$\frac{s_2 - s_1}{R} = \frac{\gamma}{\gamma - 1} \ln \left(\frac{2}{\gamma - 1} \frac{1}{M_1^2 \sin^2 \sigma} + \frac{\gamma - 1}{\gamma + 1}\right)$$

$$+ \frac{1}{\gamma - 1} \ln \left(\frac{2\gamma}{\gamma + 1} M_1^2 \sin^2 \sigma - \frac{\gamma - 1}{\gamma + 1}\right). \tag{16–2d}$$

Hence, the problem we set out to study is solved. However, in most applications, the angle σ between the oncoming velocity $\mathbf{q}_1$ and the shock wave is not known. Instead, the deflection angle δ is usually given. Therefore we must find a relation among δ, σ, and M_1. To this end we note that the component of velocity parallel to the shock remains unchanged across the shock. From Fig. 16–3, this is written as

$$q_1 \cos \sigma = q_2 \cos (\sigma - \delta),$$

or

$$\frac{q_1}{q_2} = \frac{\cos (\sigma - \delta)}{\cos \sigma}. \tag{16–3}$$

Now, the continuity equation across the shock can be written, per unit area of the shock, as

$$\rho_1 q_1 \sin \sigma = \rho_2 q_2 \sin (\sigma - \delta), \tag{16–4}$$

$$\frac{\rho_1}{\rho_2} = \frac{q_2}{q_1} \frac{\sin (\sigma - \delta)}{\sin \sigma}, \tag{16–5}$$

where $q_1 \sin \sigma$ and $q_2 \sin (\sigma - \delta)$ are the normal components of q_1 and q_2, respectively. Eliminating q_1/q_2 between Eqs. 16–3 and 16–5 and eliminating ρ_2/ρ_1 by means of Eq. 16–2b, we have

$$\tan (\sigma - \delta) = \tan \sigma \left(\frac{\gamma - 1}{\gamma + 1} + \frac{2}{\gamma + 1} \frac{1}{M_1^2 \sin^2 \sigma} \right), \tag{16–6}$$

which is an implicit transcendental equation giving σ in terms of δ and M_1. Equation 16–6 cannot be solved for σ explicitly. Figure 16–4 shows a plot of

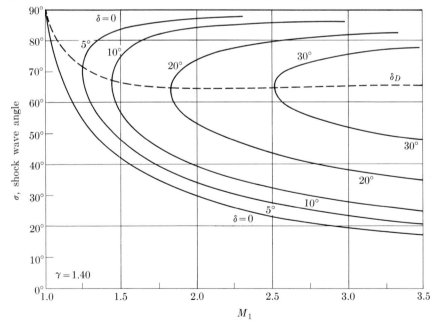

FIGURE 16–4

σ versus M_1 for several values of δ for $\gamma = 1.4$. Note that for a given value of M_1, a solution for σ exists only if δ stays below a certain value. For example, if $M_1 = 1.5$ and $\delta = 20°$, there is no solution for σ. The maximum value of δ which would still allow a solution at $M_1 = 1.5$ is about $12°$. Furthermore, it is noted that in general there are two possible solutions for a given set of M_1 and δ. These two points will be discussed in detail later in this section.

A more elegant and enlightening method of presenting the oblique shock relations is by means of the *shock polar*. A shock polar is a graph with u/a^* and v/a^* as coordinates, where u and v are the components of $\mathbf{q}$ along and perpendicular to $\mathbf{q}_1$, respectively, and a^* is the critical speed of sound of the gas (see Problem 15–5), i.e.,

$$u = q \cos \delta, \qquad v = q \sin \delta,$$

$$a^* = \sqrt{\frac{2\gamma R T^0}{\gamma + 1}} = \sqrt{\frac{2a_1^2}{\gamma + 1} \left(1 + \frac{\gamma - 1}{2} M_1^2 \right)}. \qquad (16–7)$$

On this graph, the locus of u_2/a^* and v_2/a^* for various δ is plotted for a fixed value of q_1/a^* and γ. Figure 16–5 shows one such shock polar. When a shock polar is available, then for any desired δ, one needs only to draw a line from the origin making an angle δ from the u/a^*-axis. The intersections of this line

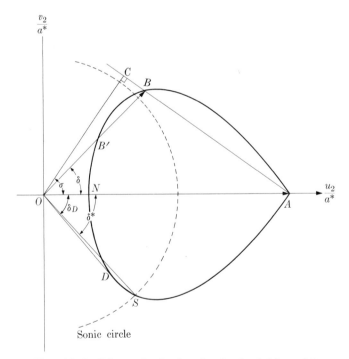

Fig. 16–5. Schematic shock polar for fixed M_1 and γ.

with the curve such as B and B' give the solution points for u_2/a^* and v_2/a^*. The shock polar can be constructed rather easily by means of the following graphical method. For the given value of M_1 of interest, the point A on the u/a^*-axis is first located so that

$$OA = \frac{q_1}{a^*} = \frac{M_1}{\sqrt{\frac{2}{\gamma+1}\left(1 + \frac{\gamma-1}{2}M_1^2\right)}}.$$

Thus OA represents the oncoming velocity vector. Now draw the line OC which makes an angle σ with respect to OA, and draw AC such that AC is perpendicular to OC. Taking σ as the shock angle as before, we see that OC and AC are the components of OA parallel and normal to the shock, respectively. Since the parallel component of the fluid velocity remains unchanged across the shock, the emerging velocity q_2/a^* must be represented by a line drawn from the origin to some point B on the line AC. The component of OB parallel to the shock is OC, as it should be; the component of OB normal to the shock is BC. To precisely locate the point B, we note from Fig. 16–5 that

$$\frac{BC}{AC} = \frac{q_2 \sin(\sigma - \delta)}{q_1 \sin \sigma}.$$

Using Eqs. 16–5 and 16–2b, we have

$$\frac{BC}{AC} = \frac{\rho_1}{\rho_2} = \frac{\gamma-1}{\gamma+1} + \frac{2}{\gamma+1}\frac{1}{M_1^2 \sin^2 \sigma}. \tag{16–8}$$

Thus with M_1 and σ known, the point B can be located by means of this equation. Repeating the same procedure for various σ, the locus of B is the desired shock polar. For a perfect gas with constant γ, the equation for the shock polar can be shown to be

$$\left(\frac{v}{a^*}\right)^2 = \left(\frac{q_1}{a^*} - \frac{u}{a^*}\right)^2 \frac{\dfrac{q_1 u}{a^{*2}} - 1}{\dfrac{2}{\gamma+1}\left(\dfrac{q_1}{a^*}\right)^2 - \dfrac{q_1 u}{a^{*2}} + 1}. \tag{16–9}$$

The derivation of Eq. 16–9 involves much algebraic manipulations and will not be presented here. It should be noted that the graphical construction method can be used for a nonperfect gas, provided that the corresponding correct relation between ρ_1/ρ_2 and the normal Mach number $M_1 \sin \sigma$ is used.

When $\delta = 0$, we see from Fig. 16–5 that there are two possible solutions, namely, the original velocity OA, or ON. The solution OA represents no disturbances at all, while the solution ON represents a normal shock. For $\delta \neq 0$, we see that again two solutions B and B' are possible; these will be referred to as *weak* and *strong solutions*, respectively. Since a^* is the critical speed of sound, the dotted unit circle shown in Fig. 16–5 is called the *sonic circle*. Solutions lying inside the circle have subsonic emerging velocities, while solutions lying outside the circle have supersonic emerging velocities. The intersection of the sonic circle with the shock polar is denoted by S, and the angle $\angle SOA$ is generally denoted by δ^*. This is the deflection angle that would make the emerging flow sonic. We also see from Fig. 16–5 that for the given M_1, the maximum deflection angle that still permits a solution is $\angle DOA$, where OD is the line drawn from the origin tangent to the shock polar. This angle is called the *detachment angle*, δ_D. It is a function of M_1 and γ. Figure 16–6 shows a plot of δ_D versus M_1 for $\gamma = 1.4$. In general, the detachment angle δ_D is slightly larger than δ^* and the difference between the two angles is less than half a degree over the entire range of M_1.

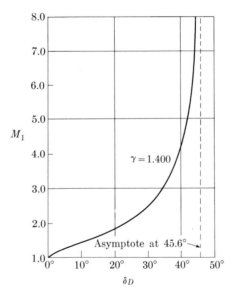

FIGURE 16–6

At this point, two pertinent questions naturally arise. Since for given M_1 and δ, there are two possible solutions when $\delta < \delta_D$, which one will actually occur in practice? Since no solution is possible when $\delta > \delta_D$, what actually happens? The answer to the first question is, unfortunately, rather involved. A somewhat vague answer would be: it depends on the conditions downstream of the shock. Note that the strong solution has a higher pressure rise than the weak solution. If the conditions downstream of the shock are such that the weak solution does not satisfy the boundary condition there, then probably the strong solution would prevail. The situation here is somewhat analogous to our analysis of nozzle flows with shock in Chapter 15. In a one-dimensional flow in a nozzle at any point where the flow is supersonic, we have two possibilities: either the flow is continuous or a normal shock may exist. They are, of course, the weak and strong solutions for the zero-deflection case. We recalled that the decision on whether a shock exists at some point in the nozzle

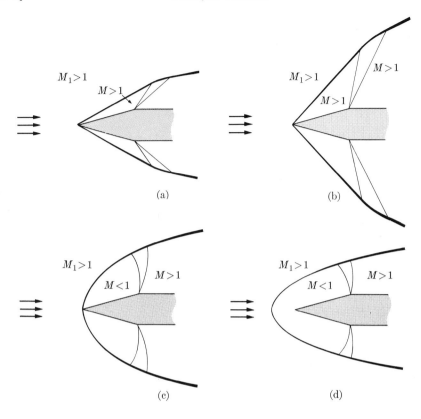

FIGURE 16–7

is decided by the downstream conditions, namely, the downstream pressure. In the oblique shock case, the same consideration prevails. It is generally a good rule, however, to consider the weak solution as more probable, unless there is strong evidence to the contrary.

To answer the question of what happens when $\delta > \delta_D$, let us consider a flow as depicted in Fig. 16–7. We consider a symmetrical wedge with a parallel afterbody in a uniform supersonic flow. In Fig. 16–7(a), M_1 is sufficiently high and $\delta < \delta^*(M_1)$. We see that the shock lies close to the wedge surface. We assume that the weak solution is valid, so that the flow behind it is supersonic. The shock is straight and is unaffected by the presence of the shoulder. At the shoulder of the body, the supersonic flow turns through a fan in a manner analogous to the flow on the upper surface of the flat plate discussed in Section 16–1. Since M_1 is lowered, but $\delta < \delta^*(M_1)$ still, Fig. 16–7(b) shows that essentially the same flow pattern remains, except that the shock angle σ becomes somewhat larger. When M_1 is lowered sufficiently that $\delta^* < \delta < \delta_D$, the flow pattern is as depicted in Fig. 16–7(c). The flow behind the shock is now subsonic; consequently, the presence of the shoulder affects the flow, and the shock wave is curved. The shock angle σ corresponding to the values of δ and M_1 obtained from the shock polar is now valid only at the tip of the wedge. At any point behind the shock, the velocity is now not parallel to the wedge surface, except at the tip and on the wedge surface. The subsonic flow behind the curved shock will accelerate and reach sonic velocity, and become supersonic upon turning at the shoulder, as shown. When M_1 is further lowered so that $\delta > \delta_D$, the resulting flow pattern is as depicted in Fig. 16–7(d). We see that the shock is now detached from the wedge. This is the reason why δ_D is called the detachment angle. The streamline at the line of symmetry will pass through a normal shock. The shock is again curved and the flow behind it is subsonic. Analogous to the previous case, it becomes supersonic upon turning at the shoulder. It may be remarked here that a quantitative analysis of flows with detached shock is very difficult, although the qualitative flow pattern is understood. The major difficulty for this type of flow is that in the flow field both subsonic and supersonic zones are present.

PROBLEMS

16–1. Show that as $M_1 \to \infty$, $q_1/a^* = \sqrt{(\gamma + 1)/(\gamma - 1)}$, the shock polar (Eq. 16–7) simplifies in this special case and gives a circle in the u_2/a^*, v_2/a^*-plane.

16–2. Show that for $\delta \ll 1$ and $M_\infty \gg 1$, the shock angle σ is given approximately by $\sigma \simeq (\gamma + 1)\delta/2$.

16–3. Given $M_1 = 2$ and $\delta = 3°$, $\gamma = 1.4$, find p_2/p_1 and ρ_2/ρ_1. [*Hint:* First solve for σ from Eq. 16–6 numerically by trial and error.]

16–3. Weak oblique shocks

The reader might have noticed that in our study of oblique shocks, no explicit formulas were given for the evaluation of shock angle, pressure ratio, etc., across an oblique shock as a function of M_1 and δ. This is because Eqs. 16–2 and 16–6 are too complicated for explicit solutions. In place of explicit analytical formulas, extensive tables and charts have been constructed and are available in the literature. However, when the deflection angle δ of the oblique shock is small, reasonably simple approximate formulas can be derived. We shall present only the formula for the pressure ratio p_2/p_1.

First of all, let us find an approximate formula for $\sin \sigma$ as a function of δ, M_1, and γ valid for small δ. Confining our attention to the weak solution, we see from Eq. 16–6 that when $\delta = 0$, $\sin \sigma = 1/M$. Let us expand $\sin \sigma$ in a Taylor series about $\delta = 0$. We have

$$\sin \sigma = \frac{1}{M} + \delta \left(\frac{d \sin \sigma}{d\delta} \right)_{\delta=0} + \frac{\delta^2}{2!} \left(\frac{d^2 \sin \sigma}{d\delta^2} \right)_{\delta=0} + \frac{\delta^3}{3!} \left(\frac{d^3 \sin \sigma}{d\delta^3} \right)_{\delta=0} + \cdots.$$

$$(16–10)$$

TABLE 16–1

$\gamma = 1.4$

M	c_1	c_2	M	c_1	c_2
1.00	∞	∞	1.80	1.336	1.618
1.02	9.950	746.293	1.90	1.238	1.529
1.04	7.001	186.333	2.00	1.155	1.467
1.06	5.689	82.987	2.50	0.873	1.320
1.08	4.903	46.943	3.00	0.707	1.269
1.20	3.015	8.307	4.00	0.516	1.232
1.30	2.408	4.300	5.00	0.408	1.219
1.40	2.041	2.919	10.00	0.201	1.204
1.50	1.789	2.288	15.00	0.134	1.202
1.60	1.601	1.949	20.00	0.100	1.201
1.70	1.455	1.748	100.00	0.020	1.200

The quantities $(d \sin \sigma/d\delta)_{\delta=0}$, $(d^2 \sin \sigma/d\delta^2)_{\delta=0}$, etc., are functions of M_1 and γ and can be obtained by successively differentiating Eq. 16–6 and solving the resulting expression with the help of trigonometric identities. The algebra involved is rather complicated. Substituting Eq. 16–10 into Eq. 16–2a, the result can be written as

$$\frac{p_2}{p_1} - 1 = \tfrac{1}{2}\gamma M_1^2[c_1\delta + c_2\delta^2 + (c_3 - D)\delta^3 + \cdots], \qquad (16\text{–}11)$$

where c_1, c_2, c_3, and D are functions of M_1 and δ and are found to be the following:

$$c_1 = \frac{2}{\sqrt{M_1^2 - 1}},$$

$$c_2 = \frac{(M_1^2 - 2)^2 + \gamma M_1^4}{2(M_1^2 - 1)^2},$$

$$(16\text{–}12)$$

$$c_3 = \frac{M_1^4}{(M_1^2 - 1)^{7/2}}$$

$$\times \left[\frac{\gamma + 1}{6} M_1^2 - \left(\frac{5 + 7\gamma - 2\gamma^2}{2(\gamma + 1)}\right)^2 - \frac{4\gamma^4 - 28\gamma^3 - 11\gamma^2 + 8\gamma + 3}{24(\gamma + 1)}\right]$$

$$+ \frac{3(M_1^2 - \tfrac{4}{3})^2}{4(M_1^2 - 1)^{7/2}},$$

$$D = \frac{\gamma + 1}{48} \frac{M_1^4}{(M_1^2 - 1)^{7/2}} [(5 - 3\gamma)M_1^4 - (12 - 4\gamma)M_1^2 + 8].$$

Although these expressions are rather complicated, for a fixed value of γ, they can be computed once and for all over the whole range of Mach numbers. For $\gamma = 1.4$, Table 16-1 gives c_1 and c_2 for $1 < M_1 < 100$. Equation 16-11 is a useful formula for the pressure rise across weak oblique shocks when M_1 is not too close to unity (transonic flow) or too large (hypersonic flow).

From Eq. 16-10, we see that for small δ, the normal Mach number $M_1 \sin \sigma$ is very close to 1.0, and σ is very close to β. From the discussion in Section 15-6, we expect the flow across weak oblique shocks to be nearly isentropic. We shall have more to say on this point in Section 16-5.

PROBLEM

16-4. Repeat Problem 16-3 using the weak shock approximation.

16-4. The Prandtl-Meyer expansion fan

Let us consider the flow in the fanlike region depicted on the leeward side of the flat plate in Fig. 16-2(b). Intuitively we expect the pressure to drop across the fan and the gas flowing through the fan to expand, in contrast to being compressed as in the case of a shock wave. This fanlike flow is called a *Prandtl-Meyer fan*.

We assume that the flow is continuous and isentropic and that the flow is again uniform after turning. The stagnation pressure p^0 and the stagnation temperature T^0 are constant over the flow field. We denote the original velocity by $\mathbf{q}$ and the speed of sound by a. The flow will be undisturbed until it reaches the Mach line oa. This Mach line is commonly referred to as a right-running Mach line, indicating that for an observer facing the stream direction, it issues forward and to the right. The angle of the Mach line relative to the oncoming velocity is simply

$$\beta = \sin^{-1} \frac{a}{q} = \sin^{-1} \frac{1}{M} \tag{16-13}$$

as is given by Eq. 16-1 (see Fig. 16-8a). We can now draw a diagram of the velocities involved. After the Mach line, the magnitude of the emerging velocity becomes $q + dq$, and the direction is deflected by $d\delta$. This is shown schematically in Fig. 16-8(b). The vector change in velocity is therefore represented by AB, the line joining the two velocities. Since it was assumed that conditions are uniform in front of and behind the Mach line oa (see Fig. 16-8a), the pressure gradient which is responsible for the change in velocity must be normal to oa. Thus, the line which represents the change of velocity must also be perpendicular to oa, or, the straight line aAB is perpendicular to oa. Now, since $\angle AOa = \angle BAC$, we have

$$\angle aOA = \angle ABC = \beta.$$

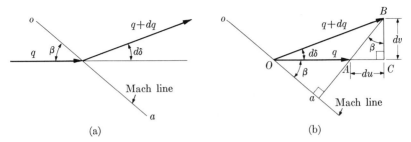

FIGURE 16–8

Now, considering the triangle OBC in Fig. 16–8(b), we have

$$\tan d\delta = \frac{dv}{q + du},$$

where du and dv are defined in the figure. Since $d\delta$ and du are infinitesimal, this can be written as

$$d\delta \cong \frac{dv}{q}. \tag{16–14}$$

But

$$du = dv \tan \angle ABC = dv \tan \beta. \tag{16–15}$$

Substituting Eq. 16–15 into Eq. 16–14, we obtain

$$d\delta \cong \frac{du}{q \tan \beta}. \tag{16–16}$$

From Eq. 16–13, we can obtain through simple trigonometry,

$$\frac{1}{\tan \beta} = \sqrt{M^2 - 1}.$$

For infinitesimal $d\delta$, we also have $OB \simeq OC$, which implies that $dq = du$. Hence Eq. 16–16 becomes

$$d\delta \cong \frac{dq}{q} \sqrt{M^2 - 1}. \tag{16–17}$$

Equation 16–17 is exact for an infinitesimal deflection $d\delta$. In order to obtain results valid for a finite deflection, we must integrate it. To do so we must first express q as a function of M. By the definition of Mach number, we have, for a perfect gas,

$$q = Ma = M\sqrt{\gamma RT}.$$

Since the stagnation temperature T^0 is constant over the flow field, we can relate T and M by the definition of T^0:

$$T^0 = T \left(1 + \frac{\gamma - 1}{2} M^2 \right).$$

TABLE 16–2

$$\gamma = 1.4$$

ω	M	ω	M	ω	M
0	1.0000	30.0	2.1339	60.0	3.5940
1.0	1.0818	31.0	2.1718	61.0	3.6600
2.0	1.1326	32.0	2.2103	62.0	3.7276
3.0	1.1769	33.0	2.2492	63.0	3.7969
4.0	1.2177	34.0	2.2887	64.0	3.8681
5.0	1.2565	35.0	2.3287	65.0	3.9412
6.0	1.2938	36.0	2.3693	66.0	4.0163
7.0	1.3300	37.0	2.4105	67.0	4.0936
8.0	1.3655	38.0	2.4523	68.0	4.1730
9.0	1.4004	39.0	2.4947	69.0	4.2548
10.0	1.4349	40.0	2.5378	70.0	4.3390
11.0	1.4692	41.0	2.5816	71.0	4.4258
12.0	1.5032	42.0	2.6261	72.0	4.5152
13.0	1.5371	43.0	2.6714	73.0	4.6076
14.0	1.5709	44.0	2.7176	74.0	4.7029
15.0	1.6747	45.0	2.7644	75.0	4.8014
16.0	1.6385	46.0	2.8122	76.0	4.9032
17.0	1.6725	47.0	2.8609	77.0	5.0085
18.0	1.7065	48.0	2.9105	78.0	5.1176
19.0	1.7406	49.0	2.9610	79.0	5.2306
20.0	1.7750	50.0	3.0126	80.0	5.3475
21.0	1.8095	51.0	3.0652	85.0	6.0064
22.0	1.8443	52.0	3.1189	90.0	6.8190
23.0	1.8793	53.0	3.1738	95.0	7.8509
24.0	1.9146	54.0	3.2298	100.0	9.2105
25.0	1.9503	55.0	3.2871	105.0	11.091
26.0	1.9862	56.0	3.3457	110.0	13.874
27.0	2.0226	57.0	3.4056	115.0	18.438
28.0	2.0593	58.0	3.4669	120.0	27.335
29.0	2.0964	59.0	3.5297	125.0	52.491
				130.0	631.03

Eliminating T, we have

$$q = M \left[\frac{\gamma R T^0}{1 + \frac{\gamma - 1}{2} M^2} \right]^{1/2}. \tag{16–18}$$

Taking the differential of the logarithm of q, we have

$$\frac{dq}{q} = \frac{M \, dM}{M^2 \left[1 + \frac{\gamma - 1}{2} M^2 \right]}. \tag{16–19}$$

Substituting Eq. 16–19 into Eq. 16–17, we have

$$d\delta = \frac{\sqrt{M^2 - 1} \, d(M^2)}{2M^2 \left[1 + \frac{\gamma - 1}{2} M^2 \right]}. \tag{16–20}$$

Integrating, we obtain the following expression:

$$\delta = \omega(M) + \text{constant}, \tag{16–21}$$

where $\omega(M)$ is given by

$$\omega(M) = \sqrt{\frac{\gamma + 1}{\gamma - 1}} \tan^{-1} \left[\sqrt{\frac{\gamma - 1}{\gamma + 1} (M^2 - 1)} \right] - \tan^{-1} \sqrt{M^2 - 1}, \tag{16–22}$$

and is tabulated in Table 16–2 for $\gamma = 1.4$. For practical reasons, the values in Table 16–2 are given in degrees instead of radians.

We recall that when we studied oblique shocks, a shock polar was constructed on a graph with u/a^* and v/a^* as coordinates. It is instructive to recast Eq. 16–22 in terms of M^* and δ, where M^* is defined as the ratio of the local speed of the gas to the critical speed of sound a^* of the gas. The relation between M and M^* can be shown with Eq. 16–7 to be

$$M^2 = \frac{\frac{2}{\gamma + 1} M^{*2}}{1 - \frac{\gamma - 1}{\gamma + 1} M^{*2}}. \tag{16–23}$$

Substituting Eq. 16–23 into Eq. 16–22, we obtain

$$\delta = \Omega(M^*) + \text{constant}, \tag{16–24}$$

where

$$\Omega(M^*) = \sqrt{\frac{\gamma + 1}{\gamma - 1}} \tan^{-1} \left[\sqrt{\frac{M^{*2} - 1}{\frac{\gamma + 1}{\gamma - 1} - M^{*2}}} \right] - \tan^{-1} \left[\sqrt{\frac{M^{*2} - 1}{1 - \frac{\gamma - 1}{\gamma + 1} M^{*2}}} \right]. \tag{16–25}$$

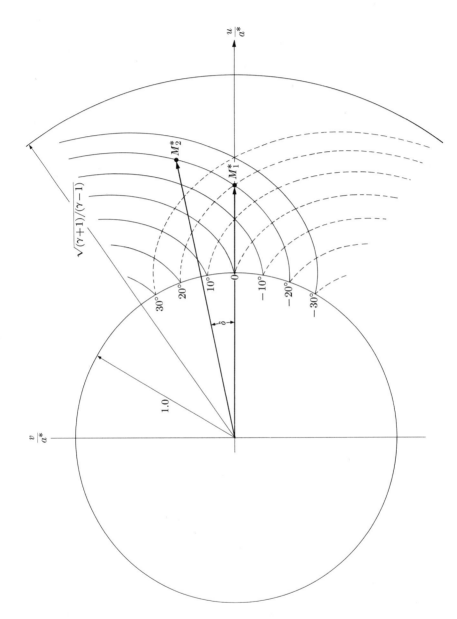

FIG. 16–9. Graphical representation of Prandtl-Meyer fan relation.

When we plot Eq. 16–24 using M^* and δ as polar coordinates, we obtain curves as shown in Fig. 16–9. Each curve corresponds to a separate value for the constant in Eq. 16–24. We see that curves exist only outside the unit circle, indicating that the flow must be supersonic. The radius of the outer circle is $\sqrt{(\gamma + 1)(\gamma - 1)}$, representing the limiting case when $M \to \infty$. For a given initial value of M_1^*, we can locate the initial velocity in Fig. 16–9 on the u/a^*-axis. For any desired value of δ, the emerging velocity M_2^* is located in Fig. 16–9 by the intersection of a line drawn from the origin making an angle δ with the u/a^*-axis and the solid curve which passes through M_1^*. Note that if δ is negative, the emerging velocity M_2^* will be smaller than M_1^*, indicating a compressive process. Strictly speaking, we can no longer use the results in this section, for in a compressive turn, oblique shock instead of Prandtl-Meyer fan is involved. In other words, we should use the shock polar (Fig. 16–5) for a compressive turn and the Prandtl-Meyer diagram (Fig. 16–9) for an expansive turn. However, it will be shown in the next section that the two analyses give almost identical results for small turning angles.

It is clear that if the flat plate shown in Fig. 16–8 is tilted in the opposite direction, a Prandtl-Meyer fan will be present above the plate and precisely the same analysis will apply. However, since the angle δ is by convention measured positive in the counterclockwise direction, we must modify Eqs. 16–21 and 16–24 to read

$$-\delta = \omega(M) + \text{constant} \tag{16–26}$$

and

$$-\delta = \Omega(M^*) + \text{constant}. \tag{16–27}$$

Equation 16–27 is also plotted on Fig. 16–9 with dashed lines. In deciding what the sign of δ should be in a practical problem, a simple memory aid is as follows. We can ignore Eqs. 16–26 and 16–27 and use only Eqs. 16–22 and 16–24 with the new convention that δ is to be considered positive when the gas expands through the Prandtl-Meyer fan. If these equations are used for weak compressive turns, then δ is a negative number.

Example 16–1. Let us suppose that the Mach number of the supersonic flow depicted in Fig. 16–1 is 2.0 and that the angle between the flat plate and the oncoming stream is 5°. Before the flow turns, we have $\delta = 0$ and $M = 2.0$. Substituting into Eq. 16–21, we find that the value of the constant is $-\omega(2.0)$. After the flow turns, we have $\delta = 5°$. Equation 16–21 now gives

$$\omega(M) = 5° + \omega(2.0) = 5° + 26° = 31°.$$

From Table 16–2, we find $M = 2.17$, which is the Mach number after the turn.

PROBLEMS

16–5. For an initial Mach number of 1.33, find (a) the final Mach numbers after a Prandtl-Meyer turn of 10°, 20°, 60°, and 90°, and (b) the corresponding pressure ratios across the Prandtl-Meyer fan. Assume $\gamma = 1.4$.

16–6. For an initial Mach number of 1.0, find the value of δ as a function of γ such that the pressure behind the Prandtl-Meyer fan is zero.

16–5. Approximate formula for small deflections

The exact Prandtl-Meyer formula, Eq. 16–21, is rather simple to use. Knowing the original Mach number and the deflection angle, the final Mach number can easily be obtained. Since the flow is isentropic, the final pressure can be obtained from the fact that the stagnation pressure remains constant through the expansion process. However, in a large number of applications, the deflection angle δ will be small, and an approximate formula valid for small δ will be very useful. For example, if the flat plate in Fig. 16–2(b) is curved, we can approximate it by many small flat segments, and analyze each small deflection by the approximate formula of this section.

Since the stagnation pressure remains constant in the Prandtl-Meyer fan, we have

$$p_0 = p\left(1 + \frac{\gamma - 1}{2} M^2\right)^{\gamma/(\gamma-1)}. \tag{16-28}$$

Taking the differential of the logarithm of p_0, we have

$$\frac{dp}{p} = -\frac{\gamma}{2} \frac{d(M^2)}{\left(1 + \frac{\gamma - 1}{2} M^2\right)}.$$

Now, since dM^2 and $d\delta$ are related by Eq. 16–20, we have

$$d\delta = \frac{\sqrt{M^2 - 1}}{2M^2} \frac{d(M^2)}{\left(1 + \frac{\gamma - 1}{2} M^2\right)}.$$

Eliminating $d(M^2)$ between the above equations, we have

$$\frac{dp}{d\delta} = -\frac{\gamma p M^2}{\sqrt{M^2 - 1}}, \tag{16-29}$$

which so far involves no approximation and is therefore still exact. Differentiating Eq. 16–29 with respect to δ once, we have

$$\frac{d^2p}{d\delta^2} = -\frac{\gamma M^2}{\sqrt{M^2 - 1}} \frac{dp}{d\delta} - \gamma p\left[\frac{1}{\sqrt{M^2 - 1}} - \frac{1}{2}\frac{M^2}{(M^2 - 1)^{3/2}}\right]\frac{dM^2}{d\delta}, \tag{16-30}$$

where $dp/d\delta$ and $dM^2/d\delta$ may be eliminated by Eqs. 16–20 and 16–29. Differentiating Eq. 16–30 once more with respect to δ, we will obtain $d^3p/d\delta^3$. Now for small δ, we can write

$$p = p_\infty + \left(\frac{dp}{d\delta}\right)_\infty \delta + \left(\frac{d^2p}{d\delta^2}\right)_\infty \frac{\delta^2}{2!} + \left(\frac{d^3p}{d\delta^3}\right)_\infty \frac{\delta^3}{3!} + \cdots , \qquad (16\text{–}31)$$

which is simply a Taylor series expansion of p for small δ. Here subscript ∞ indicates that these quantities are to be evaluated at the free-stream undisturbed condition. We define c_p as a pressure coefficient by

$$c_p = \frac{p - p_\infty}{\frac{1}{2}\rho_\infty U_\infty^2} = \frac{p - p_\infty}{\frac{1}{2}\gamma p_\infty M_\infty^2} . \qquad (16\text{–}32)$$

Hence Eq. 16–31 may be rewritten as

$$c_p = \frac{2}{\gamma p_\infty M_\infty^2} \left[\left(\frac{dp}{d\delta}\right)_\infty \delta + \left(\frac{d^2p}{d\delta^2}\right)_\infty \frac{\delta^2}{2!} + \left(\frac{d^3p}{d\delta^3}\right)_\infty \frac{\delta^3}{3!} + \cdots \right].$$

Evaluating $(dp/d\delta)_\infty$, $(d^2p/d\delta^2)_\infty$, etc., by Eqs. 16–29 and 16–30, we have

$$c_p = -c_1\delta + c_2\delta^2 - c_3\delta^3, \qquad (16\text{–}33)$$

where c_1, c_2, and c_3 are precisely the same coefficients appearing in Eqs. 16–12 which were derived for weak oblique shocks. The only difference between Eqs. 16–33 and 16–11 is that in Eq. 16–11 the coefficient of δ^3 is $c_3 - D$ instead of simply c_3 as in Eq. 16–33. The striking similarity of the two formulas is not fortuitous. One should remember that for an oblique shock, the flow is not isentropic, while the flow in a Prandtl-Meyer fan is. However, since Eq. 16–11 is valid for small deflection angle only, the entropy rise for such weak shocks are very small. In fact, it can be shown that it is proportional to δ^3. The factor D reflects the effects of entropy increase in the oblique shock wave.

As a consequence, for flow about a thin body which is only slightly inclined from the free stream, the simple approximate formula

$$c_p = -c_1\delta + c_2\delta^2 \qquad (16\text{–}34)$$

may be used regardless of whether the flow undergoes compression or expansion, provided that the maximum value of δ in the flow is sufficiently small. In most aeronautical problems, it is usually sufficient to use only the first term. This simple formula

$$c_p = -c_1\delta \qquad (16\text{–}35)$$

is commonly called the *Ackeret formula*. In applying Eqs. 16–33, 16–34, and 16–35, the sign of δ should follow the new sign convention given at the end of the last section.

16–7. Using the Ackeret formula, show that the lift generated by a two-dimensional thin wing at a small angle of attack is independent of the detailed shape of the wing, but is proportional to Δy, as shown in Fig. 16–10. [*Hint:* Locally $\delta \approx \tan \delta = dy_{\text{body}}/dx$.]

FIGURE 16–10

16-6. Applications to nozzle flows

In our discussion of one-dimensional nozzle flows, we showed that as the exit pressure is successively decreased with respect to the upstream reservoir pressure, the normal shock in the divergent section can be brought to the exit of the nozzle. When the exit pressure is further lowered, no one-dimensional solution was found possible, except for a particular exit pressure p_e when the flow is isentropic throughout. Now that we have studied oblique shocks and Prandtl-Meyer expansion fans, it will become clear what actually happens at these exit pressures.

Referring to Fig. 15–10 and using its symbols, we recall that when $p_2 = p_S$, a normal shock stands at the exit of the nozzle. As the pressure in reservoir 2 is further lowered, the flow field at the exit is no longer one-dimensional. The pressure at the exit station in the duct remains at p_e, but instead of a normal shock at the exit station, oblique shocks are issued from the lips of the nozzle, turning the flow such that the pressure rise across such oblique shocks give precisely the pressure of reservoir 2 at the outer boundary of the issuing jet (see Fig. 16–11a). As the pressure of reservoir 2 is lowered to p_e, the oblique shocks are gradually weakened until they become Mach lines (see Fig. 16–11b). For still lower pressures in reservoir 2, Prandtl-Meyer expansion fans are issued from the lips of the nozzle such that the pressure at the outer boundary of the free jet has the proper value (see Fig. 16–11c).

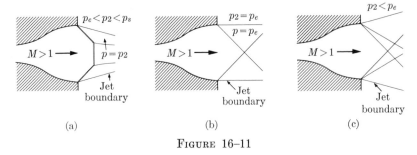

FIGURE 16–11

17 Similarity Laws in Compressible Flows

17–1. The idea of similarity

In the study of most engineering problems, sometimes the task is rendered more difficult by the fact that the solutions depend on many dimensionless parameters. A *similarity analysis* is any procedure aimed at reducing the number of such parameters; and if the process is successful, the result is called a *similarity* or a *similitude*.

There is no standard procedure available to obtain a similarity except that generally it involves transformations of both the dependent and independent variables. In practice, the transformations are often guided by physical reasonings. Very often, when a parameter takes on extreme values, a similarity can be obtained if appropriate approximations are adopted. For example, when the Reynolds number is large, we deduced the boundary-layer equations from the Navier-Stokes equations in Section 12–3. In this context, we can consider the boundary-layer theory as a large Reynolds number similarity, since its governing equations and its boundary conditions have been made independent of the Reynolds number. The transformations used in Eq. 12–11 were guided by physical reasonings given in Section 12–2.

To make the above idea more definite, let us consider a simple illustrative example. Suppose it is desired to compute and present a function $y(x; \alpha, \beta, \epsilon)$ in the range $0 \leq x \leq 1$ defined by the algebraic equation

$$x^2 - \alpha^2 y^2 = \beta x^3 y^5 + \epsilon. \tag{17-1}$$

In general, it is impossible to solve this equation explicitly for y. In practice, one must resort to some laborious numerical method for the construction of the desired function. The laborious computation must be repeated for each set of α, β, and ϵ of interest. One asks at this point if a similarity exists, so that the amount of computation labor can be reduced.

Let us consider the following transformation:

$$x = \xi_1 \alpha^m \beta^n, \qquad y = \eta_1 \alpha^p \beta^q, \qquad \epsilon = l_1 \alpha^r \beta^s, \tag{17-2}$$

where m, n, p, q, r, s, and l_1 are some unknown constants and ξ_1 and η_1 are new variables. Substituting this into Eq. 17–1, we have

$$\xi_1^2 \alpha^{2m} \beta^{2n} - \eta_1^2 \alpha^{2p+2} \beta^{2q} = \xi_1^3 \eta_1^5 \alpha^{3m+5p} \beta^{3n+5q+1} + l_1 \alpha^r \beta^s. \qquad (17\text{–}3)$$

It is obviously advantageous to cancel out as many factors of α and β as possible. Let us require the exponents of all α and β to be equal. We thus set

$$2m = 2p + 2 = 3m + 5p = r, \qquad 2n = 2q = 3n + 5q + 1 = s,$$

which yields six equations for the six unknowns m, n, p, q, r, and s. The solutions are obtained by simple algebraic manipulations.

$$m = \tfrac{5}{6}, \qquad n = -\tfrac{1}{6},$$
$$p = -\tfrac{1}{6}, \qquad q = -\tfrac{1}{6},$$
$$r = \tfrac{5}{3}, \qquad s = -\tfrac{1}{3}.$$

Substituting these values in Eq. 17–3, we obtain simply

$$\xi_1^2 - \eta_1^2 = \xi_1^3 \eta_1^5 + l_1.$$

We see now that $\eta_1(\xi_1; l_1)$ is a function of only one parameter, l_1. The computation for $\eta_1(\xi_1; l_1)$ from this equation is no easier than the computation of $y(x; \alpha, \beta, \epsilon)$, but because it contains fewer parameters, the amount of labor is considerably reduced. After $\eta_1(\xi_1; l_1)$ is carefully constructed, the desired solution is given parametrically as

$$y = \alpha^{-1/6} \beta^{-1/6} \eta_1(\xi_1, l_1), \qquad (17\text{–}4)$$

where

$$\xi_1 = \alpha^{-5/6} \beta^{-1/6} x, \qquad l_1 = \alpha^{-5/3} \beta^{-1/3} \epsilon.$$

The result in Eq. 17–4 is called a similarity.

This similarity is especially useful when the values of α, β, and ϵ of interest span only a narrow range of l_1, and the range of ξ_1 is of order unity. Such would be the case for example if ϵ is in general a very small number while α and β are of order unity, so that l_1 is always a small number. Then as an approximation we may write

$$y = \alpha^{-1/6} \beta^{-1/6} \eta_1(\xi_1; l_1 = 0); \qquad (17\text{–}5)$$

for this case only one single curve $\eta_1(\xi_1; 0)$ need be constructed.

Equation 17–4 is not the only possible similarity. For example, we may write

$$x = \xi_2 \epsilon^m \alpha^n, \qquad y = \eta_2 \epsilon^p \alpha^q, \qquad \beta = l_2 \epsilon^r \alpha^s, \qquad (17\text{–}6)$$

where m, n, p, q, r, s, and l_2 are again some unknown constants and ξ_2, η_2 are new variables. Substituting into Eq. 17–1 and equating the exponents of ϵ and α, we obtain

$$m = \tfrac{1}{2}, \qquad n = 0,$$
$$p = \tfrac{1}{2}, \qquad q = -1,$$
$$r = -3, \qquad s = +5.$$

The algebraic equation becomes

$$\xi_2^2 - \eta_2^2 = l_2 \xi_2^3 \eta_2^5 + 1. \qquad (17\text{–}7)$$

The function $\eta_2(\xi_2; l_2)$ also contains only one parameter l_2; thus the parametric representation

$$y = \epsilon^{1/2} \alpha^{-1} \eta_2(\xi_2; l_2),$$

where

$$\xi_2 = x\epsilon^{-1/2}, \qquad l_2 = \beta\epsilon^3 \alpha^{-5} \qquad (17\text{–}8)$$

is also a similarity, with equal status as Eq. 17–4. This similitude is most useful when the values of α, β, and ϵ are such that l_2 is very small and the range of interest of ξ_2 is of order unity. Then as an approximation, we may use $\eta_2(\xi_2, l_2 = 0)$. This second similitude demonstrates that for a given equation involving many parameters, there may formally exist more than one similarity. But clearly one should be guided by physical reasoning and order of magnitudes estimates of the parameters of interest in deciding on the most favorable one to use. In compressible fluid mechanics, the idea of similarity is fully exploited, as we shall see in this chapter.

PROBLEMS

17–1. Find the roots of the following algebraic equations to two significant figures for $\alpha = 0.0021$ and 0.00001.

   (a) $x^3 + 2x\alpha^2 - 3\alpha^3 = 0$
   (b) $\alpha x^2 + x - 1 = 0$

17–2. In Eq. 17–1, if $|\alpha| \ll 1$ and $\beta = 1$, $\epsilon = 2$, find a suitable similitude.

17–2. The potential equation for compressible flow

In this section we shall derive the complete governing equations for irrotational compressible flow about solid bodies. In the subsequent sections these equations will be closely scrutinized for useful similarities. For simplicity, we shall consider only two-dimensional steady flow problems, and shall assume that the gas obeys the perfect gas law and has constant C_p and C_v. Frictional and heat-conduction effects are neglected, and therefore the flow is considered isentropic. The undisturbed flow direction is in the $+x$-direction. The body in question is located on the x-axis, with its leading edge at the origin and its trailing edge at

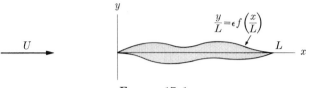

FIGURE 17–1

$x = L$, as shown in Fig. 17–1. The body shape is described by

$$\frac{y}{L} = \epsilon f \left(\frac{x}{L}\right), \tag{17–9}$$

where ϵ is the maximum value of y/L of the body. It is thus clear that $f(x/L)$ is a function of order unity. The continuity and momentum equations are

$$u \frac{\partial \rho}{\partial x} + v \frac{\partial \rho}{\partial y} + \rho \left(\frac{\partial u}{\partial x} + \frac{\partial v}{\partial y}\right) = 0, \tag{17–10a}$$

$$u \frac{\partial u}{\partial x} + v \frac{\partial u}{\partial y} = -\frac{1}{\rho} \frac{\partial p}{\partial x}, \tag{17–10b}$$

$$u \frac{\partial v}{\partial x} + v \frac{\partial v}{\partial y} = -\frac{1}{\rho} \frac{\partial p}{\partial y}, \tag{17–10c}$$

where we have assumed frictionless flow. Instead of using the energy equation, we shall use the isentropic condition that entropy remains constant along a streamline. Since the entropy of the oncoming gas is uniform, entropy must be a constant throughout the flow field. Hence we can use

$$\frac{p}{p_\infty} = \left(\frac{\rho}{\rho_\infty}\right)^\gamma, \tag{17–11}$$

where p_∞ and ρ_∞ are the free-stream pressure and density, respectively. Equations 17–10 and 17–11 give four equations and four unknowns u, v, p, and ρ. Since it was shown in Chapter 8 that an initially irrotational flow remains irrotational under the assumption of frictionless and barotropic flow, the present flow field must be irrotational everywhere, since the upstream oncoming flow is uniform and therefore irrotational. A potential function therefore exists. To obtain the equation for the velocity potential, we proceed as follows. First, we differentiate Eq. 17–11 to obtain

$$\frac{\partial p}{\partial x} = \gamma \frac{p}{\rho} \frac{\partial \rho}{\partial x}, \qquad \frac{\partial p}{\partial y} = \gamma \frac{p}{\rho} \frac{\partial \rho}{\partial y}. \tag{17–12}$$

Substituting this into Eq. 17–10b and 17–10c, we have

$$u \frac{\partial u}{\partial x} + v \frac{\partial u}{\partial y} = -\frac{a^2}{\rho} \frac{\partial \rho}{\partial x}, \qquad u \frac{\partial v}{\partial x} + v \frac{\partial v}{\partial y} = -\frac{a^2}{\rho} \frac{\partial \rho}{\partial y},$$

where $a^2 = \gamma p / \rho$ is the local speed of sound. To eliminate ρ, we multiply each of these two equations by u and v, respectively, and add. The resulting terms on the right-hand side involving $\partial \rho / \partial x$ and $\partial \rho / \partial y$ can be eliminated by the use of the continuity equation 17–10a. We then have

$$(u^2 - a^2) \frac{\partial u}{\partial x} + uv \left(\frac{\partial u}{\partial y} + \frac{\partial v}{\partial x} \right) + (v^2 - a^2) \frac{\partial v}{\partial y} = 0. \qquad (17\text{–}13)$$

This single equation now contains the velocity components u and v. The local sound speed a is related to u and v, since for this flow, stagnation enthalpy is a constant over the whole flow field:

$$\frac{a_\infty^2}{\gamma - 1} + \frac{U_\infty^2}{2} = \frac{a^2}{\gamma - 1} + \frac{u^2 + v^2}{2}, \qquad (17\text{–}14)$$

where we have written $a^2 / (\gamma - 1)$, for $C_p T$ and subscript ∞ indicates conditions in the free-stream.

Instead of the ordinary velocity potential, we use a perturbation velocity potential to describe the change of velocity from the undisturbed stream. The reason for doing this will become clear later. We write

$$u = U_\infty \left(1 + \epsilon \frac{\partial \phi}{\partial X} \right), \qquad v = \epsilon U_\infty \frac{\partial \phi}{\partial Y}, \qquad (17\text{–}15)$$

where U_∞ is the undisturbed free-stream velocity, ϵ is the maximum value of y/L of the body, and ϕ is the dimensionless perturbation velocity potential. Also, X and Y are dimensionless and are defined by

$$X = \frac{x}{L}, \qquad Y = \frac{y}{L}. \qquad (17\text{–}15a)$$

Now Eqs. 17–14 and 17–15 can be substituted into Eq. 17–13 to yield an equation for ϕ. After considerable manipulation, we obtain

$$(1 - M_\infty^2) \frac{\partial^2 \phi}{\partial X^2} + \frac{\partial^2 \phi}{\partial Y^2}$$

$$= \epsilon M_\infty^2 \left\{ (\gamma + 1) M_\infty^2 \frac{\partial \phi}{\partial X} \frac{\partial^2 \phi}{\partial X^2} + (\gamma - 1) \frac{\partial \phi}{\partial X} \left[(1 - M_\infty^2) \frac{\partial^2 \phi}{\partial X^2} + \frac{\partial^2 \phi}{\partial Y^2} \right] \right.$$

$$\left. + \frac{\partial}{\partial X} \left[(1 - M_\infty^2) \left(\frac{\partial \phi}{\partial X} \right)^2 + \left(\frac{\partial \phi}{\partial Y} \right)^2 \right] \right\}$$

$$+ \epsilon^2 M_\infty^2 \left\{ \frac{\partial \phi}{\partial X} \frac{\partial^2 \phi}{\partial X^2} + 2 \frac{\partial \phi}{\partial X} \frac{\partial \phi}{\partial Y} \frac{\partial^2 \phi}{\partial X \partial Y} \right.$$

$$\left. + \left(\frac{\partial \phi}{\partial Y} \right)^2 \frac{\partial^2 \phi}{\partial Y^2} + \frac{\gamma - 1}{2} \left(\frac{\partial^2 \phi}{\partial X^2} + \frac{\partial^2 \phi}{\partial Y^2} \right) \left[\left(\frac{\partial \phi}{\partial X} \right)^2 + \left(\frac{\partial \phi}{\partial Y} \right)^2 \right] \right\},$$

$$(17\text{–}16)$$

where $M_\infty = U_\infty/a_\infty$. Equation 17–17 is a rather formidable-looking nonlinear partial differential equation. For boundary conditions, we require (1) that the velocity of the gas at the surface of the body be parallel to the body surface, or

$$\frac{v}{u} = \left(\frac{dY}{dX}\right)_{\text{body}} = \epsilon\,\frac{df}{dX}, \qquad \text{or} \qquad \frac{\partial\phi}{\partial X} = \left(1 + \epsilon\,\frac{\partial\phi}{\partial X}\right)\frac{df}{dX} \qquad (17\text{–}17)$$

to be satisfied at $Y = \epsilon f(X)$ for $0 \le X \le 1$; and (2) that far away from the body the flow should be undisturbed. In other words, we require

$$\frac{\partial\phi}{\partial X} = 0, \qquad \frac{\partial\phi}{\partial Y} = 0 \qquad (17\text{–}18)$$

to be satisfied at large distances from the body. At this point we have completed the formal derivation of the governing equation and boundary conditions for two-dimensional compressible irrotational flows. For a given body described by Eq. 17–9, the problem is considered solved if $\phi(X, Y)$ can be found that satisfies Eqs. 17–16, 17–17, and 17–18. Note that in general we have three parameters in our problem, M_∞, γ, and ϵ, in addition to the body-shape function $f(x)$. Once $\phi(X, Y)$ is known, the velocities u and v can be obtained from Eq. 17–15 by differentiation. Once u and v are known, the corresponding value of a can be obtained from Eq. 17–14, and consequently all other state variables such as pressure can be obtained.

No general method of solving Eq. 17–16 in the present form is available. In the following sections, we shall show how the method of similarity analysis can be applied to simplify the problem.

PROBLEM

17–3. Show that

$$\frac{p}{p_\infty} = \left[1 - \epsilon\,\frac{\gamma - 1}{2}\,M_\infty^2\left(2\,\frac{\partial\phi}{\partial X} + \epsilon\,\frac{\partial\phi}{\partial X}^2 + \epsilon\,\frac{\partial\phi}{\partial Y}^2\right)\right]^{\gamma/(\gamma-1)}.$$

17–3. General comments on the potential equation

We know that in general the solution ϕ can be written as

$$\phi = \phi(X, Y; M_\infty, \epsilon, \gamma; f(X)) \qquad (17\text{–}19)$$

and also, the pressure coefficient c_p can be written as

$$c_p = \frac{p - p_\infty}{\frac{1}{2}\rho_\infty U_\infty^2} = c_p(X, Y; M_\infty, \epsilon, \gamma; f(X)). \qquad (17\text{–}20)$$

The above results, Eqs. 17–19 and 17–20, could have been obtained from dimensional analysis. We have eliminated all dimensional parameters that at first sight would seem to affect the solution: the scale of the body L, the actual velocity of the flow U_∞, and the thermodynamic state of the oncoming gas, for example, p_∞ and ρ_∞. These are already very powerful conclusions, which are valid provided that basic assumptions employed so far are satisfied. These assumptions are as follows.

(1) The flow is assumed frictionless and stagnation enthalpy is a constant over the flow field. This is equivalent to requiring that the characteristic Reynolds number of the flow $\rho_\infty U_\infty L/\mu_\infty$ be large (see Section 12–1).

(2) Body forces such as gravity are neglected.

(3) The gas is assumed to be in thermodynamic equilibrium at all times and it obeys the perfect gas law with constant C_p, C_v, and γ.

(4) The flow far upstream is uniform in velocity and thermodynamic state so that along with assumption (1) we may conclude that the flow is irrotational.

None of these assumptions is difficult to meet for most aeronautical applications. The similarity analysis which follows aims at further reducing the number of dimensionless parameters by taking advantage of approximations that are available.

17–4. Incompressible flows

Flows with $M^2 \ll 1$ are commonly referred to as *incompressible flows*, although the fluid involved is a gas which is certainly compressible in the ordinary sense. Note that for air under standard sea-level conditions, the speed of sound a is 1117 ft/sec. Thus for U_∞ up to 200 ft/sec or so, the assumption of $M_\infty^2 \ll 1$ is still very good.

When $M_\infty^2 \ll 1$, all terms involving M_∞^2 are therefore small, and in this limit Eqs. 17–16, 17–17, and 17–18 simplify into

$$\frac{\partial^2 \phi}{\partial X^2} + \frac{\partial^2 \phi}{\partial Y^2} = 0, \tag{17–21}$$

with boundary conditions

$$\frac{\partial \phi}{\partial Y} = \left(1 + \epsilon \frac{\partial \phi}{\partial X}\right) \frac{df}{dX}$$

to be satisfied at $Y = \epsilon(X)$ for $0 \le X \le 1$; and

$$\frac{\partial \phi}{\partial X} = 0, \qquad \frac{\partial \phi}{\partial Y} = 0$$

to be satisfied at infinity. Thus the solution can be written as

$$\phi = \phi(X, Y; \epsilon; f(X))$$

and

$$c_p = c_p(X, Y; \epsilon; f(X)). \tag{17–22}$$

We see clearly that γ no longer appears anywhere, and *the solution is identical with that of the flow of an incompressible fluid over the same body.* Hence if it is desired to know the pressure distribution on a certain body in a Mach number flow of helium ($\gamma = 1.667$), for example, one may either use a low-speed air tunnel or test the model in water!

PROBLEM

17–4. Show that pressure coefficient c_p defined in Eq. 17–20 becomes independent of M_∞ if $M_\infty^2 \ll 1$. Use the formula for p/p_∞ in Problem 17–3 as the starting point.

17–5. Subsonic similarity

Let us consider the *subsonic* Mach number range $1 \leq M_\infty^2 \leq 0$. For reasons to be made clear later, we assume that M_∞ is not too close to unity, so that $(M_\infty^2 - 1)^{-1}$ is of order unity. However, M_∞^2 itself may be small, so that the subsonic range actually includes the "incompressible" case.

In most applications, the bodies of interest are thin and streamlined, so that generally ϵ is a small number. We shall take advantage of this fact and assume that $\epsilon \ll 1$. From Eq. 17–9, we see that $\epsilon \ll 1$ implies a thin and flat body.

For small ϵ, the right-hand side of Eq. 17–16 can be neglected compared with each of the terms on the left-hand side. The governing equation becomes

$$(1 - M_\infty^2) \frac{\partial^2 \phi}{\partial X^2} + \frac{\partial^2 \phi}{\partial Y^2} = 0. \tag{17–23}$$

Consistent with the approximation already employed, Eq. 17–17 becomes

$$\frac{\partial \phi}{\partial Y} = \frac{df}{dX}, \tag{17–24}$$

which is to be satisfied on

$$Y = 0 \quad (0 \leq X \leq 1). \tag{17–25}$$

Note that this boundary is applied on $Y = 0$ instead of $Y = \epsilon f(X)$, since ϵ is assumed small. The other boundary condition at infinity remains unchanged. Note that the solution ϕ now depends only on M_∞^2 and $f(X)$. The following transformation is now introduced:

$$X = \bar{x}, \qquad Y = \frac{1}{\sqrt{1 - M_\infty^2}} \bar{y}, \tag{17–26}$$

$$\phi = \frac{1}{\sqrt{1 - M_\infty^2}} \bar{\Phi}(\bar{x}, \bar{y}).$$

Equations 17–23, 17–24, and 17–25 become

$$\frac{\partial^2 \bar{\Phi}}{\partial \bar{x}^2} + \frac{\partial^2 \bar{\Phi}}{\partial \bar{y}^2} = 0, \tag{17–27a}$$

with the boundary condition

$$\frac{\partial \bar{\Phi}}{\partial \bar{y}} = \frac{df}{d\bar{x}} \tag{17–27b}$$

to be satisfied on

$$\bar{y} = 0 \qquad (0 \le \bar{x} \le 1). \tag{17–27c}$$

The striking result here is that the parameter M_∞ disappears. Hence for a given body shape $f(\bar{x})$, we need only to construct a single solution $\bar{\Phi} = \bar{\Phi}(\bar{x}, \bar{y})$, from which we can generate solutions for arbitrary subsonic Mach number M_∞ and arbitrary but small ϵ:

$$\phi = \frac{1}{\sqrt{1 - M_\infty^2}} \bar{\Phi}(x, \sqrt{1 - M_\infty^2}\, y). \tag{17–28}$$

For small ϵ, the pressure coefficient is given by (see Problem 17–5)

$$c_p = \frac{p - p_\infty}{\frac{1}{2}\rho_\infty U_\infty^2} = -2\epsilon \frac{\partial \phi}{\partial X} \tag{17–29a}$$

or

$$c_p = -\frac{2\epsilon}{\sqrt{1 - M_\infty^2}} \frac{\partial \bar{\Phi}}{\partial \bar{x}}(\bar{x}, \bar{y}). \tag{17–29b}$$

And on the body surface, $y = \sqrt{1 - M_\infty^2}\, \epsilon f(\bar{x}) \ll 1$, the pressure distribution is then

$$c_{p\text{body}} = -\frac{2\epsilon}{\sqrt{1 - M_\infty^2}} \frac{\partial \bar{\Phi}}{\partial \bar{x}}(\bar{x}, \epsilon \sqrt{1 - M_\infty^2} f(\bar{x})) \cong -\frac{2\epsilon}{\sqrt{1 - M_\infty^2}} \frac{\partial \bar{\Phi}}{\partial \bar{x}}(\bar{x}, 0). \tag{17–30}$$

But, $-2\epsilon(\partial \bar{\Phi}/\partial \bar{x})$ is simply the pressure coefficient of the same thin body in an incompressible flow. Denoting it by c_{p_i}, we then have

$$c_{p\text{body}} = \frac{c_{p_i}}{\sqrt{1 - M_\infty^2}}, \tag{17–31}$$

which allows us to relate the pressure distribution on a thin, flat body to its incompressible value. Equation 17–31 is known as the *Prandtl-Glauert rule*.

We see therefore that by means of this subsonic similarity, the Prandtl-Glauert rule, any flow problem of thin bodies in inviscid uniform two-dimensional subsonic flow has been reduced to an incompressible problem.

Note that γ again does not appear anywhere. We see that data obtained from models tested in water can be used to generate information about the same model in subsonic flight!

A further important result follows from this similarity. Note that the mathematical problem posed by Eqs. 17–26 and 17–27 is now linear, and we can superpose solutions. In other words, if we are interested in a body shape described by

$$Y = \epsilon_1 f_1(X) + \epsilon_2 f_2(X), \qquad (17\text{–}32)$$

we can study this problem by first solving separately for the solutions corresponding to $Y = \epsilon_1 f(X)$ and $Y = \epsilon_2 f_2(X)$, giving us two solutions $\bar{\Phi}_1$ and $\bar{\Phi}_2$. Then the solution $\bar{\Phi}$ is simply

$$\bar{\Phi} = \epsilon_1 \bar{\Phi}_1(\bar{x}, \bar{y}) + \epsilon_2 \bar{\Phi}_2(\bar{x}, \bar{y}), \qquad (17\text{–}33)$$

and the pressure distribution c_p is

$$c_p = -\frac{2}{\sqrt{1 - M_\infty^2}} \left[\epsilon_1 \frac{\partial \bar{\Phi}_1}{\partial \bar{x}} (\bar{x}, 0) + \epsilon_2 \frac{\partial \bar{\Phi}_2}{\partial \bar{x}} (\bar{x}, 0) \right],$$

indicating that the pressure distribution produced by the body $Y = \epsilon_1 f_1(X) + \epsilon_2 f_2(X)$ is the sum of the pressure distributions produced by each of the bodies $\epsilon_1 f_1(X)$ and $\epsilon_2 f_2(X)$, as if each were alone in the flow.

PROBLEMS

17–5. Derive Eq. 17–29a from results of Problem 17–3.

17–6. On a certain thin airfoil, the minimum of c_{p_i} on the body surface is -0.05. At what free-stream Mach number M_∞ will the local Mach number at that point reach 1.0? (This Mach number is sometimes referred to as the critical Mach number.)

17–7. For a thin airfoil of arbitrary shape, show that it is always possible to write for the upper surface

$$Y_u = \epsilon_1 f_1(X) + \epsilon_2 f_2(X),$$

and for the lower surface

$$Y_l = \epsilon_1 f_1(X) - \epsilon_2 f_2(X).$$

Give an interpretation of $\epsilon_1 f_1(X)$ and $\epsilon_2 f_2(X)$. [*Hint:* Solve for $f_1(X)$ and $f_2(X)$ in terms of $Y_u(X)$ and $Y_l(X)$.]

17–6. Comments on subsonic similarity

In deriving the subsonic similarity, we took advantage of the assumption that ϵ is small, so that the right-hand side of Eq. 17–16 can be neglected. To justify this procedure, one must show that by using the approximate solution so obtained, the right-hand side of Eq. 17–16 is indeed negligible compared with each of the two terms on the left-hand side. To do this rigorously, let us perform the transformation in Eq. 17–26 on the full equation, 17–16. The resulting equation is

$$
\frac{\partial^2 \bar{\Phi}}{\partial \bar{x}^2} + \frac{\partial^2 \bar{\Phi}}{\partial \bar{y}^2}
$$

$$
= \frac{\epsilon M_\infty^2}{(1 - M_\infty^2)^{3/2}} \left\{ (\gamma + 1) M_\infty^2 \frac{\partial \bar{\Phi}}{\partial \bar{x}} \frac{\partial^2 \bar{\Phi}}{\partial \bar{x}^2} \right.
$$

$$
+ (1 - M_\infty^2) \left[(\gamma - 1) \frac{\partial \bar{\Phi}}{\partial \bar{x}} \frac{\partial^2 \bar{\Phi}}{\partial \bar{x}^2} + \frac{\partial^2 \bar{\Phi}}{\partial \bar{y}^2} \right] + \frac{\partial}{\partial \bar{x}} \left[\left(\frac{\partial \bar{\Phi}}{\partial \bar{x}} \right)^2 + \left(\frac{\partial \bar{\Phi}}{\partial \bar{y}} \right)^2 \right] \right\}
$$

$$
+ \frac{\epsilon^2 M_\infty^2}{\sqrt{1 - M_\infty^2}} \left(\frac{1}{1 - M_\infty^2} \frac{\partial \Phi}{\partial \bar{x}} \frac{\partial^2 \Phi}{\partial \bar{x}^2} + \frac{2}{\sqrt{1 - M_\infty^2}} \frac{\partial \Phi}{\partial \bar{x}} \frac{\partial \Phi}{\partial \bar{y}} \frac{\partial^2 \Phi}{\partial \bar{x} \partial \bar{y}} \right)
$$

$$
+ M_\infty^2 \epsilon^2 \left\{ \left(\frac{\partial \bar{\Phi}}{\partial \bar{y}} \right)^2 \frac{\partial^2 \bar{\Phi}}{\partial \bar{y}^2} + \frac{\gamma - 1}{2} \frac{1}{(1 - M_\infty^2)^{3/2}} \left[\left(\frac{\partial \bar{\Phi}}{\partial x} \right)^2 + (1 - M_\infty^2) \left(\frac{\partial \bar{\Phi}}{\partial y} \right)^2 \right] \right.
$$

$$
\left. \times \left[\frac{\partial^2 \bar{\Phi}}{\partial \bar{x}^2} + (1 - M_\infty^2) \frac{\partial^2 \bar{\Phi}}{\partial \bar{y}^2} \right] \right\}. \tag{17–34}
$$

Now if the right-hand side is to be neglected, then we must require not only $\epsilon \ll 1$, as was stated in the previous section, but

$$
\frac{\epsilon}{(1 - M^2)^{3/2}} \ll 1. \tag{17–35}
$$

This is the reason for our earlier statement that the subsonic similarity is applicable only if $(1 - M^2)^{-1}$ is of order unity.

17–7. Transonic similarity

Since the subsonic similarity breaks down when M_∞^2 is near unity, we shall derive in this section a different similarity for flows with M_∞ nearly equal to unity. This is called the *transonic similarity*.

As can be seen from Eq. 17–34 for a given small ϵ, as M_∞ increases toward unity, the largest term on the right-hand side is

$$
\frac{\epsilon M_\infty^2 (\gamma + 1)}{(1 - M_\infty^2)^{3/2}} \frac{\partial \bar{\Phi}}{\partial \bar{x}} \frac{\partial^2 \bar{\Phi}}{\partial \bar{x}^2}.
$$

It is clear then, in our transonic similarity analysis, that proper consideration must be given to this term. Using this consideration as a motivation, we consider the following transformation:

$$\phi = \epsilon^{-1/3} M_\infty^{-4/3} (\gamma + 1)^{-1/3} \Phi(\xi, \eta),$$

$$X = \xi, \tag{17-36}$$

$$Y = \eta \epsilon^{-1/3} M_\infty^{4/3} (\gamma + 1)^{1/3}.$$

Substituting these new variables into Eqs. 17–16 and 17–17, we have for negligibly small ϵ the approximate equation

$$\left(K - \frac{\partial \Phi}{\partial \xi}\right) \frac{\partial^2 \Phi}{\partial \xi^2} + \frac{\partial^2 \Phi}{\partial \eta^2} = 0, \tag{17-37}$$

with boundary condition

$$\frac{\partial \Phi}{\partial \eta} = \frac{df}{d\xi} \tag{17-38}$$

to be satisfied on $\eta = 0$ for $0 \leq \xi \leq 1$. The parameter K is defined by

$$K = \frac{1 - M_\infty^2}{\epsilon^{2/3} M_\infty^{8/3} (\gamma + 1)^{2/3}}, \tag{17-39}$$

and is called the *transonic similarity parameter*. The transonic similitude given by Eq. 17–36 is valid for thin bodies ($\epsilon \ll 1$), and $M_\infty^2 \doteq 1$, so that the value of K is of order unity. Note that the mathematical problem posed by Eqs. 17–37 and 17–38 is still nonlinear. Due to this fact, the superposition procedure discussed at the end of Section 17–5 cannot be applied for transonic flows.

PROBLEMS

17–8. Show that in arriving at Eq. 17–37 from Eq. 17–16, terms of order $\epsilon^{2/3}$ have been neglected compared with unity.

17–9. Show that for transonic flow, $c_p \epsilon^{-2/3} (\gamma + 1)^{1/3}$ on a body surface is a function of K only for thin bodies with identical $f(x)$ distributions.

17–10. Suppose that a transonic tunnel using helium ($\gamma = \frac{5}{3}$) as working fluid is available, and it operates at $M_\infty = 0.97$ only. It is desired to know the pressure distribution for a thin flat plate at an angle of attack of 0.027 radius in air ($\gamma = 1.4$) at $M_\infty = 0.92$. How would you perform the experiment and reduce the data?

17–8. Supersonic similarity

For *supersonic flows* with $M_\infty > 1$ about thin bodies, a similarity is again available and its derivation is analogous to that for the subsonic similarity.

Since shock waves may occur in a supersonic flow, the assumption of isentropic flow is more likely to break down. However, for Mach numbers not too high, the shock waves are weak, provided that the body is thin so that ϵ is small, and the increase of entropy across such weak shocks can be shown to be negligible.

We introduce the following transformation in analogy to Eq. 17–26:

$$\phi = \frac{1}{\sqrt{M_\infty^2 - 1}}\, \Phi(\bar{x}, \bar{y}), \qquad X = \bar{x}, \qquad Y = \frac{1}{\sqrt{M_\infty^2 - 1}}\, \bar{y}. \qquad (17\text{–}40)$$

Substituting into Eq. 17–16, we obtain an equation similar to Eq. 17–34 (except for some signs which are changed). We will see that provided

$$\frac{\epsilon M_\infty^2}{(M_\infty^2 - 1)^{3/2}} \ll 1 \qquad \text{and} \qquad M_\infty \epsilon \ll 1, \qquad (17\text{–}41)$$

the right-hand side of the resulting equation can again be neglected. Hence the resulting approximate equation becomes simply

$$\frac{\partial^2 \bar{\Phi}}{\partial \bar{x}^2} - \frac{\partial^2 \bar{\Phi}}{\partial \bar{y}^2} = 0, \qquad (17\text{–}42)$$

and the boundary condition becomes

$$\frac{\partial \bar{\Phi}}{\partial \bar{y}} = \frac{df}{d\bar{x}} \qquad (17\text{–}43)$$

to be satisfied at $\bar{y} = 0$, $0 \leq \bar{x} \leq 1$. The mathematical problem posed above is linear and hence the principle of superposition is again applicable here.

For very high Mach numbers, $M_\infty \gg 1$, the flow is called *hypersonic*. Even with reasonably small ϵ, strong shock waves will appear in the flow field, and our original assumption of isentropic flow will no longer be valid. An analysis for hypersonic similarity is beyond the scope of this book.

PROBLEM

17–11. Show that the drag coefficient c_D on a thin body in supersonic flow is given by

$$c_D\big(M_\infty; f(\xi)\big) = \frac{1}{\sqrt{M_\infty^2 - 1}}\, c_D\big(M_\infty = \sqrt{2}; f(\xi)\big),$$

where c_D is defined as $c_D = D/\tfrac{1}{2}\rho_\infty U_\infty^2$, where D is the component of the resultant pressure force on the body acting against the free-stream flow direction.

Appendix

Mechanical Properties of Some Fluids

---

(1) Density and unit weight

The density of *water* under ordinary pressure is 1 gm/cm³ or 1.94 slugs/ft³ at 4°C, and is 0.2% lower at 20°C or 68°F. With $g = 980$ cm/sec² $= 32.17$ ft/sec², the unit weight of water is 980 dynes/cm³ or 62.4 lb/ft³.

The density of *mercury* is 13.6 times that of water under ordinary conditions.

The density of *air* is given by Eq. 1–4, with $R/g = 53.35$ ft/°R or 2927 cm/°K. Under atmospheric pressure at 68°F ($=527.7$°R), the density of air is 0.00234 slug/ft³, or 0.00120 gm/cm³.

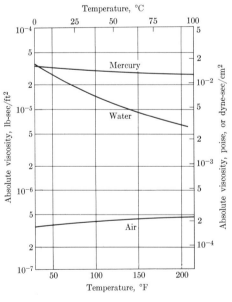

FIGURE A–1

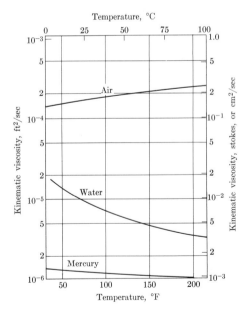

FIGURE A–2

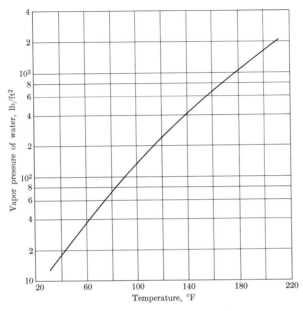

FIGURE A–3

(2) Viscosity

The values of absolute viscosity and kinematic viscosity for several fluids are shown in Figs. A–1 and A–2, respectively.

(3) Surface tension

The surface tension of water-air interface is 0.00498 lb/ft or 72.5 dynes/cm at 68°F or 20°C. The surface tension at the same temperature for mercury-air interface is 0.035 lb/ft or 500 dynes/cm.

(4) Vapor pressure

The vapor pressure of water varies with temperature, as shown in Fig. A–3.

Answers to Problems

CHAPTER 1

1-1. 1.94 slugs/ft³, 10.6 lb/ft³ $\qquad$ 1-2. $2\sigma(2R - t)/[R(R - t)]$

1-3. $T\rho^{-(\gamma-1)} = \text{constant}$ $\qquad$ 1-4. p

CHAPTER 2

2-1. $L, LT^{-1}, LT^{-2}, l, T^{-1}, T^{-2}, F, FL, FL, FL, MLT^{-1}$ (or FT), FT, ML^2 (or FLT^2), and ML^2T^{-1} (or FLT).

2-2. ML^{-3} or $FL^{-4}T^2, FL^{-2}T, FL^{-1}$, and FL^{-2}. $\qquad$ 2-3. 2

2-4. (a) 2, (b) $P/(\rho g D^3) \propto (H/D)$ $\qquad$ 2-8. $Pl^2/(EI) = \text{constant}$

2-10. (a) $fl\sqrt{m/F} = \phi(F/AE)$, (b) $fl\sqrt{m/F} = \text{constant}$

2-11. 2, 2 $\qquad$ 2-12. 2

2-13. (a) $\dfrac{\Delta p}{l} \dfrac{A}{Q} \dfrac{D^2}{\mu} = \phi\left(\dfrac{QD\rho}{A\mu}, R\right)$

2-14. (a) $\dfrac{u}{yf} = \phi\left(\dfrac{U}{yf}, \dfrac{y^2 f\rho}{\mu}\right)$ $\qquad$ 2-15. (a) $\dfrac{ut}{y} = \phi\left(\dfrac{Ut}{y}, \dfrac{y^2\rho}{\mu t}\right)$

2-16. (a) $\dfrac{\Delta p D_0^4}{\rho Q^2} = \phi\left(\dfrac{Q\rho}{D_0\mu}, \dfrac{D_0}{D}\right)$ $\qquad$ 2-17. $k = \phi\left(\dfrac{VY\rho}{\mu}, \dfrac{B}{Y}, \dfrac{e}{Y}\right)$

2-19. (a) $\dfrac{Q}{\sqrt{g}\, H^{5/2}} = \phi\left(\dfrac{L}{H}\right)$ $\qquad$ 2-20. $k = \phi\left(\theta, \dfrac{gH^2\rho}{\sigma}\right)$

2-22. $\dfrac{p}{\rho L^2 f^2} = \phi\left(\dfrac{L}{D}, \dfrac{fL^2\rho}{\mu}, fL\sqrt{\dfrac{\rho}{K}}\right)$

2-23. 5; $\tfrac{1}{5}$; 25 $\qquad$ 2-24. $\tfrac{1}{30}$ maximum

2-25. 35,200 lb with minimum $U_m = 32$ ft/sec

2-26. (a) 1.2×10^{-5} ft²/sec maximum, (b) 13.3 atmospheres minimum

2-27. 1.40, 0.0023 ft³/sec $\qquad$ 2-28. 126 lb/ft² absolute

CHAPTER 3

3-1. 16.8 slugs/ft³ minimum $\qquad$ 3-2. Yes

3-3. $L - 13.6R$ $\qquad$ 3-4. 52 lb/ft²

3-5. 100 lb $\qquad$ 3-6. 10,100 lb

3-7. 2350 lb at center $\qquad$ 3-8. 13,500 lb

3–9.   2880 lb   3–11. 688 lb, 880 lb

3–12. 750 lb, 375 lb, 375 lb

3–14. $F_H = 1120$ lb, $F_V = 560$ lb, both at 1 ft from center

3–15. 1254 lb/ft   3–16. Yes, no   3–17. No

3–22. $h - h_1 = \dfrac{R}{g}(T_1 - T)\dfrac{\log (p/p_1)}{\log (T/T_1)} = 6150$ ft

3–23. Yes   3–24. $1/\rho = (gh/K) + C$

3–25. $\dfrac{Fd}{\sigma \pi R^2} = \left(2 \cos \alpha - \dfrac{d}{R}\right)\left(1 + \dfrac{d}{R}\dfrac{1 - \sin \alpha}{\cos \alpha}\right)^2$

CHAPTER 4

4–3. $u = Uy/d$ 4–4. $u = \sqrt{2gx}$

4–6.   2°C/sec, 1°C/ft, 3°C/sec   4–7. -2 dollars/day, $+8$ dollars/day

4–10. $a_x = k\left[1 - \left(1 - \dfrac{y}{mt^2}\right)\left(1 + \dfrac{3y}{mt^2}\right)\right]$

4–11. $a = k^2/L^3$

4–13. (a) $q = U$ forward,   (b) $q = U$ backward,   (c) $q = U$ sidewise

4–14. (a) 0,   (b) $4U^2/R$,   (c) $2\sqrt{2}\,U^2/R$

4–18. (a) $v = \omega r$,   (b) $u = -\omega y$, $v = \omega x$

4–19. $A_1 + B_2 + C_3 = 0$ 4–20. $w = f(x, y, t) - yz^2$

CHAPTER 5

5–3.   (a) 0, 1,   (b) $\sin \theta$, $r \cos \theta$

5–4. $p + \rho g h = $ constant   5–5.   43 lb/ft^2

5–6. $p_B = \rho(a + g \sin \theta)L + \rho g D \cos \theta$

5–7. $\rho g R - \frac{1}{2}\rho \omega^2 R^2$ 5–8. $\frac{1}{2}b(2L + b)\omega^2/g$

5–9. $p_0/(\rho g)$ 5–10. 50.2 ft/sec, 0.165 ft^3/sec

5–11. $q_1 = \sqrt{2gH}$ 5–12. 2.54 ft/sec

5–13. 5 ft/sec, 4   5–14. (a) $U/2$, 2,   (b) $49u/60$, 1.058

5–15. 56.4 ft/sec, 0.308 ft^3/sec   5–16. 0.182 ft^3/sec

5–17. 32.1 ft/sec, $26\rho g$ vacuum   5–18. 0.40 ft and 3.46 ft

CHAPTER 6

6–3. $d\beta/dt = U/d$

6–6.   (a) 1, 0.8, 0,   (c) 1.28,   (d) $L^2 T^{-1}$, T^{-1}

6–7. $\phi = -U(x \cos \alpha + y \sin \alpha) + C$

6–8. $\phi = x^2 - y^2 + C$

6–10. $\rho g(y_1 - y_2) + \frac{1}{2}\rho k^2(x_1^2 + y_1^2 - x_2^2 - y_2^2)$

6–17. $\frac{1}{2}\rho U^2$, $\pm\pi/6$, and $\pm 5\pi/6$

CHAPTER 7

7-1. $r^2 z = C$, $r^2 z = 1$ 7-2. $r = c_1$, $\theta = c_2$

7-3. $r = \sin\theta$ 7-5. $\frac{1}{2}Ud[1 - (y^2/d^2)]$, $\frac{1}{2}Ud$

7-6. (a) $\psi = U(x\sin\alpha - y\cos\alpha) + C$,   (b) $\psi = \frac{1}{2}\omega r^2$,
      (c) $\psi = (\Gamma/2\pi)\ln r + C$

7-7. (a) $\psi = 2xy + C$, $xy = 1$

7-11. (a) $\psi = Ur\sin\theta[1 - (R^2/r^2)]$,   (b) $\sin\theta = \frac{5}{6}Rr/(r^2 - R^2)$

7-14. $\Psi = \frac{1}{2}Ur^2$

7-16. (a) $\Psi = \frac{1}{2}Ur^2 + Q[1 - (z/\sqrt{r^2 + z^2})]/4\pi$,
      (b) $2\pi Ur^2 - Q[1 + (z/\sqrt{r^2 + z^2})] = 0$

7-17. $\Psi = -\dfrac{Ur^2}{2}\left[1 - \dfrac{R^3}{(r^2 + z^2)^{3/2}}\right]$

CHAPTER 8

8-1. (a) $\xi_x = 0$, $\xi_y = 0$, $\xi_z = -a$,
      (b) $\xi_x = 0$, $\xi_y = 0$, $\xi_z = 0$,
      (c) $\xi_r = 0$, $\xi_\theta = 0$, $\xi_z = 2a$

8-3. (a) $\xi_r = -ar$, $\xi_\theta = 0$, $\xi_z = 2az$

8-5. 4   8-8. (a) $2\pi\omega R^2$,   (b) zero

CHAPTER 9

9-3. (a) 387 lb,   (b) 218 lb

9-4. (a) $F_x = \rho V_1(0.134 Q_1 + 0.159 Q_2)$, $F_y = \rho V_1(0.5 Q_1 - 1.207 Q_2)$,
      (b) $\rho Q_1 V_1\sin\theta$

9-6. 400 lb   9-7. $\frac{4}{3}$

9-8. $F_x = \frac{1}{2}\rho(v_2^2 - v_1^2)L$, $F_y = \rho u(v_1 - v_2)L$

9-9. (a) 0.00081 slug/ft^3,   (b) 4420 lb

9-11. (a) $\frac{4}{3}$,   (b) 1.020

9-12. $0.1107 V_2^2$ 9-13. $Y_1 = \frac{1}{2}Y_2(\sqrt{1 + 8F_2^2} - 1)$

9-14. 2.10 lb/ft^2 9-15. $Y_1 = Y_2\sqrt{1 + 2F_2^2}$

CHAPTER 10

10-1. $\dfrac{D}{L} = \dfrac{wL^4}{24EI}\left[\left(\dfrac{x}{L}\right)^4 - 2\left(\dfrac{x}{L}\right)^3 + \left(\dfrac{x}{L}\right)\right]$

10-2. $\dfrac{T}{T_1} = \dfrac{\log(r/a)}{\log(b/a)}$

10-4. (c) $f = c/\lambda$ 10-6. 25 ft/sec

10-8. (a) 3 ft,   (b) 2.3 ft   10-9. 2.05 ft

10-12. $\psi \doteq -a\sqrt{\dfrac{g\lambda}{2\pi}}\, e^{2\pi y/\lambda}\cos\dfrac{2\pi(x - ct)}{\lambda}$

10–13. $\phi = -a\sqrt{\dfrac{g\lambda}{2\pi}}\, e^{2\pi y/\lambda} \cos\dfrac{2\pi x}{\lambda} \cos\left(\sqrt{\dfrac{2\pi g}{\lambda}}\, t\right)$

CHAPTER 11

11–3. $Q = \dfrac{\pi \rho g s}{8\mu}\left[b^4 - a^4 - \dfrac{(b^2 - a^2)^2}{\ln(b/a)}\right]$

11–4. $w = \dfrac{\rho g a^2}{2\mu}\left[\dfrac{1}{2}\left(\dfrac{r^2}{a^2} - 1\right) - \dfrac{b^2}{a^2}\ln\left(\dfrac{r}{a}\right)\right]$

11–5. 4.598, 1.964, 0   11–8. $\delta_{jj} = 3$

11–17. (b) 3, 1, 2   11–18. (a) $2e_{12} = U/d$

11–19. (a) $e_{11} = -7/625$, $e_{22} = 7/625$, $e_{12} = -24/625$

11–20. (b) $-p$

CHAPTER 12

12–3. $1.328\sqrt{\nu L/U}$

12–4. $\tau_0 = 0.289\rho U^2\sqrt{\nu/Ux}$, $\delta = 3.46\sqrt{\nu x/U}$

12–5. 0.36, 0.14   12–6. $\sqrt{\nu/Ux}$

12–7. $Ul[(-dU/dx)/(\nu n)]^{1/2}$, $[\nu n/(-dU/dx)]^{1/2}$, $H[\nu n/(-dU/dx)]^{1/2}$

12–8. $n = 0.126[(r/r_0)^{3.5} - 1]$ 12–9. $0.664\sqrt{\nu/Ux}$

CHAPTER 13

13–6. (a) 0.38 ft/sec,   (b) 0.047 ft/sec

13–7. $l/a = 0.114$

13–9. (a) 0.0037 in., 0.0016 in., 11.96 in., 4.4 ft/sec,
  (b) 0.030 in., 0.013 in., 11.64 in., 0.54 ft/sec

13–15. (a) 0.0016 in.,   (b) 0.013 in.   13–19. (a) $6.46(\rho u_* D/\mu)^{1/7}$

13–20. 57 ft   13–21. 6.8 ft^3/sec

13–22. Use 18 in.   13–23. (a) 3.28 ft,   (b) 2.5 ft

13–24. 37.5 ft^3/sec   13–25. 0.000228   13–26. 3.44 ft

13–29. $B = 4.6$ ft, $Y = B/2$ 13–30. $Y = 8$ ft, $B = 42$ ft

CHAPTER 14

14–1. 8.314 ergs/mole-°K   14–2. 0.002378 slug/ft^3

14–3. Unchanged   14–5. 570°R

14–6. $P/T^{\gamma/(\gamma-1)} = $ constant, $p^{\gamma-1}/T = $ constant

CHAPTER 15

15–3. (a) $p_1^0 = p_1[1 + (\gamma - 1)q_1^2/2RT_1]^{\gamma/(\gamma-1)}$,
  $T_1^0 = T_1[1 + (\gamma - 1)q_1^2/2\gamma RT_1]$, $M_1 = q_1/(\gamma RT_1)^{1/2}$,
  (b) $p_2/p_2^0 = [2/(\gamma + 1)]^{\gamma/(\gamma-1)}$, $T_2/T_1^0 = 2/(\gamma + 1)$, both are independent of M_1.

15–4.   Increases 11.3-fold

15–8.   (a) 1.34, 1.70,   (b) 1.70, 1.34,   (c) 0.37.

15–9.   (a) For $M_1 = 0.25$, $M_2 = 0.62$. For $M_1 = 0.333$, no solution possible.
  If $M_2 = 1$, then $M_1 = 0.30$.
  (b) For $M_1 = 2$, no solution possible.   If $M_2 = 1$, then $M_1 = 2.2$.
  For $M_1 = 3.0$, $M_2 = 2.28$.

15–12. (a) 4640 ft/sec,   (b) 0.48,   (c) 1.03 atmosphere,   (d) 816 ft/sec

15–16. $p_c/p_1 = 0.986$, $p_e/p_1 = 0.029$, $M_c = 0.14$, $M_e = 2.94$

15–17. $\rho^0 A_2 (2C_p T^0)^{1/2} (p_2/p^0)^{1/\gamma} [1 - (p_2/p^0)^{(\gamma-1)/\gamma}]^{1/2}$

15–18. (a) 1200 lb,   (b) 3.58,   (c) 4.06

15–19. $109 A_t$ 15–20. 18.4 lb/ft

CHAPTER 16

16–3. 1.17, 1.12   16–4.   1.179, 1.125

16–5.   (a) 1.67, 2.02, 4.09, 8.39,   (b) 0.61, 0.36, 0.17, 0.00022

16–6. $\pi[\sqrt{(\gamma+1)/(\gamma-1)} - 1]/2$

CHAPTER 17

17–1.   (a) 0.0021, -0.0063, for $\alpha = 0.0021$; 0.000010, -0.000030, for
  $\alpha = 0.00001$
  (b) 1.0, -4.8×10, for $\alpha = 0.0021$; 1.0, -1.0×10^5, for $\alpha = 0.00001$

17–2. $y = \beta^{-1/5} \epsilon^{-1/10} \eta(\xi, l)$, $\xi = \epsilon^{-1/2} x$, $l = \alpha \beta^{-1/5} \epsilon^{-3/5}$

17–6. 0.93

Index

ABCDE6987654